高等职业教育计算机专业规划教材

Linux 服务器配置与管理

李治西　主编

武汉理工大学出版社
·武汉·

内 容 提 要

本书以目前被广泛使用的 Red Hat Enterprise Linux 7.4 系统编写,完全兼容 CentOS7.4 系统。采用教、学、做相结合的模式,以工作过程为导向,注重知识的实用性和可操作性,强化实践操作能力的训练。

本书依据网络工程实际工作中需要的知识和技能进行项目分解,全书共分为 14 个项目。内容涵盖:安装和使用 Linux 服务器、用户管理与文件权限、磁盘管理、网络配置与远程访问、Samba 和 NFS 服务器、DHCP 服务器、DNS 服务器、Apache 服务器配置、Vsftp 服务器、邮件服务器、代理服务器、数据库服务器、无人值守安装系统、双机热备。

本书可作为高职院校计算机网络技术专业、计算机应用技术专业及相关专业的理论与实践一体化教材,也可作为广大网络管理与维护人员搭建、配置和管理 Linux 系统的参考用书。

图书在版编目(CIP)数据

Linux 服务器配置与管理/李治西主编. —武汉:武汉理工大学出版社,2021.3
ISBN 978-7-5629-6396-7

Ⅰ.①L… Ⅱ.①李… Ⅲ.①Linux 操作系统-高等职业教育-教材 Ⅳ.①TP316.85

中国版本图书馆 CIP 数据核字(2021)第 041900 号

项目负责人:王兆国	责任编辑:黄玲玲
责任校对:李兰英	版面设计:正风图文

出版发行:武汉理工大学出版社
社　　址:武汉市洪山区珞狮路 122 号
邮　　编:430070
网　　址:http://www.wutp.com.cn
经　　销:各地新华书店
印　　刷:武汉市宏达盛印务有限公司
开　　本:787 mm×1092 mm　1/16
印　　张:21
字　　数:524 千字
版　　次:2021 年 3 月第 1 版
印　　次:2021 年 3 月第 1 次印刷
定　　价:48.00 元

凡购本书,如有缺页、倒页、脱页等印装质量问题,请向出版社发行部调换。
本社购书热线电话:027-87785758　87384729　87165708(传真)

·版权所有　盗版必究·

前　　言

Red Hat Enterprise Linux 7.4 系统作为目前广泛应用的操作系统，能提供给网络管理员多种功能和应用服务，网络管理员能根据实际网络应用环境，使用各种系统功能，提供各种网络服务。随着信息化水平的不断提高，各 IT 企业目前急需掌握该操作系统操作技能的人才。

本书注重理论与实际相结合，突出"做中教、做中学"。在知识体系上紧密结合网络管理员、网络工程师等工作岗位对 Red Hat Enterprise Linux 7.4 系统的核心技能要求，每个技能要求都进行了实际的演示，每个操作的难点、重点都进行了解释，图文并茂，详尽地向读者展示了每一个操作步骤。使读者在参照学习后，能很快地将理论知识与实际工作结合起来，并应用到具体的网络环境中。

本书共有 14 个项目：安装和使用 Linux 服务器、用户管理与文件权限、磁盘管理、网络配置与远程访问、Samba 和 NFS 服务器、DHCP 服务器、DNS 服务器、Apache 服务器、Vs-ftp 服务器、邮件服务器、代理服务器、数据库服务器、无人值守安装系统、双机热备。

本书共分为两大部分，其中第一部分以讲解 Red Hat Enterprise Linux 7.4 系统的系统管理技能为主，包含项目 1 至项目 4，通过这部分的理论学习和技能实训，使读者能够较快地掌握 Red Hat Enterprise Linux 7.4 系统的基本应用和系统管理。第二部分讲解 Red Hat Enterprise Linux 7.4 系统的主要网络服务功能，包含项目 5 至项目 14，通过该部分内容的学习，读者可以掌握 Red Hat Enterprise Linux 7.4 系统所支持的常用的网络服务功能的配置和管理。

由于编者水平有限，书中不足之处在所难免，恳请广大读者批评指正。

目　录

项目 1　安装和使用 Linux 服务器 …………………………………………………… 1

　任务 1.1　认识 Linux ………………………………………………………………… 1

　　1.1.1　Linux 系统的历史 ………………………………………………………… 1

　　1.1.2　Linux 系统的特点 ………………………………………………………… 2

　　1.1.3　Linux 系统的组成与版本 ………………………………………………… 3

　任务 1.2　安装 Linux ………………………………………………………………… 5

　　1.2.1　Linux 文件目录 …………………………………………………………… 5

　　1.2.2　Linux 物理设备 …………………………………………………………… 6

　　1.2.3　规划分区 …………………………………………………………………… 8

　　1.2.4　实现步骤 …………………………………………………………………… 8

　任务 1.3　Linux 系统基本操作 …………………………………………………… 21

　　1.3.1　命令行界面 ……………………………………………………………… 21

　　1.3.2　系统命令 ………………………………………………………………… 23

　　1.3.3　导航命令 ………………………………………………………………… 27

　　1.3.4　文件与目录操作命令 …………………………………………………… 28

　　1.3.5　文件查看命令 …………………………………………………………… 29

　　1.3.6　压缩、解压命令 ………………………………………………………… 30

　　1.3.7　VIM 编辑器 ……………………………………………………………… 32

　　1.3.8　软件包安装 ……………………………………………………………… 36

　　1.3.9　管理系统服务 …………………………………………………………… 40

项目 2　用户管理与文件权限 ……………………………………………………… 41

　任务 2.1　用户与用户组 …………………………………………………………… 41

　　2.1.1　用户与用户组的概念 …………………………………………………… 41

　　2.1.2　用户管理系统文件 ……………………………………………………… 42

　　2.1.3　用户管理命令 …………………………………………………………… 44

　　2.1.4　用户组管理命令 ………………………………………………………… 47

　　2.1.5　查看用户信息命令 ……………………………………………………… 48

　　2.1.6　查看当前登录用户命令 ………………………………………………… 49

　任务 2.2　文件权限 ………………………………………………………………… 49

　　2.2.1　查看文件权限信息 ……………………………………………………… 49

　　2.2.2　修改文件属性 …………………………………………………………… 51

　　2.2.3　文件的特殊权限 ………………………………………………………… 53

2.2.4 文件隐藏属性 ……………………………………………………………… 55
2.2.5 文件访问控制列表 …………………………………………………… 56

项目 3 磁盘管理 …………………………………………………………………… 58
任务 3.1 扩充硬盘空间 …………………………………………………………… 58
3.1.1 添加新的硬盘 ……………………………………………………… 58
3.1.2 系统使用新硬盘 …………………………………………………… 62
任务 3.2 磁盘配额 ………………………………………………………………… 66
3.2.1 磁盘配额的基本概念 …………………………………………… 66
3.2.2 配置磁盘配额 ……………………………………………………… 66
任务 3.3 动态调整硬盘分区 ……………………………………………………… 75
3.3.1 LVM 简介 …………………………………………………………… 75
3.3.2 使用 LVM 进行硬盘分区的动态调整 ………………………… 76
任务 3.4 提升数据的安全性 ……………………………………………………… 85
3.4.1 RAID 简介 …………………………………………………………… 85
3.4.2 配置 RAID …………………………………………………………… 88

项目 4 网络配置与远程访问 ……………………………………………………… 96
任务 4.1 配置网络环境 …………………………………………………………… 96
4.1.1 配置主机名 ………………………………………………………… 96
4.1.2 配置网络参数 ……………………………………………………… 98
4.1.3 网卡链路聚合 …………………………………………………… 106
任务 4.2 防火墙管理 ……………………………………………………………… 111
4.2.1 iptables 防火墙 …………………………………………………… 111
4.2.2 firewalld 防火墙 ………………………………………………… 116
任务 4.3 远程访问与连接 ……………………………………………………… 121
4.3.1 远程访问方式简介 ……………………………………………… 121
4.3.2 Telnet ……………………………………………………………… 122
4.3.3 SSH ………………………………………………………………… 123
4.3.4 远程桌面 ………………………………………………………… 127

项目 5 Samba 和 NFS 服务器 ………………………………………………… 132
任务 5.1 配置和管理 Samba 服务器 ………………………………………… 132
5.1.1 Samba 服务器简介 ……………………………………………… 132
5.1.2 Samba 服务参数 ………………………………………………… 134
5.1.3 配置 Samba 服务器 ……………………………………………… 137
任务 5.2 配置和管理 NFS 服务器 …………………………………………… 146
5.2.1 NFS 服务器简介 ………………………………………………… 146
5.2.2 NFS 服务的组件以及主要配置文件 ………………………… 147
5.2.3 配置 NFS 服务器 ………………………………………………… 148

项目 6　DHCP 服务器 …… 154
任务 6.1　DHCP 服务器简介 …… 154
6.1.1　DHCP 工作原理 …… 154
6.1.2　DHCP 的三种 IP 地址分配方式 …… 155
6.1.3　续约 IP 地址 …… 156
任务 6.2　部署 DHCP 服务器 …… 156
6.2.1　安装 DHCP 服务器 …… 156
6.2.2　认识 DHCP 服务器主配置文件 …… 156
6.2.3　配置 DHCP 服务器 …… 158

项目 7　DNS 服务器 …… 168
任务 7.1　DNS 服务器简介 …… 168
7.1.1　DNS 域名空间 …… 168
7.1.2　DNS 区域 …… 169
7.1.3　区域委派 …… 169
7.1.4　DNS 查询模式 …… 170
7.1.5　DNS 服务器的分类 …… 171
7.1.6　DNS 的区域类型 …… 172
7.1.7　资源记录 …… 172
任务 7.2　部署 DNS 服务器 …… 173
7.2.1　安装 DNS 服务器 …… 173
7.2.2　认识 BIND 配置文件 …… 173
7.2.3　配置 DNS 服务器 …… 177

项目 8　Apache 服务器 …… 190
任务 8.1　Web 服务器简介 …… 190
8.1.1　Web 服务概述 …… 190
8.1.2　Apache 服务器简介 …… 191
任务 8.2　部署 Apache 服务器 …… 192
8.2.1　安装 Apache 服务器 …… 192
8.2.2　认识 Apache 服务器的配置文件 …… 193
8.2.3　个人用户主页功能 …… 195
8.2.4　相关常规设置 …… 196
8.2.5　虚拟主机 …… 199
8.2.6　HTTPS 安全认证访问网站 …… 202
8.2.7　用户认证和授权 …… 203

项目 9　Vsftpd 服务器 …… 206
任务 9.1　FTP 服务器简介 …… 206
9.1.1　FTP 服务器基本概念 …… 206

9.1.2　FTP 服务的用户分类 208

任务 9.2　部署 Vsftpd 服务器 208

9.2.1　安装 Vsftpd 服务器 208

9.2.2　认识 Vsftpd 服务器的配置文件 209

9.2.3　配置匿名 FTP 服务器 211

9.2.4　配置普通 FTP 服务器 213

9.2.5　配置使用虚拟用户的 FTP 服务器 216

9.2.6　配置基于独立配置文件的 FTP 服务器 218

9.2.7　配置基于 SSL 的 FTP 服务器 222

项目 10　邮件服务器 227

任务 10.1　电子邮件服务器简介 227

10.1.1　电子邮件系统的组成 227

10.1.2　电子邮件协议 228

10.1.3　电子邮件的传输过程 229

任务 10.2　配置电子邮件系统 230

10.2.1　Postfix 简介 230

10.2.2　配置常规的 Postfix 服务器 232

10.2.3　配置 Dovecot 服务程序 236

10.2.4　电子邮件系统的测试 239

10.2.5　配置 Sendmail 服务程序 247

任务 10.3　部署认证功能的电子邮件系统 249

10.3.1　使用 Cyrus-SASL 开启 SMTP 的 SASL 认证 249

10.3.2　基于 TLS/SSL 的邮件服务 252

项目 11　代理服务器 257

任务 11.1　代理服务器概述 257

11.1.1　代理服务器基本概念 257

11.1.2　代理服务器的工作原理 258

11.1.3　代理服务器的分类 259

任务 11.2　配置代理服务器 261

11.2.1　安装代理服务器 261

11.2.2　认识 Squid 服务程序的配置文件 261

11.2.3　配置标准正向代理服务器 265

11.2.4　配置透明代理服务器 268

11.2.5　配置反向代理服务器 269

项目 12　数据库服务器 272

任务 12.1　MySQL 数据库服务器 272

12.1.1　MySQL 数据库简介 272

12.1.2	安装 MySQL 数据库	272
12.1.3	数据库的创建与使用	277
12.1.4	备份与恢复数据库	283
12.1.5	Apache 使用 MySQL 进行网站的认证	283

任务 12.2　MariaDB 数据库服务器 ……286
- 12.2.1　MariaDB 数据库简介 ……286
- 12.2.2　安装 MariaDB 数据库 ……286
- 12.2.3　数据库的创建与使用 ……288
- 12.2.4　备份与恢复数据库 ……292

项目 13　无人值守安装系统　295

任务 13.1　无人值守安装系统简介 ……295
- 13.1.1　无人值守安装简介 ……295
- 13.1.2　无人值守安装系统的工作流程 ……296

任务 13.2　部署无人值守安装系统 ……296
- 13.2.1　配置服务器 ……296
- 13.2.2　配置客户端 ……299

项目 14　双机热备　311

任务 14.1　双机热备概述 ……311
- 14.1.1　集群基本概念 ……311
- 14.1.2　集群的分类 ……312

任务 14.2　部署双机热备系统 ……312
- 14.2.1　使用 Pacemaker 部署双机热备系统 ……312
- 14.2.2　使用 Keepalive 部署双机热备系统 ……321

参考文献 ……325

项目 1　安装和使用 Linux 服务器

项目综述

某公司的网络管理员，根据公司的业务需求，计划使用 Linux 系统，搭建 Web、DNS、FTP、DHCP 等服务器，来满足公司员工日常办公的需求。

项目目标

- 了解 Linux 系统的发展历史、组成、版本以及系统的特点；
- 了解 Linux 系统的硬盘分区基本知识；
- 掌握 Red Hat Enterprise Linux 7.4 的安装与使用方法；
- 掌握 Red Hat Enterprise Linux 7.4 的基本操作命令；
- 掌握 VIM 编辑器的使用方法；
- 掌握 Red Hat Enterprise Linux 7.4 软件包的安装与管理方法。

◀ 任务 1.1　认识 Linux ▶

1.1.1　Linux 系统的历史

1. UNIX 系统简介

最早的 UNIX 是由美国电话电报公司（AT&T）的贝尔实验室在 20 世纪 60 年代编写的一个系统，具有多用户、多任务的特点，支持多种处理器架构。因其具有开放性、可移植性、多用户、多任务以及稳定性等特点，加上本身强大的网络通信功能，被广泛地应用在各行业中。

2. Linux 系统简介

Linux 是当前流行的一种操作系统，它是一个类似于 UNIX 的操作系统，是 UNIX 在微机上的完整实现，它的标志是一个名为 Tux 的可爱的小企鹅，如图 1-1 所示。

1990 年，芬兰人林纳斯·托瓦兹（Linus Torvalds，如图 1-2 所示）开始着手研究一个开放的、与 Minix 系统兼容的操作系统。1991 年 10 月 5 日，Linus Torvalds 在赫尔辛基大学的一台 FTP 服务器上公布了第一个 Linux 的内核版本 0.02 版。1992 年 3 月，内核 1.0 版本推出，标志着 Linux 第一个正式版本诞生。现在，Linux 凭借优秀的设计、不凡的性能，加上 IBM、Intel、AMD、DELL、Oracle、Sybase 等国际知名企业的大力支持，市场份额逐步扩大，逐渐成为主流操作系统之一。

图 1-1　Linux 的标志 Tux

图 1-2　Linux 的创始人林纳斯·托瓦兹

1.1.2　Linux 系统的特点

1. 免费开源

Linux 遵循通用公共许可证 GPL，是一款完全免费的操作系统，任何人都可以从网络上下载到它的源代码，并可以根据自己的需求进行定制化的开发，而且没有版权限制。

2. 安全、高效、稳定

Linux 继承了 UNIX 核心的设计思想，具有执行效率高、安全性高和稳定性好的特点。Linux 采取了很多安全技术措施，包括对读写文件进行权限的限制，核心程序、关键操作的授权等，这些安全举措为网络多用户环境下的用户提供了安全保障。Linux 系统可以持续运行很长一段时间而无须重启，并能提供各类服务，其安全稳定性已经在各个领域得到了广泛的认同，并得到广泛使用。

3. 支持多种硬件平台

得益于其免费开源的特点，大批程序员不断地向 Linux 社区提供代码，使得 Linux 有着异常丰富的设备驱动资源，对主流硬件的支持度极好，而且几乎能在所有主流 CPU 搭建的体系结构上运行，包括 x86、MIPS、PowerPC、SPARC、Alpha 等，可以在笔记本电脑、工作站、机顶盒、游戏机等平台上运行。

4. 多用户多任务的系统

多用户是指系统资源可以同时被不同的用户使用，每个用户对自己的资源有特定的权限，互不影响。多任务是指计算机能同一时间执行多个用户任务，且各个任务的执行相互独立。Linux 内核负责调度每个进程，使它们平等地访问处理器。

5. 丰富的网络功能

Linux 中有完善的内置网络服务，如 Apache、Sendmail、Vsftp、Samba、DNS、DHCP 等。完善的网络服务是 Linux 操作系统优于其他操作系统的一大特点。

6. 友好的用户界面

Linux 同时具有字符界面和图形界面。在字符界面用户可以通过键盘输入相应的指令来进行操作。它同时也提供了类似 Windows 图形界面的 X-Window 系统，用户可以使用鼠标方便、直观、快捷地进行操作。

1.1.3 Linux 系统的组成与版本

1. Linux 系统的组成

Linux 系统结构一般有 3 个主要部分：内核（kernel）、命令解释层（shell 或其他操作环境）、实用工具。

（1）Linux 内核

Linux 内核是系统的核心，包括基本的系统启动核心信息、各种硬件的驱动程序等。操作系统为用户提供一个操作界面，它从用户那里接收命令，并且把命令送给内核去执行，内核决定着系统的性能。

（2）命令解释层

Shell 是操作系统的用户界面，是应用程序与内核进行交互操作的接口。它接收用户输入的命令，并且把它送入内核执行。操作系统在系统内核与用户之间提供操作界面，它可以描述成一个解释器。Linux 存在多种操作环境，分别是基于图形界面的集成桌面环境和基于 Shell 命令行环境。

Shell 是一个命令解释器，它解释由用户输入的命令，并且送到内核。不仅如此，shell 还有自己的编程语言用于编辑命令，它允许用户编写由 shell 命令组成的程序。Shell 编程语言具有普通编程语言的很多特点，如它也有循环结构和分支控制结构等，用这种编程语言编写的 Shell 程序与其他应用程序具有同样的效果。

（3）实用工具

标准的 Linux 系统除系统核心程序外，都有一套名为实用工具的程序，如编辑器、浏览器、办公套件及其他系统管理工具等，用户可以自行编写需要的应用程序。

实用工具分为编辑器（用于编辑文件）、过滤器（用于接收数据并过滤数据）、交互程序（允许用户发送信息或接收来自其他用户的信息）三类。

2. Linux 版本

Linux 版本分为内核版本和发行版本。

（1）内核版本

内核是系统的心脏，其开发和规范一直由 Linus Torvalds 领导的开发小组控制，版本也是唯一的。Linux 内核使用三种不同的版本编号方式。

第一种方式用于 1.0 版本之前（包括 1.0）。第一个版本是 0.01，紧接着是 0.02、0.03、0.10、0.11、0.12、0.95、0.96、0.97、0.98、0.99 和之后的 1.0。

第二种方式用于 1.0 之后到 2.6，数字有三部分"A.B.C"，A 代表主版本号，B 代表次主版本号，C 代表修订号。只有在内核发生很大变化时 A 才变化。B 代表该版本是否稳定，偶数代表稳定版，奇数代表开发版。C 代表一些 bug 修复、安全更新、新特性和驱动的次数。以版本号 2.4.0 为例，2 代表主版本号，4 代表次版本号，0 代表修订号。在版本号中，序号的第 2 位为偶数的版本表明这是一个可以使用的稳定版本。而版本号 2.3.5 的第 2 位为奇数，表示该版本有一些新的东西加入，是一个不很稳定的测试版本。

第三种方式从 2004 年 2.6.0 版本开始，使用一种"time-based"的方式。3.0 版本之前是"A.B.C.D"的格式。七年里，前两个数字 A.B 即"2.6"保持不变，C 随着新版本的发布而增加，D 代表一些 bug 修复、安全更新、添加新特性和驱动的次数。3.0 版本之后是"A.B.C"格式，B 随着新版本的发布而增加，C 代表一些 bug 修复、安全更新、新特性和驱动的次数。同时，不再使用偶数代表稳定版、奇数代表测试版的命名方式。

(2) 发行版本

发行版本就是一些厂家或者社团将 Linux 系统内核、源代码及相关的应用程序组织构成一个完整的操作系统，从而让一般的用户可以简便地安装和使用 Linux。目前各种发行版本超过几百种，相对于内核版本而言，Linux 发行版本是由发行厂商自定义的，随发行厂商的不同而不同，与系统的内核版本没有直接关系。现在最流行的 Linux 发行版本有：Red Hat(红帽)、CentOS、Fedora、Debian、Ubuntu、Gentoo、红旗 Linux 等。

Red Hat Enterprise Linux(红帽企业版 Linux，RHEL)是目前最流行的 Linux 版本，具有极强的性能、稳定性、良好的用户界面和很好的扩展性，并且在全球范围内拥有完善的技术支持。Red Hat Enterprise Linux 的标志见图 1-3。

CentOS(社区企业操作系统)来自于 Red Hat Enterprise Linux，依照开放源代码规定释出的源代码所编译而成，去掉了很多收费的服务套件功能，不提供任何形式的技术支持，得到了广泛使用。CentOS 的标志见图 1-4。

图 1-3　Red Hat Enterprise Linux 的标志

图 1-4　CentOS 的标志

SuSE Linux 是来自德国 Novell 公司的发行版本，它吸取了 Red Hat Linux 的很多特质，性能稳定、用户界面友好。SuSE Linux 在欧洲较为流行，在我国国内也有较多应用。SuSE Linux 的标志见图 1-5。

1993 年 8 月 16 日，Debian 由美国普渡大学一名学生 Ian Murdock 首次发表，其安全性、稳定性强。Debian 不带有任何商业性质，背后也没有任何商业团体支持，因而它能够坚持其自由的风格。Debian 对 GNU 和 UNIX 精神的坚持，也获得开源社群的普遍支持。Debian 的标志见图 1-6。

图 1-5 SuSE Linux 的标志　　　　　　　　图 1-6 Debian 的标志

Ubuntu 是一个以桌面应用为主的 Linux 操作系统，基于 Debian GUN/Linux，是由全球化的专业开发团队（Canonical Ltd）打造的开源 GUN/Linux 操作系统，为桌面虚拟化提供支持平台。Ubuntu 的标志见图 1-7。

图 1-7 Ubuntu 的标志

◀ 任务 1.2　安装 Linux ▶

1.2.1　Linux 文件目录

在 Linux 系统中，并不像 Windows 系统，有人们熟知的 C 盘、D 盘等磁盘分区。Linux 系统中的一切文件都是从"根（/）"目录开始的，并按照文件系统层次标准（FHS）采用树形结构存放文件，如图 1-8 所示。

Linux 系统中的文件和目录名称是严格区分大小写的。例如，DEV、dev 代表不同的目录，并且文件名称中不能包括斜扛（/）。表 1-1 列出了 Linux 常用的目录名称以及对应内容。

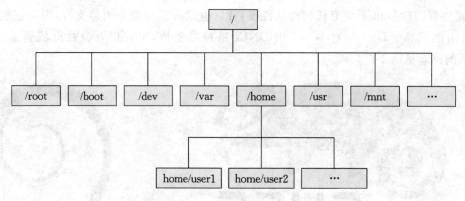

图 1-8　Linux 系统中的文件存储结构

表 1-1　Linux 系统常用的目录名称以及对应内容

目录名称	内容
/boot	保存与系统启动相关的文件,如内核文件和启动引导程序(grub)文件等
/dev	任何设备与接口都以文件形式存放在此目录
/etc	配置文件,如用户信息、服务的启动脚本、常用服务的配置文件等
/bin	存放系统命令,普通用户和 root 都可以执行,放在/bin 下的命令在单用户模式下也可以执行
/home	普通用户的主目录(也称为家目录),在创建用户时,每个用户要有一个默认登录和保存自己数据的位置,就是用户的主目录,所有普通用户的主目录是在/home/下建立一个和用户名相同的目录。如用户 user1 的主目录就是/home/user1
/lib	系统调用的函数库
/media	用于挂载设备目录,如软盘和光盘
/opt	放置第三方的软件
/root	系统管理员(root)的主目录。普通用户主目录在/home/下,root 主目录直接在"/"下
/sbin	开机过程中需要的命令
/srv	部分网络服务的数据文件目录
/tmp	临时目录。系统存放临时文件的目录,在该目录下,所有用户都可以访问和写入

在计算机上要找需要的文件或者目录,就需要知道它们所在的位置,而表示文件或目录位置的方式就是路径。表示路径的方法有绝对路径和相对路径。

① 绝对路径:以根目录为起点,完整地表示目标文件或目录的路径。

② 相对路径:以当前目录为起点,完整地表示目标文件或目录的路径。

1.2.2　Linux 物理设备

在 Linux 中一切都是文件,物理设备和硬件也不例外。既然是文件,就必须有文件名称。系统内核中的 udev 设备管理器会自动把硬件名称规范起来,目的是让用户通过设备文件的名字可以猜出设备大致的属性以及分区信息等,这对于陌生的设备来说特别方便。另外,udev 设备管理器的服务会一直以守护进程的形式运行并侦听内核发出的信号来管理/dev 目录下的设备文件。Linux 系统中常见的硬件设备及其文件名称如表 1-2 所示。

表 1-2 Linux 系统常见的硬件设备及其文件名称

硬件设备	文件名称
IDE 设备	/dev/hd[a-d]
SCSI/SATA/U 盘	/dev/sd[a-p]
软驱	/dev/fd[0-1]
打印机	/dev/lp[0-15]
光驱	/dev/cdrom
鼠标	/dev/mouse
磁带机	/dev/st0 或 /dev/ht0

现在的 IDE 设备较为少见,一般硬盘设备都是"/dev/sd"开头,一台主机上可以有多块硬盘,用 a-p 来表示 16 块硬盘(默认从 a 开始分配)。硬盘的分区编号则是:主分区或扩展分区的编号为 1~4;逻辑分区的编号从 5 开始。

值得注意的是,/dev/sda 设备编号为 a,并不表示主板上第一个插槽的存储设备,而是由系统内核的识别顺序来决定的,分区的数字编码不一定是强制顺延下来的,也有可能是手工指定的。

图 1-9 说明了/dev/sda3 这个设备文件包含的信息。

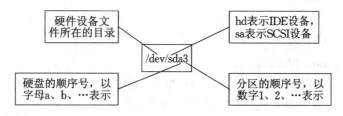

图 1-9 设备文件名称

图 1-10 给出了硬盘主分区、扩展分区、逻辑分区之间的关系。

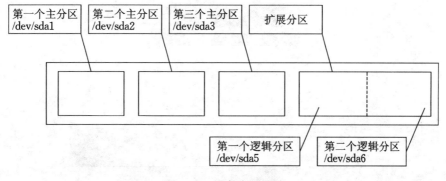

图 1-10 硬盘分区规划

1.2.3 规划分区

在安装 Linux 系统时,需要根据实际情况,进行硬盘分区,比较简单的分区方案就是为 Linux 提供 2 个分区,一个是根分区(/,根据 Linux 系统安装后占用资源的大小和实际需要保存的数据量来拟定该分区的容量大小),一个是交换分区(SWAP,类似于 Windows 系统的虚拟内存,将一部分硬盘空间虚拟成内存来使用,从而解决内存容量不足的问题,一般设置为与物理内存相同大小)。当系统安装成功后,也可以使用分区扩容的方法,动态调整分区大小,如使用 LVM(逻辑卷管理器)。

要进行分区,如 Windows 操作系统一样,需要进行硬盘的格式化操作,这就需要使用文件系统(操作系统中负责管理和存储文件信息的软件机构)。Linux 系统有几种常见的文件系统:

① Ext2:Linux 早期的发现版本中默认的文件系统类型。

② Ext3:一款日志文件系统,能够在系统出现意外掉电重启后,避免文件系统资料丢失,并能自动修复数据的不一致与错误。

③ Ext4:Ext3 的改进版本,作为 RHEL 6 系统中的默认文件管理系统,它支持的存储容量高达 1EB,且有无限多的子目录。另外,Ext4 文件系统能够批量分配 block 块,从而极大地提高了读写效率。

④ XFS:一种高性能的日志文件系统,是 RHEL 7 中默认的文件管理系统,擅长处理大文件,提供平滑的数据传输。在发生意外宕机后,可以快速地恢复可能被破坏的文件。

1.2.4 实现步骤

1. 配置 VM 虚拟机

(1) 安装 VMware Workstation。

(2) 在 VMware Workstation 管理界面,单击"创建新的虚拟机"选项(图 1-11),在"新建虚拟机向导"界面保持默认选项,单击"下一步"按钮(图 1-12)。

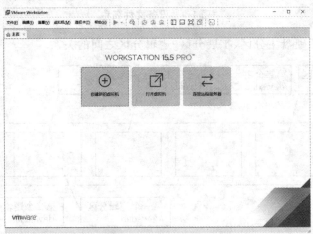

图 1-11 虚拟机管理界面

图 1-12　选择虚拟机配置类型

(3) 如图 1-13 所示，选择"稍后安装操作系统"，单击"下一步"按钮。

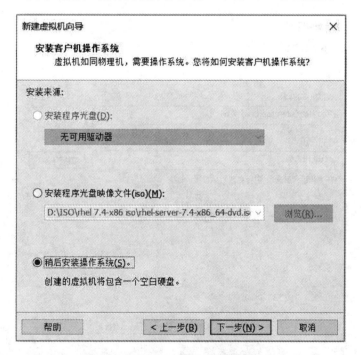

图 1-13　指定虚拟机操作系统的安装来源

(4) 如图 1-14 所示，选择客户端操作系统类型、版本，单击"下一步"按钮。

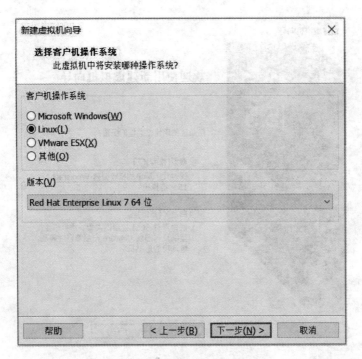

图 1-14 选择操作系统的版本

（5）如图 1-15 所示，输入虚拟机的名称，并选择安装位置后，单击"下一步"按钮。

图 1-15 命名虚拟机以及指定安装路径

（6）如图 1-16 所示，设置虚拟机系统的"最大磁盘大小"为 20 GB，单击"下一步"按钮。

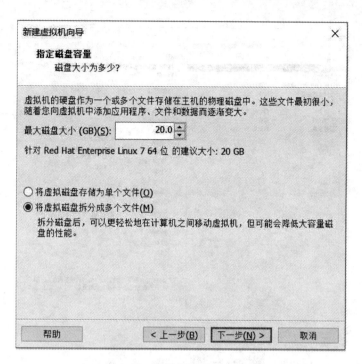

图 1-16 设置虚拟机磁盘大小

(7) 如图 1-17 所示，单击"自定义硬件"按钮。

图 1-17 虚拟机的配置界面

(8) 如图 1-18、图 1-19 所示，可以更改虚拟机系统的内存大小（虚拟机已根据真实计算机的情况，给出了内存分配的建议）和 CPU 处理器的相关配置。

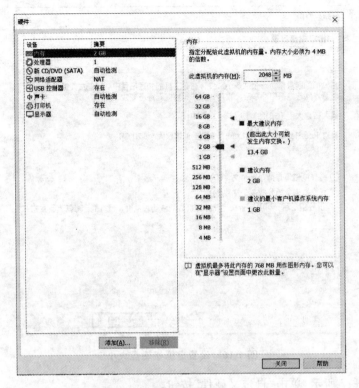

图 1-18 设置虚拟机的内存大小

图 1-19 设置虚拟机的 CPU 参数

（9）如图1-20所示，在"使用ISO映像文件"中选择下载好的RHEL系统镜像文件。

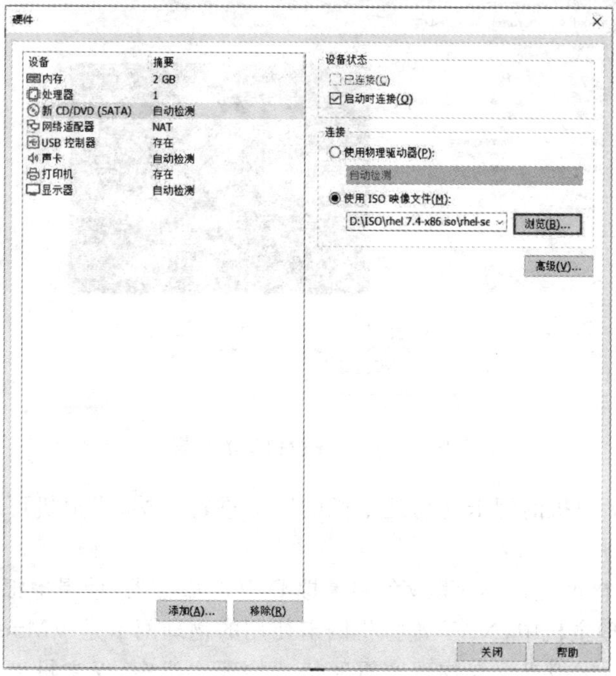

图1-20 设置虚拟机的光驱设备

（10）如图1-21所示，选择"网络连接"类型为"桥接模式"，完成配置，单击"关闭"按钮，返回到"新建虚拟机向导"界面，单击"完成"按钮，完成虚拟机配置，如图1-22所示。

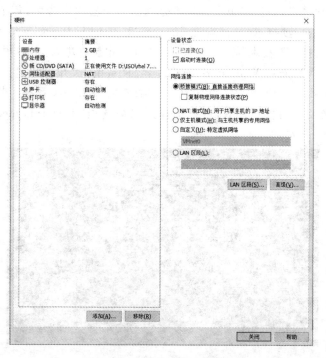

图1-21 设置虚拟机网络模式

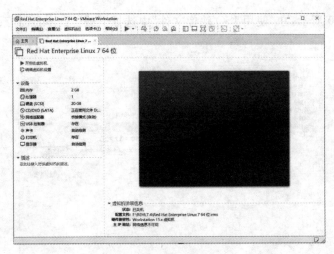

图 1-22 完成虚拟机的配置

① 桥接模式：虚拟机的网卡与物理主机的网卡进行交接，虚拟机可以通过物理主机的网卡访问外网。

② NAT 模式：借助虚拟 NAT 设备和虚拟 DHCP 服务器，使得虚拟机可以通过物理主机访问外网。在物理主机中，NAT 虚拟机网卡使用的物理网卡是 VMnet8。

③ 仅主机模式：虚拟机仅与物理主机通信。物理主机中，仅主机模式模拟网卡使用的物理网卡是 VMnet1。

2. 安装 Linux 系统

(1) 安装 RHEL7.4 或者 Centos7.4 时，计算机的 CPU 需要支持 VT（Virtualization Technology，虚拟化技术，是指将单台计算机分割出多个独立资源区，每个资源区均可以按照需要模拟出系统的一项技术）。目前购买的计算机，其 CPU 几乎都支持 VT，如果启动虚拟机后提示"CPU 不支持 VT 技术"等错误信息，需要重启电脑并进入 BIOS 中开启 VT 功能。

(2) 在虚拟机(VMware Workstation)管理界面中，单击"开启此虚拟机"，如图 1-23 所示，选择"Install Red Hat Enterprise Linux 7.4"直接安装 Linux 系统。若要先检查光盘是否完整有效再安装及启动救援模式，可以选择第二个选项"Test this media & install Red Hat Enterprise Linux 7.4"。

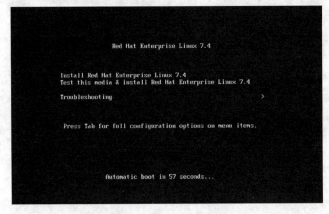

图 1-23 RHEL7 系统安装界面

项目 1　安装和使用 Linux 服务器

(3) 选择直接安装 Linux 系统后,需要等待系统加载安装镜像,如图 1-24 所示。

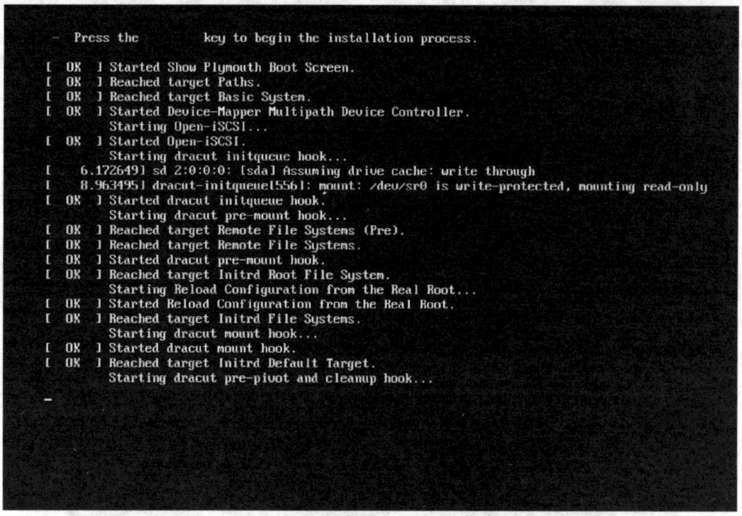

图 1-24　安装向导的初始化

(4) 如图 1-25 所示,选择系统的安装语言后,单击"继续"按钮。

图 1-25　设置系统的安装语言

(5) 如图 1-26 所示,"软件"界面的"软件选择",可以根据用户的需求来调整系统的基本环境,如要安装文件服务器、Web 服务器等,可以单击选中"带 GUI 的服务器"单选按钮,如图 1-27 所示。

(6) 如图 1-28 所示,单击"安装位置"选项后,可以选择"自动配置分区"或者"我要配置分区",这里默认选择"自动配置分区",如图 1-29 所示。回到安装主界面,单击"开始安装"按钮,进行系统安装,如图 1-30 所示。

图 1-26　安装系统界面

图 1-27　指定系统软件类型

图 1-28　配置系统的安装位置

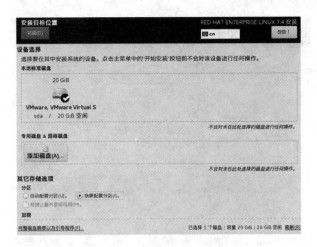

图 1-29 指定系统安装媒介

图 1-30 完成系统安装媒介的选择

(7) 在图 1-31 中，为 Linux 系统设置系统管理员密码，如图 1-32 所示。同时，也可以创建其他系统用户。

图 1-31 设置系统用户

图 1-32　设置 root 管理员的密码

（8）系统安装完毕后，单击"重启"按钮，计算机重启后，就可以看到系统的初始化界面，如图 1-33 所示，单击"LICENSE INFORMATION"选项，接受许可协议，单击"完成"按钮，如图 1-34 所示，回到"初始设置"主界面，单击"完成配置"按钮。

图 1-33　系统初始化界面

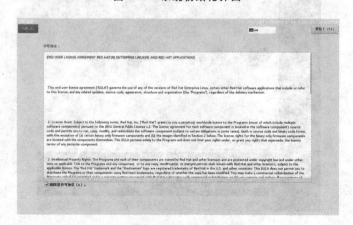

图 1-34　同意许可协议

(9) 在图 1-35 中,选择默认语言"汉语",单击"前进"按钮。

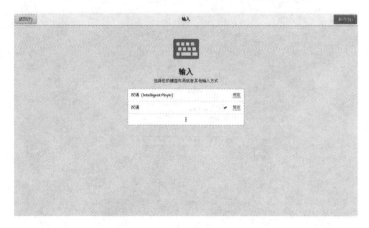

图 1-35　选择系统语言

(10) 在图 1-36 中,选择默认的键盘输入方式,单击"前进"按钮。

图 1-36　设置系统的输入方式

(11) 在图 1-37 中,选择系统时区(北京,中国),单击"前进"按钮。

图 1-37　设置系统时区

(12) 在图 1-38、图 1-39 中，为系统创建一个本地的普通用户(jack)，并设置密码，单击"前进"按钮，完成系统设置，如图 1-40 所示。

图 1-38　创建本地普通用户

图 1-39　设置本地普通用户的密码

图 1-40　系统初始化配置完成

（13）现在，就可以使用 Linux 系统了，如图 1-41 所示。

图 1-41 系统界面

任务 1.3 Linux 系统基本操作

1.3.1 命令行界面

1. 文本界面

Linux 系统可以通过文本界面或者图形界面登录。在图形界面下，输入正确的用户名和密码就可以进入系统。在文本界面下，输入正确的用户名和密码就可以进入提示符状态，如图 1-42 所示。在图形界面中，可以按组合键 Ctrl+Alt+F2，切换到文本界面。也可以使用命令进行切换，切换图形界面命令：systemctl set-default graphical.target，切换文本界面命令：systemctl set-default multi-user.target，输入命令后，重启系统。

需要注意：只有用 root 用户登录系统，才能具有系统管理员的权限，相当于 Windows 系统的 administrator 用户。

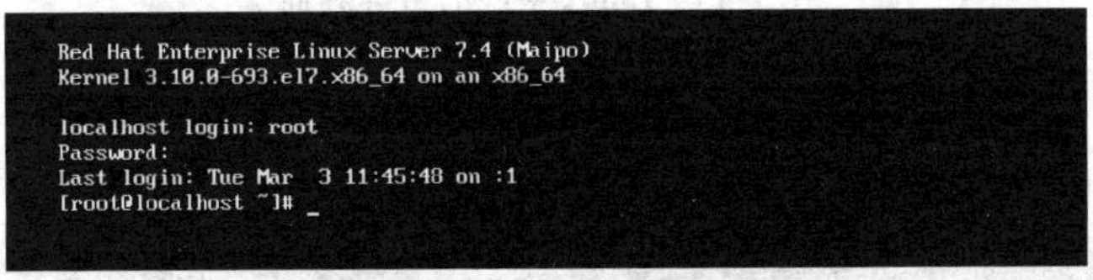

图 1-42 Linux 文本界面

如果是 root 用户登录 Linux 系统，在登录文本界面后，会出现"#"命令提示符；如果是用普通用户登录系统，则出现"$"命令提示符。如[root@localhost~]#的意思为：root 为当前用户名，localhost 为当前 Linux 主机名，~为当前目录名。可以在"#"命令提示符下，输入"shutdown-h now"，关闭系统，如图 1-43 所示，使用命令 su，切换普通用户 jack、系统用户 root 登录。

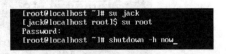

图 1-43 关闭系统

同时，可以使用命令 exit 来注销当前用户。

2. 虚拟控制台

虚拟控制台也称虚拟终端(tty)。Linux 系统允许一个用户多次登录,提供了虚拟控制台的方式,允许用户在同一时间从控制台进行多次登录。可以通过组合键 Alt+功能键 (F1～F6)来访问 6 个虚拟控制台。虚拟控制台使得用户同时在多个控制台上工作,完美体现了 Linux 系统多用户的特性。

3. 重置 root 系统用户密码

系统管理人员如果将操作系统的密码忘记了,对于 Windows 系统,可以通过第三方软件来初始化(或清除)系统密码,对于 Linux 系统,同样也可以进行密码重置。

(1) 在 Linux 系统桌面空白处,单击鼠标右键,单击"打开终端"命令,然后输入如图 1-44 所示命令,确认 Linux 系统是否是 RHEL7.4。

```
[root@localhost ~]# cat /etc/redhat-release
Red Hat Enterprise Linux Server release 7.4 (Maipo)
[root@localhost ~]#
```

图 1-44 查看 Linux 系统版本

(2) 重启系统,在出现引导界面时,按键盘 e 键进入内核编辑界面,如图 1-45 所示。

图 1-45 Linux 系统的引导界面

(3) 在 linux16 参数这行的最后添加"rd.break"参数,然后按下组合键 Ctrl+X,执行修改后的内核程序,如图 1-46 所示。

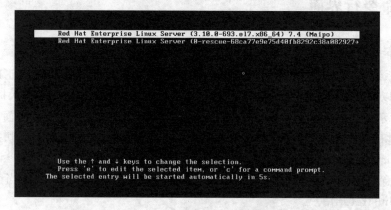

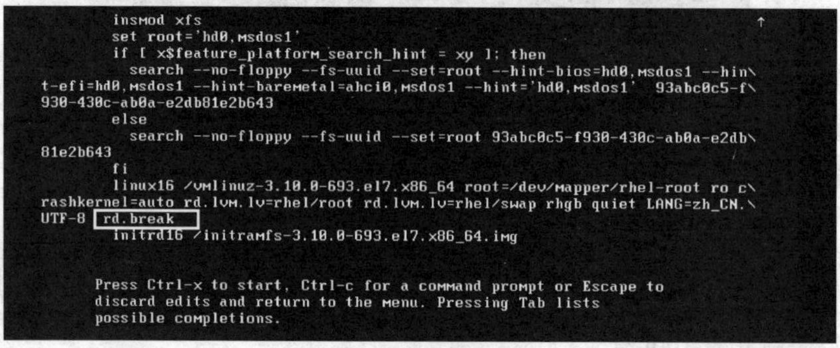

图 1-46 内核信息编辑界面

(4) 待系统进入救援模式后,依次输入如图 1-47 所示命令,系统重启后,就可以使用重置后的密码登录系统。注意,在输入 passwd 后,输入密码和确认密码是不显示的。

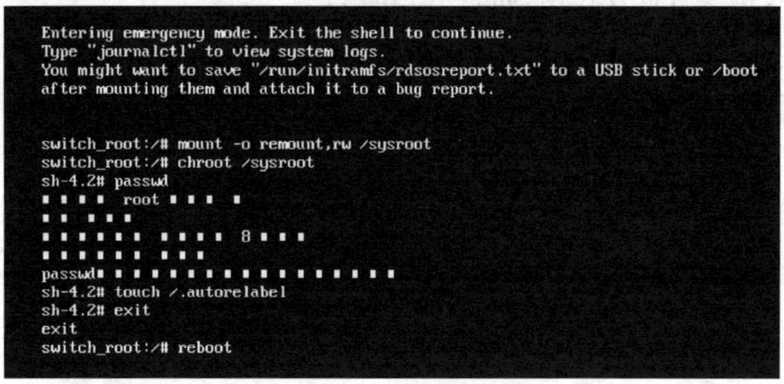

图 1-47 重置 Linux 系统的 root 管理员密码

命令如下:

```
mount -o remount,rw /sysroot    //让路径文件系统为读写模式,实现自由修改
chroot /sysroot                 //切换根目录
passwd                          //重置密码
touch /.autorelabel             //重启系统会自动标记 SELinux 的文件系统
exit
reboot
```

4. 命令行帮助

Linux 提供了丰富的命令,同时,为了帮助用户快速、便捷地使用这些目录,Linux 提供了 4 种帮助命令。

① 帮助选项:shell 命令使用 -help 的选项获得帮助,如 useradd-help(可以简写为-h)。
② man 命令:如使用 man systemctl 查看 systemctl 命令的相关使用方法。
③ info 命令:如使用 info ifconfig 查看 ifconfig 命令的相关使用方法。
④ help 命令:如使用 help cd 查看 cd 命令的相关使用方法。

1.3.2 系统命令

1. echo 命令

echo 命令用于在终端输出字符串或者变量提取后的值。其用法为:echo[字符串|$变量]。
例如执行下面的命令后,终端屏幕上显示信息为 xx。

```
[root@localhost ~]# echo xx
xx
```

2. date 命令

date 命令用于显示、设置系统的时间、日期。其用法为:date -[dsu] + [指定的格式]。
参数说明如下:
① -d:显示字符串所指的日期与时间,字符串前后必须加上双引号。

② -s：根据字符串来设置日期与时间，字符串前后必须加上双引号。
③ -u：显示 GMT。
日期格式字符串参数以及作用，如表 1-3 所示。

表 1-3　日期格式字符串参数以及作用

参数	作用
%H	小时（以 00-23 来表示）
%I	小时（以 01-12 来表示）
%M	分钟（以 00-59 来表示）
%m	月份（以 01-12 来表示）
%s	总秒数
%S	秒（以 00-59 表示）
%T	时间（含时、分、秒，以 24 小时制来表示）
%y	年份（以年份的最后两位数字来表示）
%Y	年份（以完整的数字来表示）
%b	月份（以 Jan-Dec 来表示）
%B	月份（以 January-December 来表示）
%d	日期（以 01-31 来表示）
%D	日期（以月/日/年来表示）
%j	今年中的第几天

例如使用 date 查看当前系统时间，命令如下：

[root@localhost~]# date
2020 年 03 月 04 日 星期三 20:09:10 CST

例如输出昨天的日期，命令如下：

[root@localhost~]# date -d "1 day ago" +"%Y-%m-%d"
2020-03-03

3. ps 命令

ps 命令用于查看系统的进程状态。其用法为：ps -[Aaux]。
参数说明如下：
① -A：所有的进程均显示出来，与-e 具有同样的效用。
② -a：显示所有进程，包括其他用户的进程。
③ -u：显示用户及其他信息。
④ -x：显示没有控制终端的进程，通常与 a 参数一起使用，可列出较完整信息。
例如，查看系统进程状态，如图 1-48 所示。

```
[root@localhost ~]# ps -aux
USER        PID %CPU %MEM    VSZ   RSS TTY      STAT START   TIME COMMAND
root          1  0.0  0.3 128164  6816 ?        Ss   19:55   0:01 /usr/lib/syst
root          2  0.0  0.0      0     0 ?        S    19:55   0:00 [kthreadd]
root          3  0.0  0.0      0     0 ?        S    19:55   0:00 [ksoftirqd/0]
root          5  0.0  0.0      0     0 ?        S<   19:55   0:00 [kworker/0:0H]
```

图 1-48　查看系统进程状态

图 1-48 中所输入信息的含义，如表 1-4 所示。

表 1-4 系统进程状态信息含义

名称	含义
USER	进程属于哪个用户账户
PID	进程的 ID
%CPU	进程占用 CPU 资源的百分比
%MEM	进程占用物理内存的百分比
VSZ	进程占用虚拟内存的大小，单位为 KB
RSS	进程占用实际物理内存的大小，单位为 KB
TTY	进程是在哪个终端运行的。其中，tty1～tty7 代表本地控制台终端（可以通过 Alt＋F1～F7 快捷键切换不同的终端），若与终端机无关，则显示"?"，若为 pts/0 等，则表示为由网络连接进主机的程序
STAT	该进程当前的状态。常见的状态有以下几种： (1) R:进程目前正在运行或在运行队列中等待 (2) S:进程目前正在睡眠当中,但可被某些信号唤醒 (3) T:进程目前正在侦测或者是停止了 (4) Z:进程已经终止,但是其父进程却无法正常终止,直到父进程调用 wait4()系统函数后,进程释放 (5) D:进程不可中断
START	进程被触发启动的时间
TIME	进程实际使用 CPU 运行的时间
COMMAND	进程的实际指令

4. ifconfig 命令

ifconfig 命令用于查看网卡配置、网络状态等信息，同时，也可以配置网卡信息。其语法为：ifconfig [网络设备] [参数]。

常见的参数说明如下：

① add:设置网络设备 IPv6 的 IP 地址。
② del:删除网络设备 IPv6 的 IP 地址。
③ down:关闭指定的网络设备。
④ up:启动指定的网络设备。
⑤ netmask:设置网络设备的子网掩码。

例如配置网卡 ens33 的 IP 地址为 192.168.2.1,并查看配置结果(注意,该配置是临时性的,当系统重启后,配置清除)。

```
[root@localhost ~]# ifconfig ens33 192.168.2.1 netmask 255.255.255.0
[root@localhost ~]# ifconfig
ens33: flags=4163<UP,BROADCAST,RUNNING,MULTICAST>  mtu 1500
       inet 192.168.2.1   netmask 255.255.255.0   broadcast 192.168.2.255
       inet6 fe80::20c:29ff:fec4:e7d1   prefixlen 64   scopeid 0x20<link>
```

```
        ether 00:0c:29:c4:e7:d1  txqueuelen 1000  (Ethernet)
        RX packets 954   bytes 63154 (61.6 KiB)
        RX errors 0  dropped 0  overruns 0  frame 0
        TX packets 49   bytes 9110 (8.8 KiB)
        TX errors 0  dropped 0 overruns 0   carrier 0  collisions 0

lo: flags=73<UP,LOOPBACK,RUNNING>  mtu 65536
        inet 127.0.0.1  netmask 255.0.0.0
        inet6 ::1  prefixlen 128   scopeid 0x10<host>
        loop  txqueuelen 1  (Local Loopback)
        RX packets 664   bytes 56360 (55.0 KiB)
        RX errors 0  dropped 0  overruns 0   frame 0
        TX packets 664   bytes 56360 (55.0 KiB)
        TX errors 0  dropped 0 overruns 0   carrier 0  collisions 0

virbr0: flags=4099<UP,BROADCAST,MULTICAST>  mtu 1500
        inet 192.168.122.1   netmask 255.255.255.0  broadcast 192.168.122.255
        ether 52:54:00:e4:73:dc  txqueuelen 1000  (Ethernet)
        RX packets 0   bytes 0 (0.0 B)
        RX errors 0  dropped 0  overruns 0   frame 0
        TX packets 0   bytes 0 (0.0 B)
        TX errors 0  dropped 0 overruns 0   carrier 0  collisions 0
```

5. mount 命令

mount 命令用来挂载文件夹、光盘等。其用法为：mount[选项][设备名][挂载点]。常用的选项及功能如表 1-5 所示。

表 1-5 mount 命令的选项及功能

选项	功能
-t vfstype	指定文件系统的类型，如 nfs、iso9660（光盘）等。通常不用指定，mount 会自动选择正确的类型
-o	设置挂载方式： loop：用来把一个文件当成硬盘分区挂接上系统； ro：采用只读方式挂接设备； rw：采用读写方式挂接设备
-a	挂载/etc/fstab 文件中记录的设备

例如，将/dev/cdrom 挂载到/media 之下：

```
[root@localhost ~]# mount /dev/cdrom /media/
```

6. df 命令

df 命令用来检查文件系统的分区、挂载、空间占用情况。其用法为：df -[ah]。
参数说明如下：

① -a：显示所有文件系统的磁盘使用情况。

② -h:以容易理解的格式输出文件系统大小。

df 命令的使用,如图 1-49 所示。

```
[root@localhost ~]# df -h
文件系统                   容量    已用    可用   已用%  挂载点
/dev/mapper/rhel-root       17G   3.2G    14G    19%   /
devtmpfs                  897M      0   897M     0%   /dev
tmpfs                     912M      0   912M     0%   /dev/shm
tmpfs                     912M   9.0M   903M     1%   /run
tmpfs                     912M      0   912M     0%   /sys/fs/cgroup
/dev/sda1                1014M   179M   836M    18%   /boot
tmpfs                     183M    24K   183M     1%   /run/user/0
/dev/sr0                   3.8G   3.8G      0   100%   /media
```

图 1-49 系统的分区、挂载、空间占用情况

1.3.3 导航命令

1. pwd 命令

pwd 命令用于显示当前目录的路径,如图 1-50 所示。

```
[root@localhost ~]# pwd
/root
```

图 1-50 显示当前目录的路径

2. cd 命令

cd 命令用于改变当前工作目录,它的用法与 DOS 下的 cd 目录基本一样。

cd..进入上一层目录;cd-进入上一个进入的目录;cd~进入用户的 home 目录。如图 1-51 所示。

```
[root@localhost ~]# cd /var
[root@localhost var]#
```

图 1-51 返回/var 目录

3. ls 命令

ls 命令用于浏览目录的内容。其用法为:ls -[al][目录]。

参数说明如下:

① -a:列出所有文件,包括隐藏文件(以"."开头)。

② -l:使用长格式显示文件详细信息,包括文件的属性、大小等。

如图 1-52 所示,查看/home 的内容。

```
[root@localhost ~]# ls -al /home/
总用量 4
drwxr-xr-x.   3 root root   18 3月   3 01:01 .
dr-xr-xr-x.  17 root root  245 3月   3 23:50 ..
drwx------.  14 jack jack 4096 3月   3 22:33 jack
```

图 1-52 查看/home 的内容

1.3.4 文件与目录操作命令

1. touch 命令

touch 命令用于创建空文件。其用法为:touch 文件名。

2. mkdir 命令

mkdir 命令用于创建一个目录或多个目录。其用法为:mkdir -[mp]目录名。

参数说明如下:

① -m:在创建目录时,设置权限模式。

② -p:创建目录和它的子目录。

例如,创建目录/123,并在该目录下创建文件 1.txt。

```
[root@localhost ~]# mkdir /123
[root@localhost ~]# cd /123
[root@localhost 123]# touch 1.txt
[root@localhost 123]# ls -al
总用量 0
drwxr-xr-x.  2 root root  19 3月   5 00:38 .
dr-xr-xr-x. 18 root root 256 3月   5 00:38 ..
-rw-r--r--.  1 root root   0 3月   5 00:38 1.txt
```

3. cp 命令

cp 命令用于复制文件或目录。其用法为:cp -[pdr] 源文件或目录 目标文件或目录。

参数说明如下:

① -p:保留原文件或目录的属性。

② -d:若对象为"链接文件",则保留该"链接文件"的属性。

③ -r:递归处理,将指定目录下的文件与子目录一并处理。

④ -f:覆盖已经存在的目标文件而不给出提示。

例如,复制目录/123 内的 1.txt 文件到/home 目录下。

```
[root@localhost ~]# cp /123/1.txt /home/
[root@localhost ~]# ls /home/
1.txt  jack
```

4. mv 命令

mv 命令用于移动文件或重命名文件。其用法为:mv -[bfiuv]源文件或目录 目标文件或目录。

参数说明如下:

① -b:若需覆盖文件,则覆盖前先行备份。

② -f:若目标文件或目录与现有的文件或目录重复,则直接覆盖现有的文件或目录。

③ -i:覆盖前先行询问用户。

④ -u:在移动或更改文件名时,若目标文件已存在,且其文件日期比源文件新,则不覆盖目标文件。

⑤ -v:执行时显示详细的信息。

例如复制目录/123 内的 1.txt 文件到/home 目录下,并将文件名称重命名为 2.txt。

> [root@localhost~]# mv /123/1.txt /home/2.txt

5. rm 命令

rm 命令用于删除文件或目录。其用法为:rm -[fir]目标文件或目录。
参数说明如下:
① -f:强制删除文件或目录。
② -i:删除文件或目录前,向用户提示是否删除,用户确认后,才进行删除。
③ -r:删除目录,将指定目录下的所有文件以及子目录全部删除。
例如,删除/home 目录下的文件 2.txt。

> [root@localhost~]# rm /home/2.txt
> rm:是否删除普通空文件 "/home/2.txt"? y

例如,强制删除/home 目录下的文件 1.txt。

> [root@localhost~]# rm -f /home/1.txt

6. rmdir 命令

rmdir 命令用于删除一个空目录。其用法为:rmdir 目录名。

1.3.5 文件查看命令

1. file 命令

file 命令可辨识文件和目录的类型。其用法为:file [-Lvz] 文件或目录。
参数说明如下:
① -L:查看软连接所对应的文件的文件类型。
② -v:直接显示版本信息。
③ -z:探测压缩后的文件类型。
例如,查看/home 目录的类型。

> [root@localhost~]# file /home/
> /home/: directory

2. cat 命令

cat 命令用于将文件的内容打印并输出到显示器或终端窗口。其用法为:cat 文件名。

3. head 命令

head 命令用于显示文件的前几行。其用法为:head -n 文件名。
例如查看/etc 目录的 cgconfig.conf 文件的前 3 行内容。

> [root@localhost~]# head -3 /etc/cgconfig.conf

4. less 与 more 命令

less 命令类似于 more 命令,用来按页显示文件,more 命令用于逐页阅读文本,但是 less 命令可以前后翻页,more 命令只能使用空格键向后翻页。
其用法为:less 文件名,more 文件名。

5. grep 命令

grep 命令用来在指定的文本文件中匹配字符串,输出匹配字符串所在行的全部内容。

其用法为：grep 关键字 文件名。

例如，查看/etc 目录的 cgconfig.conf 文件含有 2007 字符串的信息。

```
[root@localhost ~]# grep 2007 /etc/cgconfig.conf
# Copyright IBM Corporation. 2007
```

1.3.6 压缩、解压命令

1. zip 命令

zip 命令用于将当前目录下的所有文件和文件夹全部压缩成 zip 格式的文件。其用法为：zip [选项] 压缩文件名 源文件或源文件夹。

常用选项参数说明如下：

① -A：调整可执行的自动解压缩文件。
② -b＜工作目录＞：指定暂时存放文件的目录。
③ -c：替每个被压缩的文件加上注释。
④ -d：从压缩文件内删除指定的文件。
⑤ -D：压缩文件内不建立目录名称。
⑥ -e：压缩时指定加密。
⑦ -F：尝试修复已损坏的压缩文件。
⑧ -g：将文件压缩后附加在已有的压缩文件之后，而非另行建立新的压缩文件。
⑨ -r：递归处理，将指定目录下的所有文件和子目录一并处理。
⑩ -v：显示指令执行过程或显示版本信息。
⑪ -z：替压缩文件加上注释。

例如，将/home/web 目录下的所有文件和文件夹压缩成 web.zip 文件。

```
[root@localhost ~]# cd /home/web
[root@localhost web]# zip -r -q web.zip /home/web/
[root@localhost web]# ls -al
总用量 4
drwxr-xr-x. 2 root root   21 3月   5 02:20 .
drwxr-xr-x. 4 root root   29 3月   5 02:17 ..
-rw-r--r--. 1 root root  168 3月   5 02:20 web.zip
```

2. unzip 命令

unzip 命令用于解压 zip 格式的文件。其用法为：unzip [选项] 压缩文件名 目标文件或目标文件夹。

常用选项参数说明如下：

① -c：将解压缩的结果显示到屏幕上，并对字符做适当的转换。
② -f：更新现有的文件。
③ -l：显示压缩文件内所包含的文件。
④ -t：检查压缩文件是否正确。
⑤ -C：压缩文件中的文件名称区分大小写。
⑥ -o：不提示，直接覆盖原有文件。

⑦ -n:解压缩时不要覆盖原有的文件。
⑧ -P<密码>:使用 zip 的密码选项。
⑨ -d<目录>:指定文件解压缩后所要存储的目录。

例如,将/home/web 的 web.zip 文件解压到当前目录。

```
[root@localhost web]# unzip web.zip
Archive:  web.zip
 creating: home/web/
```

也可以指定将文件解压到指定的目录,例如,将/home/web 的 web.zip 文件解压到/home/jack 目录下。

```
[root@localhost web]# unzip web.zip -d /home/jack
Archive:  web.zip
 creating: /home/jack/home/web/
```

3. tar 命令

tar 命令用来为文件和目录创建档案,是 Linux 系统最常用的备份工具之一。tar 可用于建立、还原、查看、管理文件,也可方便地追加新文件到备份文件中,或仅更新部分备份文件,以及解压、删除指定的文件。其用法为:tar[主选项+辅选项]文件或目录。

主选项(必选项)参数说明如下:
① -c:创建压缩档案文件。
② -r:向压缩归档文件末尾追加文件。
③ -t:显示压缩文件的内容。
④ -u:更新原压缩包中的文件。
⑤ -x:解压。

辅选项(可选项)参数说明如下:
① -f:使用档案名字,注意,这个参数是最后一个参数,后面只能接档案名。
② -v:显示详细信息。
③ -j:支持 bzip2 解压文件。
④ -z:支持 gzip 解压文件。
⑤ -Z:支持 compress 解压文件。
⑥ -C :指定解压到的目录。

例如将/home/web 目录下的所有.txt 文件打包压缩成 mytxt.tar.gz 文件。

```
[root@localhost web]# tar -czvf mytxt.tar.gz ./*.txt
./1.txt
./2.txt
./3.txt
```

然后将 mytxt.tar.gz 文件解压到/home/jack 目录中。

```
[root@localhost web]# tar -xzvf mytxt.tar.gz -C /home/jack
./1.txt
./2.txt
./3.txt
```

例如将/home/photo 目录下的所有.jpg 文件打包压缩成 myjpg.tar.biz 文件。

```
[root@localhost photo]# tar -cjvf myjpg.tar.biz ./*.jpg
./1.jpg
./2.jpg
./3.jpg
```

然后将 myjpg.tar.biz 文件解压到/home/jack 目录中。

```
[root@localhost photo]# tar -xjvf myjpg.tar.biz -C /home/jack
./1.jpg
./2.jpg
./3.jpg
```

解压总结

- *.tar 用 tar -xvf 解压。
- *.gz 用 gzip -d 或者 gunzip 解压。
- *.tar.bz2 用 tar -xjf 解压。
- *.Z 用 uncompress 解压。
- *.tar.gz 和 *.tgz 用 tar -xzf 解压。
- *.bz2 用 bzip2 -d 或者用 bunzip2 解压。
- *.tar.Z 用 tar -xZf 解压。
- *.zip 用 unzip 解压。
- *.rar 用 unrar e 解压。

1.3.7 VIM 编辑器

VIM 是一个类似于 VI 的文本编辑器,且在 VI 基础上改进和增加了很多特性。它是一个基于 shell 的全屏幕文本编辑器,没有菜单,全部操作都基于命令,能够在任何 shell、字符终端、基于字符的网络连接中使用,能高效地在文件中进行编辑、删除、复制、查找、替换、移动等操作。

1. VIM 工作模式

VIM 有命令模式、输入模式、末行模式三种工作模式。三种工作模式之间的切换方式如图 1-53 所示。

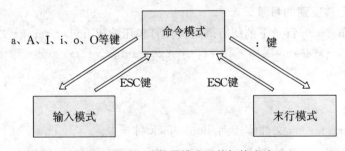

图 1-53　VIM 编辑器模式间的切换方法

在使用 VIM 编辑器时,默认是使用命令模式,如需要进行文本编辑,就要切换到输入模式,编辑完成后,要先回到命令模式,再切换到末行模式,才能进行文本的保存,以及退出 VIM 编辑器。

2. 常用的 VIM 命令

命令模式下的命令及其作用如表 1-6 所示。

表 1-6 命令模式下的命令及其作用

命令	作用
i	在当前光标位置的左侧添加文本
I	在当前行的开始处添加文本
a	在当前光标位置的右侧添加文本
A	在当前行的末尾位置添加文本
o	在当前行的下面新建一行
O	在当前行的上面新建一行
dd	删除(剪切)光标所在整行
3dd	删除(剪切)从光标处开始的 3 行
yy	复制光标所在整行
3yy	复制从光标处开始的 3 行
p	将之前删除(dd)或复制(yy)过的数据粘贴到光标后
/字符串	在文本中从上至下搜索该字符串
?字符串	在文本中从下至上搜索该字符串
:s/x/y	将当前光标所在行的第一个 x 替换成 y
:s/x/y/g	将当前光标所在行所有 x 替换成 y
:%s/x/y/g	将全文中的所有 x 替换成 y
n	显示搜索命令定位到的下一个字符串
N	显示搜索命令定位到的上一个字符串
u	撤销上一步的操作

末行模式下的命令及其作用如表 1-7 所示。

表 1-7 末行模式下的命令及其作用

命令	作用
:w	保存
:q	退出
:q!	强制退出(放弃对文本的修改内容)
:wq	保存并退出
:wq!	强制保存并退出
:set nu	显示行号
:set nonu	不显示行号
:命令	执行该命令
:整数	跳转到该行

例如，使用 VIM 编辑器编辑/home 目录下的 hello.txt 文件。

```
[root@localhost ~]# touch /home/hello.txt
[root@localhost ~]# vim /home/hello.txt
```

打开该文档后，按下"i"键，从"命令模式"（图 1-54）切换到"输入模式"（图 1-55）。

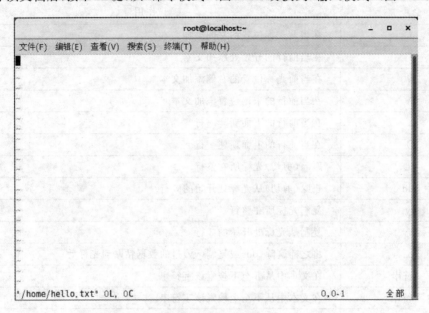

图 1-54　命令模式

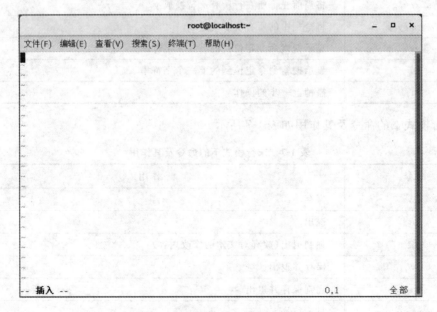

图 1-55　输入模式

在"输入模式"下，就可以编辑文本内容，如图 1-56 所示。

可以按下"ESC"键，从"输入模式"切换到"命令模式"，然后输入"yy"对文本进行复制，再按"p"键进行粘贴，如图 1-57 所示。

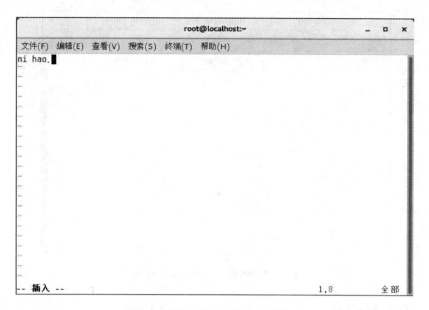

图 1-56　输入模式下输入文本内容

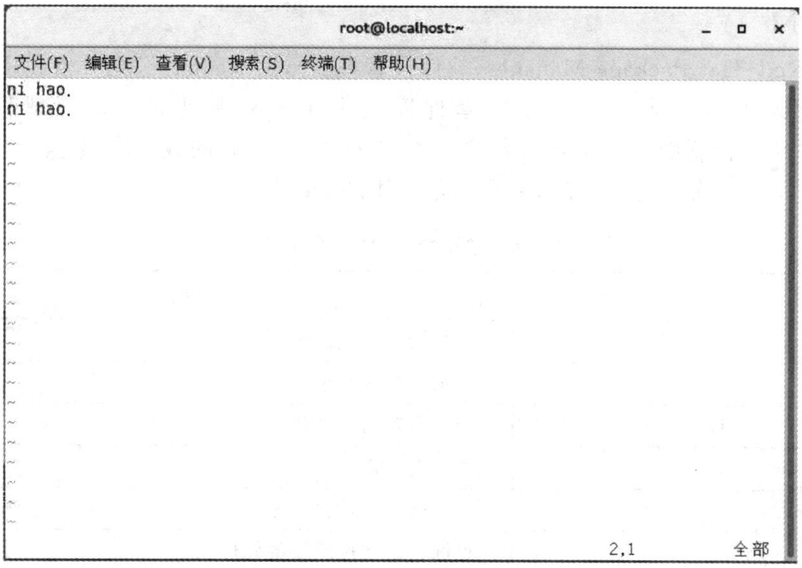

图 1-57　复制后进行粘贴

编辑完成后，按下"ESC"键，从"输入模式"切换到"命令模式"，然后输入":wq"切换到"末行模式"，保存并退出，如图 1-58 所示。

编辑完成后，通过 cat 命令查看该文件内容。

```
[root@localhost~]# cat /home/hello.txt
ni hao.-
ni hao.
```

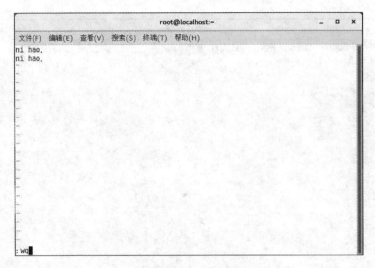

图 1-58 保存并退出文本编辑器

1.3.8 软件包安装

1. RPM

RPM(Red-Hat Package Manager,软件包管理器),由 Redhat 公司开发。RPM 是以一种数据库记录的方式来将所需要的套件安装在 Linux 主机的一套管理程序。就如 Windows 系统的控制面板,Linux 系统中存在一个关于 RPM 的数据库,它详细记录了安装的包与包之间的依赖相关性。表 1-8 列出了常用的 RPM 命令。

表 1-8 常用的 RPM 命令及功能

命令	功能
-i	安装指定的软件包
-v	显示指令执行过程
-U	升级指定的软件包
-h	显示安装进度
-a	显示所有软件包
-f	查询指定文件所属的软件包
-q	查询软件包
-l	显示软件包的文件列表
-e	删除指定的软件包
rpm -ivh 软件包名称	安装软件包的命令格式
rpm -uvh 软件包名称	升级软件包的命令格式
rpm -e 软件包名称	删除指定的软件包命令格式
rpm -qa | grep 软件包名称	查询已经安装的(包含关键字)软件包命令格式
rpm -ql 软件包名称	查询软件包安装的位置命令格式
rpm -qf 软件包名称	查询文件属于哪个 RPM 包的命令格式

例如,查询系统已经安装的所有 RPM 软件包,如图 1-59 所示(显示部分 RPM 软件包)。

```
root@localhost:~
文件(F) 编辑(E) 查看(V) 搜索(S) 终端(T) 帮助(H)
[root@localhost ~]# rpm -qa
xorg-x11-drv-qxl-0.1.5-3.el7.x86_64
iw-4.3-1.el7.x86_64
cups-filesystem-1.6.3-29.el7.noarch
freerdp-libs-1.0.2-10.el7.x86_64
iwl3160-firmware-22.0.7.0-56.el7.noarch
liberation-fonts-common-1.07.2-15.el7.noarch
xorg-x11-drv-void-1.4.1-2.el7.x86_64
ipset-libs-6.29-1.el7.x86_64
ibus-gtk3-1.5.3-13.el7.x86_64
sound-theme-freedesktop-0.8-3.el7.noarch
iwl6050-firmware-41.28.5.1-56.el7.noarch
poppler-data-0.4.6-3.el7.noarch
python-hwdata-1.7.3-4.el7.noarch
perl-Pod-Perldoc-3.20-4.el7.noarch
gnome-abrt-0.3.4-8.el7.x86_64
nss-tools-3.28.4-8.el7.x86_64
grub2-pc-modules-2.02-0.64.el7.noarch
oddjob-mkhomedir-0.31.5-4.el7.x86_64
perl-Time-HiRes-1.9725-3.el7.x86_64
gnome-shell-extension-user-theme-3.22.2-10.el7.noarch
libreport-2.1.11-38.el7.x86_64
tzdata-java-2017b-1.el7.noarch
```

图 1-59　查询已经安装的软件包

例如,在已安装的软件包中,查询包含关键字“hy”的软件包名称,如图 1-60 所示。

```
[root@localhost ~]# rpm -qa |grep hy
empathy-3.12.12-4.el7.x86_64
hyperv-daemons-license-0-0.30.20161211git.el7.noarch
hypervvssd-0-0.30.20161211git.el7.x86_64
telepathy-salut-0.8.1-6.el7.x86_64
telepathy-farstream-0.6.0-5.el7.x86_64
telepathy-haze-0.8.0-1.el7.x86_64
telepathy-glib-0.24.0-1.el7.x86_64
telepathy-mission-control-5.16.3-3.el7.x86_64
hypervfcopyd-0-0.30.20161211git.el7.x86_64
hyperv-daemons-0-0.30.20161211git.el7.x86_64
telepathy-gabble-0.18.1-4.el7.x86_64
hypervkvpd-0-0.30.20161211git.el7.x86_64
telepathy-filesystem-0.0.2-6.el7.noarch
telepathy-logger-0.8.0-5.el7.x86_64
hyphen-2.8.6-5.el7.x86_64
```

图 1-60　查询包含关键字“hy”的软件包

例如,安装软件包 telnet-0.17-64.el7.x86_64.rpm,如图 1-61 所示。注意,要先进入软件包所在的目录里面,否则会出现安装错误,如图 1-62 所示。

```
[root@localhost ~]# mount /dev/cdrom /media/
mount: /dev/sr0 写保护,将以只读方式挂载
[root@localhost ~]# cd /media/Packages/
[root@localhost Packages]# rpm -ivh telnet-0.17-64.el7.x86_64.rpm
警告:telnet-0.17-64.el7.x86_64.rpm: 头V3 RSA/SHA256 Signature, 密钥 ID fd431d51
: NOKEY
准备中...                          ################################# [100%]
        软件包 telnet-1:0.17-64.el7.x86 64 已经安装
```

图 1-61　安装软件包

```
[root@localhost ~]# rpm -ivh telnet-0.17-64.el7.x86_64.rpm
错误:打开 telnet-0.17-64.el7.x86_64.rpm 失败: 没有那个文件或目录
```

图 1-62　出现安装错误

例如,删除刚安装的软件包,注意,如果卸载正常,系统是没有任何提示信息的,如图 1-63所示。

```
[root@localhost Packages]# rpm -e telnet
[root@localhost Packages]# rpm -qa |grep telnet
```

图 1-63　卸载软件包

2. YUM

RPM 安装软件的一个最大弊端,就是软件包之间有一定的依赖性,如果一个软件与很多个软件包都有依赖关系,安装这个软件就会非常麻烦。YUM 可以自动解决依赖性问题,从而降低了软件安装的难度和复杂度,减轻了系统用户的负担。使用 YUM 安装软件时,需要至少一个 YUM 源,YUM 源就是存放很多 RPM 软件的目录。

(1) 配置 YUM 源

YUM 源定义文件存放在/etc/yum.repos.d/目录下,在 RHEL7.4 系统里,默认没有 YUM 源定义文件,用户可以自行定义任何可以使用的 YUM 源文件,该文件的扩展名为 repo。表 1-9 列出了 YUM 源文件格式及功能。

表 1-9　YUM 源文件格式及功能

选项	功能
[ ]	YUM 源唯一的 ID 号
name	YUM 源名称
baseurl	YUM 源的 URL 地址 (可以是 ftp:// 网络路径,http://网络路径,或 file:///本地路径)
mirrorlist	指定镜像站点目录
enabled	是否启用该 YUM 源(0 代表禁用,1 代表启用,默认为启用)
gpgcheck	安装软件时是否检查签名(0 代表禁用,1 代表启用)
gpqkey	如果检查软件包的签名,该选项为检查签名的密钥文件

例如,编辑一个 YUM 源文件 mydvd.repo,首先创建该文件。

```
[root@localhost ~]# cd /etc/yum.repos.d/
[root@localhost yum.repos.d]# touch mydvd.repo
[root@localhost yum.repos.d]# vim mydvd.repo
```

然后编辑该文件,文件内容如图 1-64 所示,这里使用本地路径作为 YUM 源的 URL 地址,需要注意的是,要记得先挂载光盘到/media 目录(mount /dev/cdrom /media)。

```
[mydvd]
name=mydvd
baseurl=file:///media
enabled=1
gpgcheck=0
```

图 1-64　配置 YUM 源的 URL 地址

(2) YUM 工具使用

YUM 工具主要有查询、安装、卸载、升级软件包等功能,这些功能都要使用 YUM 指令来实现,其用法为:yum[参数][命令][软件包名]。

参数说明如下:
① -y:安装过程中,若有提示选择,全部都选择"yes"。
② -q:不显示安装过程。

表 1-10 列出了常用的 YUM 命令。

表 1-10 常用的 YUM 命令及功能

命令	功能
yum repolist	列出所有仓库
yum list	列出仓库中所有软件包
yum info 软件包名称	查看软件包信息
yum install 软件包名称	安装软件包
yum remove 软件包名称	卸载软件包
yum reinstall 软件包名称	重新安装软件包
yum update 软件包名称	升级软件包
yum check-update	检查可更新的软件包
yum clean all	清除所有仓库缓存
yum groups list	检查系统中已安装的软件包组
yumgroupinfo 软件包组	查询指定的软件包组信息
yumgroupinstall 软件包组	安装指定的软件包组
yumgroupremove 软件包组	删除指定的软件包组

例如安装 dhcp 软件,如图 1-65 所示。

```
[root@localhost yum.repos.d]# yum -y install dhcp
已加载插件:langpacks, product-id, search-disabled-repos, subscription-manager
This system is not registered with an entitlement server. You can use subscription-manager to register.
mydvd                                                     | 4.1 kB   00:00
(1/2): mydvd/group_gz                                     | 137 kB   00:00
(2/2): mydvd/primary_db                                   | 4.0 MB   00:00
正在解决依赖关系
--> 正在检查事务
---> 软件包 dhcp.x86_64.12.4.2.5-58.el7 将被 安装
--> 解决依赖关系完成

依赖关系解决

================================================================
 Package      架构         版本                 源         大小
================================================================
正在安装:
 dhcp         x86_64       12:4.2.5-58.el7      mydvd      513 k

事务概要
================================================================
安装  1 软件包

总下载量:513 k
安装大小:1.4 M
Downloading packages:
Running transaction check
Running transaction test
Transaction test succeeded
Running transaction
  正在安装    : 12:dhcp-4.2.5-58.el7.x86_64                    1/1
mydvd/productid                                   | 1.6 kB  00:00:00
  验证中      : 12:dhcp-4.2.5-58.el7.x86_64                    1/1

已安装:
  dhcp.x86_64 12:4.2.5-58.el7

完毕!
```

图 1-65 安装 dhcp 软件

1.3.9 管理系统服务

在 RHEL6 系统中使用 service、chkconfig 等命令来管理系统服务,在 RHEL7 系统中则使用 systemctl 命令来管理系统服务,它们的命令对比如表 1-11、表 1-12 所示。

表 1-11 启动、重启、停止、重新加载、查看系统服务

RHEL 6	RHEL7	功能
service [服务] start	systemctl start [服务]	启动服务
service [服务] restart	systemctl restart [服务]	重启服务
service [服务] stop	systemctl stop [服务]	停止服务
service [服务] reload	systemctl reload [服务]	重新加载服务
service [服务] status	systemctl status [服务]	查看服务状态

表 1-12 系统服务开机启动或不启动、查看各级别下服务状态

RHEL 6	RHEL7	功能
chkconfig [服务] on	systemctl enable [服务]	开机自动启动服务
chkconfig [服务] off	systemctl disable [服务]	开机不自动启动服务
chkconfig [服务]	systemctl is-enabled [服务]	查看服务是否开机自动启动
chkconfig --list	systemctl list-unit-files --type=service	查看各级别下服务状态

自我测试

一、填空题

1. Linux 是当前流行的一种操作系统,它是一个类似_____的操作系统。
2. Linux 系统结构一般有 3 个主要部分:_____、_____、_____。
3. Linux 版本分为_____和_____。
4. Linux 系统,可以通过_____或者_____登录。
5. ifconfig 命令用于查看_____、_____等信息,同时,也可以配置网卡信息。

二、简答题

1. 简述 Linux 系统的特点。
2. 简述 Linux 常见的文件系统。
3. 简述命令 date、pwd、touch、file 的功能。
4. 简述 RPM 与 YUM 软件仓库的作用。

三、实训题

1. 使用虚拟机安装 Red Hat Enterprise Linux 7.4,并配置系统的 IPv4。
2. 配置 YUM 软件仓库,安装 dhcp 软件。

项目 2　用户管理与文件权限

📚 项目综述

根据公司的部门划分，开发部有三个项目组，分别为设计组、实施组、运维组，都需要通过网络远程使用服务器。为此，公司的网络管理员需要使用 Linux 系统给三个项目组分配用户账户，以及设置文件权限，以满足开发部员工的需求。

📚 项目目标

- 了解用户、用户组基本知识；
- 了解文件、目录权限基本知识；
- 掌握用户、用户组的管理方法；
- 掌握文件、目录的属性及其权限的设置方法。

◀ 任务 2.1　用户与用户组 ▶

2.1.1　用户与用户组的概念

Linux 作为多用户、多任务的操作系统，其所提供的账户类型主要有用户账户和组账户。每个用户在使用系统前，都必须有一个账户，用户通过该账户登录到系统，访问其资源，每一个账户都有一个唯一的身份号码（UID），在 Linux 系统中，分为三种用户账户：

① 超级用户：系统的管理员用户，拥有系统的最高管理权限，默认是 root 用户，其 UID 为 0。

② 系统用户：系统安装后默认就会存在，主要是便于系统管理，默认情况下是不能登录系统的。由系统创建的系统用户的 UID 为 0~200，由用户创建的系统用户的 UID 为 201~999。

③ 普通用户：具有登录系统的权限，只能对自己目录下的文件进行访问和修改，其 UID 从 1000 开始。

为了有效、方便地管理具有相同权限的用户，就需要使用到用户组，每一个用户组都有一个唯一的用户组号码（GID）。普通用户组 GID 为 1000~60000。由操作系统创建的系统用户组 GID 为 0~200，由用户创建的系统用户组 GID 为 201~999。

用户和用户组的对应关系可以是：一对一、一对多、多对一和多对多。Linux 系统中，每一个用户至少要属于一个用户组,可以通过设置用户组的权限来指定用户的权限。在使用 useradd 命令创建用户时,系统默认会自动创建一个同名的用户组,作为用户的用户组,同时,会在/home 目录下创建一个同名的目录,作为该用户的主目录。如果一个用户属于多个用户组,那么该用户的主组记录在/etc/passwd 文件中,其他组为该用户的附属组。

注意:每个用户有且只有一个主组,但是其附加组可以有多个或者没有。

2.1.2 用户管理系统文件

在 Linux 系统中,与用户管理有关的系统文件有:/etc/passwd、/etc/shadow、/etc/group、/etc/gshadow。

1. 用户的配置文件/etc/passwd

系统中的每一个合法用户对应该文件中的每一行记录,有多少行就表示多少个用户。在该文件中,每一行都由七个字段构成,各字段之间通过":"号分隔,如图 2-1 所示。

```
[root@localhost ~]# cat /etc/passwd
root:x:0:0:root:/root:/bin/bash
bin:x:1:1:bin:/bin:/sbin/nologin
daemon:x:2:2:daemon:/sbin:/sbin/nologin
adm:x:3:4:adm:/var/adm:/sbin/nologin
lp:x:4:7:lp:/var/spool/lpd:/sbin/nologin
sync:x:5:0:sync:/sbin:/bin/sync
shutdown:x:6:0:shutdown:/sbin:/sbin/shutdown
halt:x:7:0:halt:/sbin:/sbin/halt
mail:x:8:12:mail:/var/spool/mail:/sbin/nologin
operator:x:11:0:operator:/root:/sbin/nologin
games:x:12:100:games:/usr/games:/sbin/nologin
ftp:x:14:50:FTP User:/var/ftp:/sbin/nologin
nobody:x:99:99:Nobody:/:/sbin/nologin
```

图 2-1 查看 passwd 文件

passwd 文件的参数含义如表 2-1 所示。

表 2-1 passwd 文件的参数含义

名称	含义
第一个字段	用户登录时使用的账户名,在系统中是唯一的
第二个字段	存放加密的口令,x 表示该用户的密码已被映射到/etc/shadow 文件中
第三个字段	系统内部用它来标识用户(UID)
第四个字段	系统内部用它来标识用户所属的组(GID)
第五个字段	用户说明,对这个账户的说明
第六个字段	用户组目录(家目录),用户登录系统后进入的目录
第七个字段	当前用户登录后所使用的 shell。在 RHEL 系统中,默认的 shell 是 bash;如果不希望用户登录系统,可以通过 usermod 命令,或者修改 passwd 文件将该字段设置为/sbin/nologin,大多数内置系统账户都设为/sbin/nologin,表示禁止登录系统

2. 用户密码配置文件/etc/shadow

/etc/shadow 文件是/etc/passwd 的影子文件,但它并不是由/etc/passwd 产生的,几乎所有的 Linux 系统都对密码进行了加密,密文就保存在/etc/shadow 文件中,只有 root 用户才具有读取权限。在该文件中,每一行都由九个字段构成,各字段之间通过":"号分隔,如图 2-2 所示。

```
[root@localhost ~]# cat /etc/shadow
root:$6$qFW3TT2d$In3i1eHKYmpq6WbE.mPUmlwSlrHDfZKtKrqA4lMxqZiWLq7R2GFilPhEA5e8m5GMq
rH1pVzBs6qkWboxaOLjK1:18324:0:99999:7:::
bin:*:16925:0:99999:7:::
daemon:*:16925:0:99999:7:::
adm:*:16925:0:99999:7:::
lp:*:16925:0:99999:7:::
sync:*:16925:0:99999:7:::
shutdown:*:16925:0:99999:7:::
halt:*:16925:0:99999:7:::
mail:*:16925:0:99999:7:::
operator:*:16925:0:99999:7:::
games:*:16925:0:99999:7:::
ftp:*:16925:0:99999:7:::
```

图 2-2 查看 shadow 文件

shadow 文件的参数含义如表 2-2 所示。

表 2-2 shadow 文件的参数含义

名称	含义
第一个字段	用户账户名
第二个字段	用户的加密密码
第三个字段	最后一次修改时间,从 1970 年 1 月 1 日到用户最后一次更改密码的天数
第四个字段	最小时间间隔,从 1970 年 1 月 1 日到用户可以更改密码的天数
第五个字段	最大时间间隔,从 1970 年 1 月 1 日到必须更改密码的天数,否则密码将过期
第六个字段	警告时间,在密码过期之前多少天提醒用户更新密码,默认值为 7 天
第七个字段	不活动时间,在用户密码过期之后到禁用账号的天数
第八个字段	失效时间,从 1970 年 1 月 1 日起,到账号被禁用的天数
第九个字段	保留位

3. 用户组配置文件/etc/group

/etc/group 文件存放了使用系统用户组信息,在该文件中,每一行都由四个字段构成,各字段之间通过":"号分隔,如图 2-3 所示。

```
[root@localhost ~]# cat /etc/group
root:x:0:
bin:x:1:
daemon:x:2:
sys:x:3:
adm:x:4:
```

图 2-3 查看 group 文件

group 文件的参数含义如表 2-3 所示。

表 2-3 group 文件的参数含义

名称	含义
第一个字段	组名
第二个字段	用户组的口令，用 x 表示口令是被/etc/gshadow 文件保护的
第三个字段	用户组号码(GID)
第四个字段	该组的成员，成员之间使用","隔开

4. 用户组密码配置文件/etc/gshadow

/etc/gshadow 文件用于存放加密的用户组的口令、组管理员等信息，该文件只有 root 用户可读。在该文件中，每一行都由四个字段构成，各字段之间通过":"号分隔，如图 2-4 所示。

```
[root@localhost ~]# cat /etc/gshadow
root:::
bin:::
daemon:::
sys:::
adm:::
tty:::
disk:::
lp:::
mem:::
kmem:::
wheel:::jack
cdrom:::
mail:::postfix
```

图 2-4 查看 gshadow 文件

gshadow 文件的参数含义如表 2-4 所示。

表 2-4 gshadow 文件的参数含义

名称	含义
第一个字段	组名
第二个字段	用户组的口令，保存已加密的口令
第三个字段	用户组管理员，管理员有权对该组添加、删除账号
第四个字段	该组的成员，成员之间使用","隔开

2.1.3 用户管理命令

1. useradd 命令

useradd 命令用于创建新的用户，其用法为：useradd [选项] 用户名。

useradd 命令常用的选项及功能如表 2-5 所示。

表 2-5 useradd 命令常用的选项及功能

选项	功能
-d	指定用户主目录
-g	指定用户组（主组）
-G	指定用户一个或多个附加组
-m	若主目录不存在则自动创建
-M	不创建主目录
-s	指定登录时所用的 shell 类型，默认是 /bin/bash
-u	手动创建新用户的 UID
-e	指定账号的到期时间

例如创建用户 lucy，由于没有指定该用户的用户组和主目录，系统会自动创建同名的用户组以及在 /home 目录下创建同名的目录，作为该用户的主目录，如图 2-5 所示。

```
[root@localhost ~]# useradd lucy
[root@localhost ~]# ls /home/
hello.txt  jack  lucy  photo  web
[root@localhost ~]# groups lucy
lucy : lucy
[root@localhost ~]# tail -l /etc/passwd
gdm:x:42:42::/var/lib/gdm:/sbin/nologin
gnome-initial-setup:x:992:987::/run/gnome-initial-setup/:/sbin/nologin
sshd:x:74:74:Privilege-separated SSH:/var/empty/sshd:/sbin/nologin
avahi:x:70:70:Avahi mDNS/DNS-SD Stack:/var/run/avahi-daemon:/sbin/nologin
postfix:x:89:89::/var/spool/postfix:/sbin/nologin
ntp:x:38:38::/etc/ntp:/sbin/nologin
tcpdump:x:72:72::/:/sbin/nologin
jack:x:1000:1000:jack:/home/jack:/bin/bash
dhcpd:x:177:177:DHCP server:/:/sbin/nologin
lucy:x:1001:1001::/home/lucy:/bin/bash
```

图 2-5 创建用户 lucy

例如创建用户 tom，作为 root 组的成员，设置主目录为 /home/mytom，并且该用户不能登录系统，如图 2-6 所示。

```
[root@localhost ~]# useradd -d /home/mytom -g root -s /sbin/nologin tom
[root@localhost ~]# tail -l /etc/passwd
gnome-initial-setup:x:992:987::/run/gnome-initial-setup/:/sbin/nologin
sshd:x:74:74:Privilege-separated SSH:/var/empty/sshd:/sbin/nologin
avahi:x:70:70:Avahi mDNS/DNS-SD Stack:/var/run/avahi-daemon:/sbin/nologin
postfix:x:89:89::/var/spool/postfix:/sbin/nologin
ntp:x:38:38::/etc/ntp:/sbin/nologin
tcpdump:x:72:72::/:/sbin/nologin
jack:x:1000:1000:jack:/home/jack:/bin/bash
dhcpd:x:177:177:DHCP server:/:/sbin/nologin
lucy:x:1001:1001::/home/lucy:/bin/bash
tom:x:1002:0::/home/mytom:/sbin/nologin
```

图 2-6 创建用户 tom

2. usermod 命令

usermod 命令用于修改用户的属性,其用法为:usermod[选项]用户名。

usermod 命令常用的选项及功能如表 2-6 所示。

表 2-6 usermod 命令常用的选项及功能

选项	功能
-l	修改用户名
-c	修改用户备注信息
-d	修改用户主目录
-L	锁定账号,临时禁止用户登录
-U	对账号解锁
-g	修改用户所属组
-G	修改用户所属的附属组
-u	修改用户的 UID
-e	修改用户有效期
-f	修改用户密码在多少天后过期
-s	修改用户登录后所使用的 shell 类型

例如,修改用户 tom 的主目录为/var/mytom,如图 2-7 所示。

```
[root@localhost ~]# usermod -d /var/mytom tom
[root@localhost ~]# tail -l /etc/passwd
gnome-initial-setup:x:992:987::/run/gnome-initial-setup/:/sbin/nologin
sshd:x:74:74:Privilege-separated SSH:/var/empty/sshd:/sbin/nologin
avahi:x:70:70:Avahi mDNS/DNS-SD Stack:/var/run/avahi-daemon:/sbin/nologin
postfix:x:89:89::/var/spool/postfix:/sbin/nologin
ntp:x:38:38::/etc/ntp:/sbin/nologin
tcpdump:x:72:72::/:/sbin/nologin
jack:x:1000:1000:jack:/home/jack:/bin/bash
dhcpd:x:177:177:DHCP server:/:/sbin/nologin
lucy:x:1001:1001::/home/lucy:/bin/bash
tom:x:1002:0::/var/mytom:/sbin/nologin
```

图 2-7 修改用户 tom 的主目录

3. userdel 命令

userdel 命令用于删除用户,其用法为:useradd[-rf]用户名。

参数说明如下:

① -r:同时删除用户及用户主目录。

② -f:强制删除用户。

例如,删除用户 tom 及其主目录。

```
[root@localhost~]# usedel-r tom
```

4. passwd

passwd 命令用于修改用户密码、过期时间等信息,其用法为:passwd[选项]用户名。

passwd 命令常用的选项及功能如表 2-7 所示。

表 2-7 passwd 命令常用的选项及功能

选项	功能
-d	删除密码，使用户可以以空密码登录系统
-u	解除锁定，允许用户登录
-l	锁定用户，禁止用户登录
--stdin	允许用户通过标准输入修改密码
-e	强制用户在下次登录系统时修改密码
-S	显示用户的密码信息

例如，修改用户 lucy 的密码为"123456"。

```
[root@localhost ~]# passwd lucy
更改用户 lucy 的密码。
新的密码：
无效的密码：密码少于 8 个字符
重新输入新的密码：
passwd：所有的身份验证令牌已经成功更新。
```

也可以使用下面这样的方式，将 lucy 用户的密码修改为"123456789"。

```
[root@localhost ~]# echo "123456789" | passwd --stdin lucy
更改用户 lucy 的密码。
passwd：所有的身份验证令牌已经成功更新。
```

2.1.4 用户组管理命令

1. groupadd 命令

groupadd 命令用于创建用户组，其用法为：groupadd［选项］用户组名。
groupadd 命令常用的选项及功能如表 2-8 所示。

表 2-8 groupadd 命令常用的选项及功能

选项	功能
-r	指定新建用户组的 GID
-g	创建系统用户组

例如，创建一个用户组 teacher。

```
[root@localhost ~]# groupadd teacher
```

2. groupmod 命令

groupmod 命令用于修改用户组的属性，其用法为：groupmod［选项］用户组名。
groupmod 命令常用的选项及功能如表 2-9 所示。

表 2-9 groupmod 命令常用的选项及功能

选项	功能
-n	修改用户组名
-g	修改用户组的 GID

例如，修改用户组 teacher 的名称为 tea。

```
[root@localhost ~]# groupmod -n tea teacher
```

3. groupdel 命令

groupdel 命令用于删除用户组，其用法为：groupdel 用户组名。

例如，删除用户组 tea。

```
[root@localhost ~]# groupdel tea
```

4. gpasswd 命令

gpasswd 命令用于把用户添加到组，把用户设置为组管理员，把用户从组中删除，其用法为：gpasswd [选项] 用户名 用户组名。

gpasswd 命令常用的选项及功能如表 2-10 所示。

表 2-10 gpasswd 命令常用的选项及功能

选项	功能
-a	添加用户到组
-d	将用户从组中删除
-A	设置用户为组管理员

例如，将用户 jack 添加到 lucy 组。

```
[root@localhost ~]# gpasswd -a jack lucy
正在将用户"jack"加入"lucy"组中
```

下面创建公司开发部的三个项目组（设计组、实施组、运维组），并创建各项目组的账户。

```
[root@localhost ~]# groupadd shejizu
[root@localhost ~]# useradd -g shejizu shejizuuser1
[root@localhost ~]# echo "123456" | passwd --stdin shejizuuser1
[root@localhost ~]# groupadd shishizu
[root@localhost ~]# useradd -g shishizu shishizuuser1
[root@localhost ~]# echo "123456" | passwd --stdin shishizuuser1
[root@localhost ~]# groupadd yunweizu
[root@localhost ~]# useradd -g yunweizu yunweizuuser1
[root@localhost ~]# echo "123456" | passwd --stdin yunweizuuser1
```

2.1.5 查看用户信息命令

id 命令可以用于显示用户的 UID、GID 及所属群组的组列表，其用法为：id [选项] 用户名。

id 命令常用的选项及功能如表 2-11 所示。

表 2-11　id 命令常用的选项及功能

选项	功能
-g	显示用户所属群组的 ID
-G	显示用户所属附加群组的 ID
-u	显示用户 ID

例如,查看用户 jack 的 ID 信息。

```
[root@localhost~]# id jack
uid=1000(jack) gid=1000(jack)组=1000(jack),10(wheel),1001(lucy)
```

可以看到用户 jack 的 UID 为 1000,GID 为 1000,组名为 jack,并且它还属于 whell 组、lucy 组。

2.1.6　查看当前登录用户命令

whoami 命令用于查看当前登录者的信息。

例如,查看当前登录者的信息。

```
[root@localhost~]# whoami
root
```

任务 2.2　文件权限

2.2.1　查看文件权限信息

在 Linux 系统中,使用 ls 命令加上参数"-l"可以查看当前目录下的文件或目录的详细信息,如图 2-8 所示。

```
[root@localhost ~]# ls -l /home/
总用量 12
-rw-r--r--.  1 root          root          16 3月   5 16:09 hello.txt
drwx------. 14 jack          jack        4096 3月   5 14:09 jack
drwx------. 14 lucy          lucy        4096 3月   8 01:21 lucy
drwx------.  3 shejizuuser1  root          78 3月   7 16:40 mytom
drwxr-xr-x.  2 root          root          66 3月   5 14:08 photo
drwx------.  3 shejizuuser1  shejizu       78 3月   8 15:50 shejizuuser1
drwx------.  3 shishizuuser1 shishizu      78 3月   8 15:51 shishizuuser1
drwxr-xr-x.  2 root          root          65 3月   5 03:02 web
drwx------.  3 yunweizuuser1 yunweizu      78 3月   8 15:52 yunweizuuser1
```

图 2-8　查看/home 目录下文件(目录)详细信息

这些详细信息,从左到右的含义,如表 2-12 所示。

表 2-12　信息参数的含义

名称	含义
文件类型	d:目录文件；l:链接文件；-:普通文件；b:块设备文件；c:字符设备文件；p:管道文件
文件权限	每 3 个字符为一组，左边 3 个字符代表所有者权限，中间 3 个字符代表与所有者同一组用户的权限，右边 3 个字符代表其他用户的权限
链接数	文件的链接数
所有者	文件或目录的所有者
所属组	文件或目录的所属组
容量大小	文件或目录的大小
日期和时间	文件或目录最后修改的时间
名称	文件或目录的名称

Linux 系统的软连接，相当于 Windows 系统中的快捷方式，使用命令 ln 来创建。其用法为：ln -s 源文件或目录 目标文件或目录。

文件和目录的读、写、执行权限，使用 r、w、x 来表示，若需要去除对应的权限，则使用"-"表示，也可以使用数字 4、2、1 来表示读、写、执行权限。表 2-13 列出了三种权限的含义。

表 2-13　文件与目录的权限含义

权限	文件	目录
r	查看文件内容	查看目录内容（显示子目录、文件列表）
w	修改文件内容	修改目录内容（在目录中创建、移动、删除文件或子目录）
x	执行该文件	执行 cd 命令，加入或退出该目录

表 2-14 给出两种权限表示方式的比较。

表 2-14　权限的两种表示方式

权限	字母	数字
只读	r--	4
只写	-w-	2
只执行	--x	1
读和写	rw-	6
读和执行	r-x	5
读、写、执行	rwx	7

2.2.2 修改文件属性

1. chmod 命令

chmod 命令用于设置文件或目录的权限,可以采用两种形式的权限表示方法:字符形式和数字形式。其用法为:chmod[操作对象][操作符号][权限][文件|目录],参数说明如表 2-15 所示。

表 2-15 chmod 命令的参数及作用

名称	选项	作用
操作对象	u	表示用户所有者,即文件或目录的所有者
	g	表示组群的所有者,即与文件的用户所有者有相同组群 GID 的所有用户
	o	表示其他用户
	a	表示所有用户,它是系统默认值
操作符号	+	添加某个权限
	-	取消某个权限
	=	赋予给定权限并取消原先权限
权限	r	读取权限
	w	写入权限
	x	可执行权限

(1) 字符形式改变文件或目录的权限

格式为:chmod [ugoa][＋-=][rwx][文件|目录]

例如,给/home 目录下的 hello.txt 文件的所有者添加执行权限,组内用户添加写权限。

```
[root@localhost~]# ls -al /home/
-rw-r--r--. 1 root      root      16 3月  5 16:09 hello.txt
[root@localhost~]# chmod u+x,g+w /home/hello.txt
[root@localhost~]# ls -al /home/
-rwxrw-r--. 1 root      root      16 3月  5 16:09 hello.txt
```

(2) 数字形式改变文件或目录的权限

格式为:chmod [-R][权限][文件|目录],-R 表示递归修改指定目录下所有子项的权限。

例如给/home 目录下的 jack 目录的所有者读、写、执行权限,组内用户读、写权限,其他用户读权限。

```
[root@localhost~]# ls -al /home/
drwx------. 14 jack      jack      4096 3月  5 14:09 jack
[root@localhost~]# chmod 764 /home/jack/
[root@localhost~]# ls -al /home/
drwxrw-r--. 14 jack      jack      4096 3月  5 14:09 jack
```

2. chown 命令

chown 命令用来更改文件和目录的用户所有者和组群的所有者。其用法为：chown [-R][用户.用户组][文件|目录],-R 表示将下级子目录下的所有文件和目录的所有权一起更改。

例如，将/home 目录下的 hello.txt 文件的所有者修改为 jack，所属用户组修改为 jack。

```
[root@localhost~]# ls-al /home/
-rwxrw-r--.  1 root         root         16 3月   5 16:09 hello.txt
[root@localhost~]# chown jack.jack /home/hello.txt
[root@localhost~]# ls-al /home/
-rwxrw-r--.  1 jack         jack         16 3月   5 16:09 hello.txt
```

现在，就可以使用文件权限的知识以及相关命令，给公司的三个项目组创建属于各组用户账户的文件，从而实现各组文件的权限控制。

```
[root@localhost~]# mkdir /shejizu
[root@localhost~]# chown shejizuuser1.shejizu /shejizu
[root@localhost~]# mkdir /shishizu
[root@localhost~]# chown shishizuuser1.shishizu /shishizu
[root@localhost~]# mkdir /yunweizu
[root@localhost~]# chown yunweizuuser1.yunweizu /yunweizu
[root@localhost~]# ls-al /shejizu/
drwxr-xr-x.  2 shejizuuser1 shejizu    6 3月  10 00:12 .
[root@localhost~]# ls-al /shishizu/
drwxr-xr-x.  2 shishizuuser1 shishizu   6 3月  10 00:12 .
[root@localhost~]# ls-al /yunweizu/
drwxr-xr-x.  2 yunweizuuser1 yunweizu   6 3月  10 00:12 .
```

通过查看三个项目组的目录文件，可以看出目录文件的所有者具有读、写、执行权限，组内用户以及其他用户具有读、执行权限。

现在来测试一下，当实施组用户登录系统，访问设计组文件目录内的文件时，是否能进行写操作。

```
[root@localhost~]# su shejizuuser1              //使用 su 命令，切换到设计组用户
[shejizuuser1@localhost root]$ touch /shejizu/1.txt
                                                //在设计组的文件目录内创建 1.txt 文件
[shejizuuser1@localhost root]$ su shishizuuser1  //切换到实施组用户
密码：
[shishizuuser1@localhost root]$ cd /shejizu      //访问设计组的文件目录
[shishizuuser1@localhost shejizu]$ ls            //可以进行读操作
1.txt
[shishizuuser1@localhost shejizu]$ vi 1.txt      //进行写操作
```

如图 2-9 所示，实施组用户是没有该文件的写权限的。

图 2-9 实施组用户没有文件的写权限

2.2.3 文件的特殊权限

在实际的应用中,只有读取、写入、执行权限,并不一定能满足系统用户对于文件目录的安全性的需要,这时可以使用特殊权限扩展系统基础权限的功能。

1. SUID

只有可执行的二进制程序才可以设置 SUID,它可以让二进制程序的执行者临时拥有文件所有者的权限。其用法为:chmod u+s 文件名。

例如,以/example/text 目录为例,进行 SUID 权限的设置。

```
[root@localhost ~]# mkdir -p /example/test
[root@localhost ~]# chmod 755 /example/test/
[root@localhost ~]# ls -l /example/
drwxr-xr-x. 2 root root 6 3月  10 01:08 test
[root@localhost ~]# chmod u+s /example/test/
[root@localhost ~]# ls -l /example/
drwsr-xr-x. 2 root root 6 3月  10 01:08 test
```

通过查看文件属性,发现文件所有者的权限由 rwx 变成了 rws,其中 x 改变成 s,这是因为该文件被赋予了 SUID 权限。如果原本没有 x 执行权限(rw-),那么被赋予 SUID 权限后 x 执行权限位置上,将变成大写的 S(rwS)。

2. SGID

同 SUID 一样,可以通过设置二进制程序的 SGID,从而让执行者临时拥有文件所属组的权限,或者在某个目录中创建的文件自动继承该目录的用户组(只可以对目录进行设置)。

与 SUID 一样,如果执行者原来有执行权限,赋予 SGID 权限后,原有的 x 变为 s,若没有执行权限则变为 S。其用法为:chmod g+s 文件名。

例如,以/example/exam 目录为例,进行 SGID 权限的设置。

```
[root@localhost ~]# mkdir -p /example/exam
[root@localhost ~]# chmod g+s /example/exam/
[root@localhost ~]# ls -dl /example/exam
drwxr-sr-x. 2 root root 6 3月  10 01:20 /example/exam
```

例如,设计组中的用户在使用/shejizu 目录时,不同的用户在该目录下创建文件时,会自动继承该目录的用户组。

```
[root@localhost~]# ls-dl /shejizu/
drwxr-xr-x. 2 shejizuuser1 shejizu 19 3月   10 00:21 /shejizu/
[root@localhost~]# chmod g+s /shejizu
[root@localhost~]# ls-dl /shejizu/
drwxr-sr-x. 2 shejizuuser1 shejizu 19 3月   10 00:21 /shejizu/
[root@localhost~]# chmod 777 /shejizu         //为了让所有用户都能具有 rwx 权限
[root@localhost~]# ls-dl /shejizu/
drwxrwsrwx. 3 shejizuuser1 shejizu 65 3月   10 04:10 /shejizu/
0[root@localhost~]# useradd-g shejizu shejizuuser2
[root@localhost~]# echo "123456" | passwd--stdin shejizuuser2
[shejizuuser2@localhost root]$ mkdir /shejizu/shejuzuuser2
[shejizuuser2@localhost root]$ touch /shejizu/shejuzuuser2/123.txt
[shejizuuser2@localhost root]$ su root
密码:
[root@localhost~]# ls-al /shejizu/shejuzuuser2/123.txt
-rw-r--r--. 1 shejizuuser2 shejizu 0 3月   10 04:15 /shejizu/shejuzuuser2/123.txt
```

可以看到,当切换到用户 shejizuuser2 后,在该目录下创建文件时,该文件所属用户组为 shejizu。即使切换到用户 yunweizuuser1,再在该目录下创建文件,该文件所属用户组仍然为 shijizu。

```
[root@localhost~]# su yunweizuuser1
[yunweizuuser1@localhost root]$ mkdir /shejizu/yunweizu/
[yunweizuuser1@localhost root]$ touch /shejizu/yunweizu/555.txt
[yunweizuuser1@localhost root]$ su root
密码:
[root@localhost~]# ls-al /shejizu/yunweizu/
-rw-r--r--. 1 yunweizuuser1 shejizu  0 3月   10 04:21 555.txt
```

3. SBIT

SBIT 特殊权限位可确保用户只能删除自己的文件,而不能删除其他用户的文件。换句话说,当对某个目录设置了 SBIT 黏滞位权限后,那么该目录中的文件就只能被其所有者执行删除操作了。

当给目录设置 SBIT 特殊权限位后,文件的其他人权限部分的 x 执行权限就会被替换成 t(原本有 x 执行权限)或者 T(原本没有 x 执行权限)。

例如,使用 shishizuuser1 用户创建目录/111,设置该目录下的 SBIT 特殊权限位,并在该目录下创建文件/ceshi.txt,同时赋予该文件 777 权限(让所有人都具有 rwx 权限),这时切换到 yunweizuuser1 用户,看看是否可以删除该文件。

```
[root@localhost~]# mkdir /111
[root@localhost~]# chown-R shishizuuser1.shishizu /111
[root@localhost~]# su shishizuuser1
```

```
[shishizuuser1@localhost root]$ touch /111/ceshi.txt
[shishizuuser1@localhost root]$ chmod 777 /111/ceshi.txt
[shishizuuser1@localhost root]$ chmod-R o+t /111
[shishizuuser1@localhost root]$ ls-dl /111/ceshi.txt
-rwxrwxrwt. 1 shishizuuser1 shishizu 0 3月  10 04:57 /111/ceshi.txt
[shishizuuser1@localhost root]$ su yunweizuuser1
密码:
[yunweizuuser1@localhost root]$ rm-f /111/ceshi.txt
rm:无法删除"/111/ceshi.txt": 权限不够
```

需要注意的是,当给目录设置 SBIT 特殊权限位后,root 用户是可以进行文件删除的。

2.2.4 文件隐藏属性

chattr 命令用于设置文件的隐藏权限,其用法为:chattr [＋/-参数] 文件;lsattr 用于显示文件的隐藏权限,其用法为:lsattr [参数] 文件。

chattr 命令常见的参数及作用如表 2-16 所示。

表 2-16 chattr 命令常见的参数及作用

参数	功能
i	无法对文件进行修改,若对目录进行设置,则仅能修改其中的子文件内容而不能新建或删除文件
a	仅允许补充内容,无法覆盖/删除内容,也不能删除文件
u	文件删除后仍保存在硬盘,方便恢复

例如,在目录/ceshi 下,创建文件 111.txt,在没有设置隐藏权限"a"时,可以删除文件,但是设置了该权限后,将不能删除文件,如果取消该隐藏权限,将可以删除该文件。

```
[root@localhost ~]# mkdir /ceshi
[root@localhost ~]# touch /ceshi/111.txt
[root@localhost ~]# rm /ceshi/111.txt
rm:是否删除普通空文件 "/ceshi/111.txt"? y
[root@localhost ~]# touch /ceshi/111.txt
[root@localhost ~]# chattr+a /ceshi/111.txt
[root@localhost ~]# rm /ceshi/111.txt
rm:是否删除普通空文件 "/ceshi/111.txt"? y
rm:无法删除"/ceshi/111.txt": 不允许的操作
[root@localhost ~]# lsattr /ceshi/111.txt
-----a---------- /ceshi/111.txt
[root@localhost ~]#
[root@localhost ~]# chattr-a /ceshi/111.txt
[root@localhost ~]# rm /ceshi/111.txt
rm:是否删除普通空文件 "/ceshi/111.txt"? y
```

2.2.5 文件访问控制列表

访问控制列表(ACL)在网络设备上,如交换机、路由器、防火墙上,通常用来对网络数据进行控制,从而提升网络的安全性。在 Linux 系统中,如果对某个指定的用户进行单独的权限控制,就需要用到 ACL。基于普通文件或目录设置 ACL 就是针对指定的用户或用户组设置文件或目录的操作权限。ACL 对目录、文件的权限控制如下:

① 针对某个目录设置了 ACL,则目录中的文件会继承其 ACL。

② 针对文件设置了 ACL,则文件不再继承其所在目录的 ACL。

1. setfacl 命令

setfacl 命令用于对单一用户或用户组、单一文件或目录进行读/写/执行权限的控制,其用法为:setfacl [参数] 文件名称。

其中,对于目录使用 -R 参数,对于普通文件使用 -m 参数,删除某个文件的 ACL 使用 -b 参数。

2. getfacl 命令

getfacl 命令用于显示文件上设置的 ACL 信息,其用法为:getfacl 文件名称。

例如,使用 setfacl 命令设置用户 shishizuuser1 具有目录/etc 的读取、写入、执行权限,然后使用该用户在目录/etc 下创建目录/shishizu,并在该目录下新建文件 111.txt。同时,使用 getfacl 命令查看目录/etc 上所设置的 ACL 信息。

```
[root@localhost ~]# setfacl -Rm u:shishizuuser1:rwx /etc
[root@localhost ~]# su shishizuuser1
[shishizuuser1@localhost root]$ mkdir /etc/shishizu
[shishizuuser1@localhost root]$ touch /etc/shishizu/111.txt
[shishizuuser1@localhost root]$ su root
密码:
[root@localhost ~]# ls -l /etc/
drwxr-xr-x.  2 shishizuuser1 shishizu       21 3月  10 05:36 shishizu
[root@localhost ~]# getfacl /etc/
getfacl: Removing leading '/' from absolute path names
# file: etc/
# owner: root
# group: root
user::rwx
user:shishizuuser1:rwx
group::r-x
mask::rwx
other::r-x
```

自我测试

一、填空题

1. Linux 提供的账户类型主要有_____和_____。
2. 在 Linux 系统中,与用户管理有关的系统文件有:_____、/etc/shadow、/etc/group、_____。
3. useradd 命令,用于创建新的_____。
4. chattr 命令用于设置文件的_____权限。

二、简答题

1. 简述 Linux 系统中的三种用户账户。
2. 简述 id 命令的功能。
3. 简述 SUID、SGID 权限的功能。

三、实训题

1. 创建用户,用户名为 tom,uid＝222,gid＝222,用户家目录为/home/tomdir。
2. 新建用户组,组名为 share,id 为 888,将 jack 添加到 share 组里面。
3. 创建文件 ceshi.txt,将文件的所有者修改为 jack,属组为 share。

项目 3　磁盘管理

📚 项目综述

随着公司发展，服务器上的数据越来越多，占用的服务器空间也逐渐增加，服务器的硬盘容量已经不能满足办公需求。公司的网络管理员，准备通过在服务器上添加新的硬盘，扩充分区容量，帮助公司解决办公难的问题。同时，使用磁盘阵列提高公司数据的安全性、稳定性。

📚 项目目标

- 了解扩充硬盘的基本方法；
- 了解磁盘配额基本知识；
- 了解 LVM 的基本知识；
- 了解 RAID 的基本知识；
- 掌握扩充硬盘的方法；
- 掌握磁盘配额的配置；
- 能正确使用 LVM 技术对硬盘分区进行动态调整；
- 能正确使用 RAID 技术配置磁盘阵列。

◀ 任务 3.1　扩充硬盘空间 ▶

3.1.1　添加新的硬盘

可以使用 VMware Workstation 为虚拟机添加新的硬件设备，这里为 Linux 系统添加一块新的硬盘。为确保硬盘添加后，系统能正常识别，建议先将系统关机，再添加硬盘。

实现步骤如下：

（1）在虚拟机管理主界面，依次单击菜单栏上的"虚拟机"→"设置"，打开如图 3-1 所示界面。

（2）在图 3-1 中，单击"添加"按钮，选择"硬盘"，单击"下一步"按钮，如图 3-2 所示。

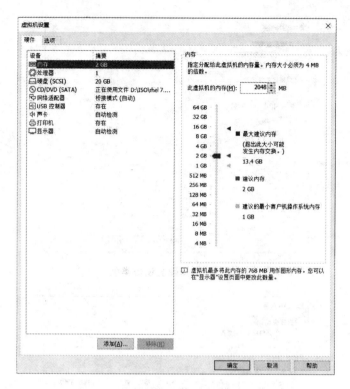

图 3-1　虚拟机设置界面

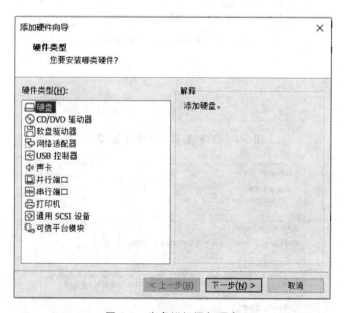

图 3-2　为虚拟机添加硬盘

（3）在图 3-3 中，选择默认的"SCSI"虚拟硬盘类型，单击"下一步"按钮。

（4）在图 3-4 中，选择"创建新虚拟磁盘"，单击"下一步"按钮。

（5）在图 3-5 中，设置新硬盘的容量，单击"下一步"按钮。

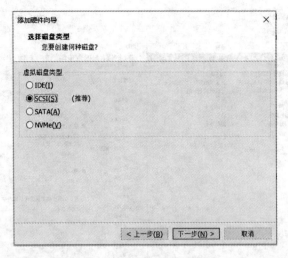

图 3-3 选择硬盘设备类型

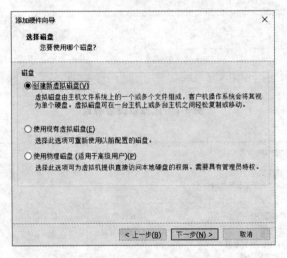

图 3-4 选择"创建新虚拟磁盘"选项

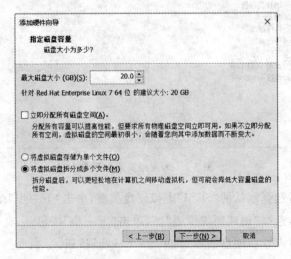

图 3-5 设定硬盘容量

(6) 在图 3-6 中,单击"完成"按钮,完成硬盘的添加,如图 3-7 所示,这时虚拟机已经添加了一块新硬盘。

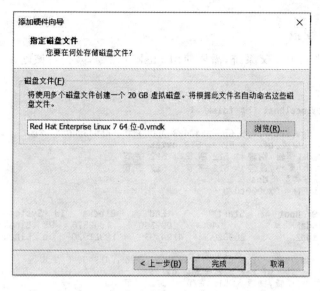

图 3-6　设置硬盘的文件名和保存的位置

图 3-7　查看虚拟机硬件的设置信息

3.1.2 系统使用新硬盘

1. 查看系统硬盘

添加硬盘后,进入 Linux 系统,需要使用 fdisk -l 命令,来查看该硬盘是否已经被系统识别,如图 3-8 所示。

```
[root@localhost ~]# fdisk -l
磁盘 /dev/sda: 21.5 GB, 21474836480 字节, 41943040 个扇区
Units = 扇区 of 1 * 512 = 512 bytes
扇区大小(逻辑/物理): 512 字节 / 512 字节
I/O 大小(最小/最佳): 512 字节 / 512 字节
磁盘标签类型: dos
磁盘标识符: 0x0000c070

   设备 Boot      Start         End      Blocks   Id  System
/dev/sda1   *      2048     2099199     1048576   83  Linux
/dev/sda2       2099200    41943039    19921920   8e  Linux LVM

磁盘 /dev/sdb: 21.5 GB, 21474836480 字节, 41943040 个扇区
Units = 扇区 of 1 * 512 = 512 bytes
扇区大小(逻辑/物理): 512 字节 / 512 字节
I/O 大小(最小/最佳): 512 字节 / 512 字节
```

图 3-8 列出系统所有分区类型

从图 3-8 中可以看到两块硬盘,分别是第一块硬盘/dev/sda,第二块硬盘/dev/sdb。第一块硬盘划分了两个分区,分别是/dev/sda1、/dev/sda2;第二块硬盘容量 21.5 GB,目前在进行磁盘分区,无法使用。

2. 硬盘分区

在 Linux 系统中,可以使用 fdisk 命令进行硬盘分区,其用法为:fdisk [硬盘名称]。表 3-1 列出了 fdisk 命令常用的参数及功能。

表 3-1 fdisk 命令常用的参数及功能

参数	功能
n	添加硬盘新的分区
m	列出所有命令
d	删除一个硬盘分区
l	列出所有分区类型
t	更改分区类型
p	查看硬盘分区信息
q	不保存直接退出
w	保存并退出

实现步骤如下:

（1）在硬盘/dev/sdb 上，使用 fdisk 命令，创建第一个主分区，容量为 4 GB，创建后，查看硬盘分区信息，如图 3-9 所示。

```
[root@localhost ~]# fdisk /dev/sdb
欢迎使用 fdisk (util-linux 2.23.2)。

更改将停留在内存中，直到您决定将更改写入磁盘。
使用写入命令前请三思。

Device does not contain a recognized partition table
使用磁盘标识符 0x88e70219 创建新的 DOS 磁盘标签。

命令(输入 m 获取帮助)：n
Partition type:
   p   primary (0 primary, 0 extended, 4 free)
   e   extended
Select (default p): p
分区号 (1-4，默认 1)：1
起始 扇区 (2048-41943039，默认为 2048)：
将使用默认值 2048
Last 扇区，+扇区 or +size{K,M,G} (2048-41943039，默认为 41943039)：+4G
分区 1 已设置为 Linux 类型，大小设为 4 GiB

命令(输入 m 获取帮助)：p

磁盘 /dev/sdb：21.5 GB, 21474836480 字节，41943040 个扇区
Units = 扇区 of 1 * 512 = 512 bytes
扇区大小(逻辑/物理)：512 字节 / 512 字节
I/O 大小(最小/最佳)：512 字节 / 512 字节
磁盘标签类型：dos
磁盘标识符：0x88e70219

   设备 Boot      Start         End      Blocks   Id  System
/dev/sdb1          2048     8390655     4194304   83  Linux
```

图 3-9 创建第一个主分区

需要注意的是，主分区或扩展分区的编号为 1～4；逻辑分区的编号从 5 开始。在设置起始扇区时，不需要进行任何操作，直接按"回车"键，系统会自动进行扇区的计算。只需要在 Last 扇区(结束扇区)指定所要创建的分区大小即可。

（2）在硬盘/dev/sdb 上，创建第二个主分区，容量为 6 GB，创建后，查看硬盘分区信息，如图 3-10 所示。

```
命令(输入 m 获取帮助)：n
Partition type:
   p   primary (1 primary, 0 extended, 3 free)
   e   extended
Select (default p): p
分区号 (2-4，默认 2)：2
起始 扇区 (8390656-41943039，默认为 8390656)：
将使用默认值 8390656
Last 扇区，+扇区 or +size{K,M,G} (8390656-41943039，默认为 41943039)：+6G
分区 2 已设置为 Linux 类型，大小设为 6 GiB

命令(输入 m 获取帮助)：p

磁盘 /dev/sdb：21.5 GB, 21474836480 字节，41943040 个扇区
Units = 扇区 of 1 * 512 = 512 bytes
扇区大小(逻辑/物理)：512 字节 / 512 字节
I/O 大小(最小/最佳)：512 字节 / 512 字节
磁盘标签类型：dos
磁盘标识符：0x88e70219

   设备 Boot      Start         End      Blocks   Id  System
/dev/sdb1          2048     8390655     4194304   83  Linux
/dev/sdb2       8390656    20973567     6291456   83  Linux
```

图 3-10 创建第二个主分区

（3）保存硬盘分区信息，并退出，使用 fdisk -l 命令查看系统分区情况，如图 3-11 所示。

```
命令(输入 m 获取帮助): w
The partition table has been altered!
Calling ioctl() to re-read partition table.
正在同步磁盘。
[root@localhost ~]# fdisk -l
磁盘 /dev/sda：21.5 GB，21474836480 字节，41943040 个扇区
Units = 扇区 of 1 * 512 = 512 bytes
扇区大小(逻辑/物理)：512 字节 / 512 字节
I/O 大小(最小/最佳)：512 字节 / 512 字节
磁盘标签类型：dos
磁盘标识符：0x0000c070

   设备 Boot      Start         End      Blocks   Id  System
/dev/sda1   *      2048     2099199     1048576   83  Linux
/dev/sda2        2099200    41943039    19921920   8e  Linux LVM

磁盘 /dev/sdb：21.5 GB，21474836480 字节，41943040 个扇区
Units = 扇区 of 1 * 512 = 512 bytes
扇区大小(逻辑/物理)：512 字节 / 512 字节
I/O 大小(最小/最佳)：512 字节 / 512 字节
磁盘标签类型：dos
磁盘标识符：0x88e70219

   设备 Boot      Start         End      Blocks   Id  System
/dev/sdb1          2048     8390655     4194304   83  Linux
/dev/sdb2       8390656    20973567     6291456   83  Linux
```

图 3-11 查看系统分区情况

3. 创建文件系统

硬盘分区后，需要在分区上创建文件系统，该分区才能使用。在 Linux 系统中，使用 mkfs 命令来创建文件系统，其用法为：mkfs［选项］分区设备文件名。

常用的选项为"-t 文件系统"，主要是指定格式化的文件系统，如 xfs、ext3、ext4 等。

如图 3-12 所示，使用 mkfs 命令在/dev/sdb1、/dev/sdb2 两个分区上创建 xfs 类型的文件系统。

```
[root@localhost ~]# mkfs.xfs /dev/sdb1
meta-data=/dev/sdb1              isize=512    agcount=4, agsize=262144 blks
         =                       sectsz=512   attr=2, projid32bit=1
         =                       crc=1        finobt=0, sparse=0
data     =                       bsize=4096   blocks=1048576, imaxpct=25
         =                       sunit=0      swidth=0 blks
naming   =version 2              bsize=4096   ascii-ci=0 ftype=1
log      =internal log           bsize=4096   blocks=2560, version=2
         =                       sectsz=512   sunit=0 blks, lazy-count=1
realtime =none                   extsz=4096   blocks=0, rtextents=0
[root@localhost ~]# mkfs.xfs /dev/sdb2
meta-data=/dev/sdb2              isize=512    agcount=4, agsize=393216 blks
         =                       sectsz=512   attr=2, projid32bit=1
         =                       crc=1        finobt=0, sparse=0
data     =                       bsize=4096   blocks=1572864, imaxpct=25
         =                       sunit=0      swidth=0 blks
naming   =version 2              bsize=4096   ascii-ci=0 ftype=1
log      =internal log           bsize=4096   blocks=2560, version=2
         =                       sectsz=512   sunit=0 blks, lazy-count=1
realtime =none                   extsz=4096   blocks=0, rtextents=0
```

图 3-12 格式化系统分区

4. 挂载使用

如图 3-13 所示，将分区/dev/sdb1 挂载到/web 目录下，分区/dev/sdb2 挂载到/ftp 目录下，挂载完成后，使用命令 df -h 查看系统挂载信息以及硬盘的使用情况。

```
[root@localhost ~]# mkdir /web
[root@localhost ~]# mkdir /ftp
[root@localhost ~]# mount /dev/sdb1 /web/
[root@localhost ~]# mount /dev/sdb2 /ftp/
[root@localhost ~]# df -h
文件系统                容量    已用    可用 已用% 挂载点
/dev/mapper/rhel-root   17G    3.2G    14G   19%  /
devtmpfs               897M      0    897M    0%  /dev
tmpfs                  912M      0    912M    0%  /dev/shm
tmpfs                  912M   9.0M    903M    1%  /run
tmpfs                  912M      0    912M    0%  /sys/fs/cgroup
/dev/sda1             1014M   179M    836M   18%  /boot
tmpfs                  183M    20K    183M    1%  /run/user/0
/dev/sr0               3.8G   3.8G      0   100%  /run/media/root/RHEL-7.4 Server.x86_64
/dev/sdb1              4.0G    33M    4.0G    1%  /web
/dev/sdb2              6.0G    33M    6.0G    1%  /ftp
```

图 3-13　挂载系统分区并查看系统挂载信息

5. 添加交换分区

可以使用 mkswap 命令，对硬盘分区进行格式操作，如图 3-14 所示。然后，创建交换分区（SWAP），可以使用 free -m 命令查看当前系统的交换分区情况（大小为 2 GB），如图 3-15 所示。

```
[root@localhost ~]# fdisk /dev/sdb
欢迎使用 fdisk (util-linux 2.23.2)。

更改将停留在内存中，直到您决定将更改写入磁盘。
使用写入命令前请三思。

命令(输入 m 获取帮助)：n
Partition type:
   p   primary (2 primary, 0 extended, 2 free)
   e   extended
Select (default p): p
分区号 (3,4, 默认 3)：3
起始 扇区 (20973568-41943039, 默认为 20973568)：
将使用默认值 20973568
Last 扇区, +扇区 or +size{K,M,G} (20973568-41943039, 默认为 41943039)：+3G
分区 3 已设置为 Linux 类型, 大小设为 3 GiB

命令(输入 m 获取帮助)：p

磁盘 /dev/sdb：21.5 GB, 21474836480 字节，41943040 个扇区
Units = 扇区 of 1 * 512 = 512 bytes
扇区大小(逻辑/物理)：512 字节 / 512 字节
I/O 大小(最小/最佳)：512 字节 / 512 字节
磁盘标签类型：dos
磁盘标识符：0x88e70219

   设备 Boot      Start         End      Blocks   Id  System
/dev/sdb1            2048     8390655     4194304   83  Linux
/dev/sdb2         8390656    20973567     6291456   83  Linux
/dev/sdb3        20973568    27265023     3145728   83  Linux

命令(输入 m 获取帮助)：w
The partition table has been altered!

Calling ioctl() to re-read partition table.
```

图 3-14　创建第三个主分区

```
[root@localhost ~]# mkswap /dev/sdb3
正在设置交换空间版本 1，大小 = 3145724 KiB
无标签，UUID=af2347a6-175d-428c-af88-c4d7d9778e37
[root@localhost ~]# free -m
              total        used        free      shared  buff/cache   available
Mem:           1823         671         651           9         500         945
Swap:          2047           0        2047
```

图 3-15　创建交换分区

接下来使用 swapon 命令激活交换分区,再次使用 free -m 命令[该命令用于显示系统内存的使用情况,包括物理内存、交换内存(swap)和内核缓冲区内存]查看当前系统的交换分区情况(大小已经变为 5 GB),如图 3-16 所示。

```
[root@localhost ~]# swapon /dev/sdb3
[root@localhost ~]# free -m
              total        used        free      shared  buff/cache   available
Mem:           1823         688         628           9         506         927
Swap:          5119           0        5119
```

图 3-16 激活交换分区

可以使用 swapoff 命令关闭指定的交换空间,如图 3-17 所示,将系统交换分区的大小还原为 2 GB。

```
[root@localhost ~]# swapoff /dev/sdb3
[root@localhost ~]# free -m
              total        used        free      shared  buff/cache   available
Mem:           1823         687         629           9         506         928
Swap:          2047           0        2047
```

图 3-17 关闭指定的交换空间

◀ 任务 3.2 磁盘配额 ▶

3.2.1 磁盘配额的基本概念

磁盘配额对于系统管理员尤其重要,管理员可以通过磁盘配额功能来限制用户或用户组在磁盘内的存储空间,从而避免用户过度使用磁盘空间,导致系统磁盘空间不够用。

磁盘配额,就是对用户或者用户组能够使用的磁盘空间、文件数量进行分配和限制,以防止个别用户占用大量的磁盘空间或创建大量的文件,从而确保系统的稳定性和持续性。其中磁盘空间是限制用户能够使用的磁盘数据块大小,也就是限制磁盘空间大小,默认单位为 KB;文件数量是限制用户能够拥有的文件个数(inode 用量)。

在 Linux 系统中使用 quota 命令进行磁盘配额的配置与管理,磁盘配额的限制方法分为两种:

① 软限制:指定一个软性的配额数值(当前用户在这个文件系统上的最低限制容量),用户在宽限期内,允许暂时超出 soft 规定的数值,但是必须在宽限期内将磁盘容量降低到 soft 的容量限制之下,同时系统会发出警告信息。

② 硬限制:指定一个硬性的配额数值(当前用户在这个文件系统上的最高限制容量),是绝对禁止用户超过的,当达到硬限制时,系统也会发出警告并强制终止用户的操作。硬限制的配额数值应大于相应的软限制的配额数值,否则软限制将失效。

3.2.2 配置磁盘配额

1. 启用磁盘配额功能

在 Linux 系统中,启用磁盘配额功能,就需要通过编辑/etc/fstab 文件,在该磁盘上添加 usrquota(用户配额)、grpquota(用户组配额)选项。

/etc/fstab 是系统自动挂载的配置文件,该文件记录了系统启动过程中需要自动挂载的文件系统、挂载点、文件系统类型等信息,如图 3-18 所示。

```
# /etc/fstab
# Created by anaconda on Tue Mar  3 00:32:39 2020
#
# Accessible filesystems, by reference, are maintained under '/dev/disk'
# See man pages fstab(5), findfs(8), mount(8) and/or blkid(8) for more info
#
/dev/mapper/rhel-root   /           xfs     defaults    0 0
UUID=93abc0c5-f930-430c-ab0a-e2db81e2b643 /boot  xfs  defaults  0 0
/dev/mapper/rhel-swap   swap        swap    defaults    0 0
```

图 3-18 /etc/fstab 文件内容

fstab 文件参数含义如表 3-2 所示。

表 3-2 fstab 文件参数含义

名称	含义
第一列	磁盘设备文件或者该设备的 Label 或者 UUID(通用唯一标识符,由系统自动生成、管理)
第二列	设备的挂载点
第三列	文件系统的类型,包括 ext3、ext4、xfs 等
第四列	文件系统的参数,常用文件系统参数见表 3-3
第五列	使用 dump 命令备份文件系统的频率。0 表示不备份,1 表示每天进行备份,2 表示不定期备份
第六列	系统开机时是否使用 fsck 命令对文件系统进行检查。0 表示不检查;1 表示最早检查;2 表示当 1 检查结束后,就进行检查

文件系统参数的含义见表 3-3。

表 3-3 文件系统参数的含义

参数	说明
async/sync	设置是否为同步方式运行,默认为 async(异步)
auto/noauto	当下载 mount -a 的命令时,此文件系统是否被主动挂载。默认为 auto(自动)
rw/ro	是否以只读(ro)或者读写(rw)模式挂载
exec/noexec	此文件系统是否能够进行"执行"操作
user/nouser	是否允许用户使用 mount 命令挂载
suid/nosuid	是否允许 SUID 存在
usrquota	启动文件系统支持磁盘配额模式
grpquota	启动文件系统对群组磁盘配额模式的支持
defaults	同时具有 rw、suid、exec、auto、nouser、async 等默认参数的设置

例如为系统添加第三块硬盘/dev/sdc,并划分两个分区/dev/sdc1、/dev/sdc2,容量大小分别为 4 GB、6 GB,/dev/sdc1 分区使用 xfs 文件系统,/dev/sdc2 使用 ext4 文件系统,并配置/ets/fstab 文件,启用分区的磁盘配额功能,并分别挂载到/gx1、/gx2 目录下。

实现步骤如下:

(1) 为系统添加第三块硬盘,容量 30 GB,如图 3-19 所示。

```
[root@localhost ~]# fdisk -l

磁盘 /dev/sda:21.5 GB,21474836480 字节,41943040 个扇区
Units = 扇区 of 1 * 512 = 512 bytes
扇区大小(逻辑/物理):512 字节 / 512 字节
I/O 大小(最小/最佳):512 字节 / 512 字节
磁盘标签类型:dos
磁盘标识符:0x0000c070

   设备 Boot      Start        End      Blocks   Id  System
/dev/sda1   *      2048     2099199    1048576   83  Linux
/dev/sda2       2099200    41943039   19921920   8e  Linux LVM

磁盘 /dev/sdb:21.5 GB,21474836480 字节,41943040 个扇区
Units = 扇区 of 1 * 512 = 512 bytes
扇区大小(逻辑/物理):512 字节 / 512 字节
I/O 大小(最小/最佳):512 字节 / 512 字节
磁盘标签类型:dos
磁盘标识符:0x88e70219

   设备 Boot      Start        End      Blocks   Id  System
/dev/sdb1          2048     8390655    4194304   83  Linux
/dev/sdb2       8390656    20973567    6291456   83  Linux
/dev/sdb3      20973568    27265023    3145728   83  Linux

磁盘 /dev/sdc:32.2 GB,32212254720 字节,62914560 个扇区
Units = 扇区 of 1 * 512 = 512 bytes
扇区大小(逻辑/物理):512 字节 / 512 字节
I/O 大小(最小/最佳):512 字节 / 512 字节
```

图 3-19 查看系统所有分区类型

(2) 将硬盘/dev/sdc 划分两个分区/dev/sdc1、/dev/sdc2,容量大小分别为 4 GB、6 GB,分区结果如图 3-20 所示。

```
命令(输入 m 获取帮助):p

磁盘 /dev/sdc:32.2 GB,32212254720 字节,62914560 个扇区
Units = 扇区 of 1 * 512 = 512 bytes
扇区大小(逻辑/物理):512 字节 / 512 字节
I/O 大小(最小/最佳):512 字节 / 512 字节
磁盘标签类型:dos
磁盘标识符:0xd07e132c

   设备 Boot      Start        End      Blocks   Id  System
/dev/sdc1          2048     8390655    4194304   83  Linux
/dev/sdc2       8390656    20973567    6291456   83  Linux
```

图 3-20 划分两个分区/dev/sdc1、/dev/sdc2

(3) 创建文件系统,dev/sdc1 分区使用 xfs 文件系统,/dev/sdc2 使用 ext4 文件系统,如图 3-21 所示。

(4) 挂载使用,首先创建挂载目录,然后配置/etc/fstab 文件,如图 3-22 所示,完成后重启系统。

```
[root@localhost~]# mkdir /gx1
[root@localhost~]# mkdir /gx2
```

```
[root@localhost ~]# mkfs.xfs /dev/sdc1
meta-data=/dev/sdc1             isize=512    agcount=4, agsize=262144 blks
         =                      sectsz=512   attr=2, projid32bit=1
         =                      crc=1        finobt=0, sparse=0
data     =                      bsize=4096   blocks=1048576, imaxpct=25
         =                      sunit=0      swidth=0 blks
naming   =version 2             bsize=4096   ascii-ci=0 ftype=1
log      =internal log          bsize=4096   blocks=2560, version=2
         =                      sectsz=512   sunit=0 blks, lazy-count=1
realtime =none                  extsz=4096   blocks=0, rtextents=0
[root@localhost ~]# mkfs.ext4 /dev/sdc2
mke2fs 1.42.9 (28-Dec-2013)
文件系统标签=
OS type: Linux
块大小=4096 (log=2)
分块大小=4096 (log=2)
Stride=0 blocks, Stripe width=0 blocks
393216 inodes, 1572864 blocks
78643 blocks (5.00%) reserved for the super user
第一个数据块=0
Maximum filesystem blocks=1610612736
48 block groups
32768 blocks per group, 32768 fragments per group
3192 inodes per group
Superblock backups stored on blocks:
        32768, 98304, 163840, 229376, 294912, 819200, 884736

Allocating group tables: 完成
正在写入inode表: 完成
Creating journal (32768 blocks): 完成
Writing superblocks and filesystem accounting information: 完成
```

图 3-21 创建文件系统

```
#
# /etc/fstab
# Created by anaconda on Tue Mar  3 00:32:39 2020
#
# Accessible filesystems, by reference, are maintained under '/dev/disk'
# See man pages fstab(5), findfs(8), mount(8) and/or blkid(8) for more info
#
/dev/mapper/rhel-root   /                       xfs     defaults        0 0
UUID=93abc0c5-f930-430c-ab0a-e2db81e2b543 /boot                   xfs     defaults        0 0
/dev/mapper/rhel-swap   swap                    swap    defaults        0 0

/dev/sdc1       /gx1    xfs     defaults,usrquota,grpquota 0 0
/dev/sdc2       /gx2    ext4    defaults,usrquota,grpquota 0 0
```

图 3-22 挂载目录

系统重启后,进入系统。使用命令 df -h 查看系统挂载信息,可以看到两个分区已经自动挂载到了/gx1、/gx2 目录,如图 3-23 所示。

```
[root@localhost ~]# df -h
文件系统                 容量    已用   可用  已用% 挂载点
/dev/mapper/rhel-root    17G    3.2G   14G    19%  /
devtmpfs                897M      0  897M     0%  /dev
tmpfs                   912M      0  912M     0%  /dev/shm
tmpfs                   912M   9.0M  903M     1%  /run
tmpfs                   912M      0  912M     0%  /sys/fs/cgroup
/dev/sdc1               4.0G    33M  4.0G     1%  /gx1
/dev/sdc2               5.8G    24M  5.5G     1%  /gx2
```

图 3-23 查看系统挂载信息

2. 配置/gx1 磁盘配额

/dev/sdc1 分区使用 xfs 文件系统,挂载到/gx1 目录,需要使用针对 xfs 文件系统的 xfs_quota 磁盘配额管理工具来进行配置。

实现步骤如下：

（1）安装 xfsprogs 软件包。

```
[root@localhost~]# yum -y install xfsprogs
```

（2）使用 xfs_quota 命令，为用户 jack 配置磁盘配额，该命令语法为：xfs_quota -x -c ′limit[-ug] b[soft|hard]=N i[soft|hard]=N name′ 目录名。

参数说明如下：

① -x：启用专家模式。
② -c：直接调用管理命令。
③ limit：实际限制的内容，可以设置磁盘容量的软、硬限制以及文件数的软、硬限制。
④ -u：表示指定用户（如果指定组使用 -g）。
⑤ bsoft：设置磁盘容量的软限制数值。
⑥ bhard：设置磁盘容量的硬限制数值。
⑦ isoft：设置磁盘文件数的软限制数值。
⑧ ihard：设置磁盘文件数的硬限制数值。
⑨ name：用户名或用户组名。

这里，设置用户 jack 对目录/gx1 的磁盘配额限制为：磁盘容量的软限制和硬限制分别为 3 MB、5 MB，创建文件数的软限制和硬限制分别为 3 个、6 个。同时设置所有用户对目录/gx1 的权限为 777。

```
[root@localhost~]# xfs_quota -x -c 'limit -u bsoft=3m bhard=5m isoft=3 ihard=6 jack ' /gx1
[root@localhost~]# chmod 777 /gx1
```

可以使用以下命令，查看系统磁盘配额的有关信息，如图 3-24 所示。

```
[root@localhost ~]# xfs_quota -x -c 'print' /gx1
Filesystem          Pathname
/gx1                /dev/sdc1 (uquota, gquota)
[root@localhost ~]# xfs_quota -x -c 'report -ubih' /gx1
User quota on /gx1 (/dev/sdc1)
                        Blocks                              Inodes
User ID      Used   Soft   Hard Warn/Grace    Used   Soft   Hard Warn/Grace
---------- --------------------------------- ---------------------------------
root            0      0      0  00 [------]     3      0      0  00 [------]
jack            0     3M     5M  00 [------]     0      3      6  00 [------]

[root@localhost ~]# xfs_quota -x -c 'state' /gx1
User quota state on /gx1 (/dev/sdc1)
  Accounting: ON
  Enforcement: ON
  Inode: #67 (2 blocks, 2 extents)
Group quota state on /gx1 (/dev/sdc1)
  Accounting: ON
  Enforcement: ON
  Inode: #68 (1 blocks, 1 extents)
Project quota state on /gx1 (/dev/sdc1)
  Accounting: OFF
  Enforcement: OFF
  Inode: #68 (1 blocks, 1 extents)
Blocks grace time: [7 days]
Inodes grace time: [7 days]
Realtime Blocks grace time: [7 days]
```

图 3-24　查看系统磁盘配额信息

① xfs_quota -x -c ′print′ 目录名：列出目前目录的文件系统参数信息。
② xfs_quota -x -c ′report -ubih′目录名：列出目前目录的 quota 设置(-u：查看用户；-b：

查看 block；-i：查看 inode；-h：人性化）。

③ xfs_quota -x -c ´state 目录名´：列出目前目录支持 quota 的文件系统的信息，有没有启动 quota 功能。

(3) 切换到用户 jack，进行测试，如图 3-25 所示。

```
[root@localhost ~]# su jack
[jack@localhost root]$ cd /gx1
[jack@localhost gx1]$ touch cs{1..6}.txt
[jack@localhost gx1]$ touch cs7.txt
touch: 无法创建"cs7.txt": 超出磁盘限额
[jack@localhost gx1]$ dd if=/dev/zero of=/gx1/cs1.txt bs=3M count=1
记录了1+0 的读入
记录了1+0 的写出
3145728字节(3.1 MB)已复制, 0.0392918 秒, 80.1 MB/秒
[jack@localhost gx1]$ dd if=/dev/zero of=/gx1/cs2.txt bs=6M count=1
dd: 写入"/gx1/cs2.txt" 出错: 超出磁盘限额
记录了1+0 的读入
记录了0+0 的写出
2097152字节(2.1 MB)已复制, 0.0189426 秒, 111 MB/秒
```

图 3-25 磁盘配额限制已生效

可以看出，创建文件超过 6 个，文件容量超过 5 MB 时，将受到系统的限制，无法创建。这里 dd 命令用于读取、转换并输出数据，其中：

① if＝文件名：输入文件名，指定源文件。其中/dev/zero 是一个输入设备，可用来初始化文件。

② of＝文件名：输出文件名，指定目的文件。

③ ibs＝bytes：一次读入 bytes 个字节，即指定一个块大小为 bytes 个字节；

obs＝bytes：一次输出 bytes 个字节，即指定一个块大小为 bytes 个字节；

bs＝bytes：同时设置 ibs 和 obs 的块大小为 bytes 个字节。

④ count＝blocks：仅拷贝 blocks 个块，块大小等于 ibs 指定的字节数。

3. 配置/gx2 磁盘配额

/dev/sdc2 分区使用 ext4 文件系统，挂载到/gx2 目录，需要使用针对 ext4 文件系统的 quota 磁盘配额管理工具来进行配置。

实现步骤如下：

(1) 安装 quota 软件包。

```
[root@localhost ~]# yum -y install quota
```

(2) 使用 quotacheck 命令，检测磁盘配额并创建配额文件，其用法为 quotacheck [-augv] 目录名。

参数说明如下：

① -a：扫描已添加 usrquota 或者 grpquota 的分区。

② -u：扫描时，计算每个用户所占用的目录和文件数目。

③ -g：扫描时，计算每个用户组所占用的目录和文件数目。

④ -v：显示扫描过程。

⑤ -m：强制进行扫描。

⑥ -c：创建配额数据文件。

```
[root@localhost ~]# quotacheck -cmvug /gx2
```

注意,命令执行后,需要查看下/gx2 目录下,是否已经创建了配额文件 aquota.user、aquota.group。

```
[root@localhost~]# ls -al /gx2
总用量 40
drwxr-xr-x.  3 root root  4096 3月  12 02:37 .
dr-xr-xr-x. 28 root root  4096 3月  12 01:06 ..
-rw-------.  1 root root  6144 3月  12 02:37 aquota.group
-rw-------.  1 root root  6144 3月  12 02:37 aquota.user
drwx------.  2 root root 16384 3月  12 01:02 lost+found
```

(3) 使用 equota 命令,设置用户和用户组的配额,其用法为:edquota [-u 用户名] [-g 用户组名] [-t 设置宽限时间]。

这里,设置用户 lucy 对目录/gx2 的磁盘配额限制为:磁盘容量的软限制和硬限制分别为 30KB、50KB,创建文件数的软限制和硬限制分别为 3 个、5 个,如图 3-26 所示。同时设置所有用户对目录/gx2 的权限为 777。

```
[root@localhost~]# chmod 777 /gx2
[root@localhost~]# edquota -u lucy
```

```
Disk quotas for user lucy (uid 1001):
  Filesystem           blocks       soft       hard     inodes     soft     hard
  /dev/sdc1                 0          0          0          0        0        0
  /dev/sdc2                 0         30         50          0        3        5
```

图 3-26　设置用户对目录/gx2 的磁盘配额限制

图 3-26 中,从左到右,每一列的含义如表 3-4 所示。

表 3-4　文件的参数含义

名称	含义
Filesystem	进行配额的文件系统名称
blocks	当前用户在这个文件系统上所使用的磁盘空间,单位为 kbytes
soft	磁盘容量软限制
hard	磁盘容量硬限制
inodes	当前用户使用的 inodes
soft	文件个数软限制
hard	文件个数硬限制

设置完成后,quotaon 命令开启目录/gx2 的磁盘配额功能,同时,使用命令 repquota -auvs、quota -uv,显示指定文件系统的使用和配额的摘要、磁盘已使用的空间与限制(单一用户),如图 3-27 所示。其中:

① -a:列出在/etc/fstab 文件里,有加入 quota 设置的分区的使用状况,包括用户和群组。

② -g:列出所有群组的磁盘空间限制。

③ -u:列出所有用户的磁盘空间限制。

④ -v:显示该用户或群组的所有空间限制。

⑤ -s：以 MB、GB 等方式显示。

```
[root@localhost~] # quotaon -aguv
/dev/sdc2[/gx2]: group quotas turned on
/dev/sdc2[/gx2]: user quotas turned on
```

```
[root@localhost ~]# repquota -auvs
*** Report for user quotas on device /dev/sdc1
Block grace time: 7days; Inode grace time: 7days
                        Space limits              File limits
User            used    soft    hard    grace    used    soft    hard    grace
----------------------------------------------------------------
root      --    0K      0K      0K               3       0       0
jack      --    0K      3072K   5120K            0       3       6

*** Status for user quotas on device /dev/sdc1
Accounting: ON; Enforcement: ON
Inode: #67 (2 blocks, 2 extents)

*** Report for user quotas on device /dev/sdc2
Block grace time: 7days; Inode grace time: 7days
                        Space limits              File limits
User            used    soft    hard    grace    used    soft    hard    grace
----------------------------------------------------------------
root      --    20K     0K      0K               2       0       0
jack      --    0K      10K     20K              0       3       5
lucy      --    0K      30K     50K              0       3       5

Statistics:
Total blocks: 7
Data blocks: 1
Entries: 3
Used average: 3.000000

[root@localhost ~]# quota -uv lucy
Disk quotas for user lucy (uid 1001):
    Filesystem  blocks  quota   limit   grace   files   quota   limit   grace
    /dev/sdc1   0       0       0               0       0       0
    /dev/sdc2   0       30      50              0       3       5
```

图 3-27　显示指定文件系统的使用和配额的摘要、磁盘已使用的空间与限制

（4）切换到用户 lucy，进行测试，如图 3-28、图 3-29 所示。

```
[root@localhost ~]# su lucy
[lucy@localhost root]$ cd /gx2
[lucy@localhost gx2]$ touch cs{1..5}.txt
[lucy@localhost gx2]$ touch cs6.txt
touch: 无法创建"cs6.txt": 超出磁盘限额
[lucy@localhost gx2]$ dd if=/dev/zero of=/gx2/cs1.txt bs=20K count=1
记录了1+0 的读入
记录了1+0 的写出
20480字节(20 kB)已复制，0.000668081 秒，30.7 MB/秒
[lucy@localhost gx2]$ dd if=/dev/zero of=/gx2/cs2.txt bs=60K count=1
sdc2: warning, user block quota exceeded.
sdc2: write failed, user block limit reached.
dd: 写入"/gx2/cs2.txt" 出错：超出磁盘限额
记录了1+0 的读入
记录了0+0 的写出
28672字节(29 kB)已复制，0.000487753 秒，58.8 MB/秒
```

图 3-28　磁盘配额限制已生效

这里，可以使用命令 edquota [-p <源用户>] [-ug] [用户或群组名称]，实现用户之间的配额限制信息的套用。例如，使用 edquota -p lucy -u jack 命令，将用户 lucy 的配额限制信息套用给用户 jack，如图 3-30 所示。

也可以创建用户 alex，设置用户组 alex 对目录/gx2 的磁盘配额进行限制，配置为：磁盘容量的软限制和硬限制分别为 10 KB、30 KB，创建文件数的软限制和硬限制分别为 3 个、5 个，如图 3-31 所示，查看配置的结果，如图 3-32 所示。

```
[root@localhost gx2]# repquota -auvs
*** Report for user quotas on device /dev/sdc1
Block grace time: 7days; Inode grace time: 7days
                   Space limits                  File limits
User         used    soft    hard  grace   used  soft  hard  grace
----------------------------------------------------------------
root    --     0K      0K      0K              3     0     0
jack    --     0K   3072K   5120K              0     3     6

*** Status for user quotas on device /dev/sdc1
Accounting: ON; Enforcement: ON
Inode: #67 (2 blocks, 2 extents)

*** Report for user quotas on device /dev/sdc2
Block grace time: 7days; Inode grace time: 7days
                   Space limits                  File limits
User         used    soft    hard  grace   used  soft  hard  grace
----------------------------------------------------------------
root    --    20K      0K      0K              2     0     0
jack    --     0K     10K     20K              0     3     5
lucy    ++    48K     30K     50K  6days       5     3     5  6days

Statistics:
Total blocks: 7
Data blocks: 1
Entries: 3
Used average: 3.000000
```

图 3-29　显示指定文件系统的使用和配额的摘要

```
[root@localhost ~]# edquota -p lucy -u jack
[root@localhost ~]# quota -uv jack
Disk quotas for user jack (uid 1000):
   Filesystem  blocks   quota   limit   grace   files   quota   limit   grace
     /dev/sdc1      0       0       0                0       0       0
     /dev/sdc2      0      30      50                0       3       5
```

图 3-30　将用户 lucy 的配额限制信息套用给用户 jack

```
Disk quotas for group alex (gid 1003):
   Filesystem           blocks    soft    hard   inodes    soft    hard
   /dev/sdc2                 0      10      30        0       3       5
   /dev/sdc1                 0       0       0        0       0       0
```

图 3-31　设置磁盘配额限制

```
[root@localhost ~]# quota -vg alex
Disk quotas for group alex (gid 1003):
   Filesystem  blocks   quota   limit   grace   files   quota   limit   grace
     /dev/sdc2      0      10      30                0       3       5
     /dev/sdc1      0       0       0                0       0       0
```

图 3-32　查看配置的结果

测试该用户是否继承了用户组 alex 的磁盘配额设置信息，如图 3-33 所示。

```
[root@localhost ~]# useradd alex
[root@localhost ~]# quotaoff -aguv
/dev/sdc2 [/gx2]: group quotas turned off
/dev/sdc2 [/gx2]: user quotas turned off
[root@localhost ~]# edquota -g alex
[root@localhost ~]# quotaon -aguv
/dev/sdc2 [/gx2]: group quotas turned on
/dev/sdc2 [/gx2]: user quotas turned on
```

```
[root@localhost ~]# su alex
[alex@localhost root]$ cd /gx2
[alex@localhost gx2]$ touch cc{1..5}.txt
sdc2: warning, group file quota exceeded.
[alex@localhost gx2]$ touch cc6.txt
sdc2: write failed, group file limit reached.
touch: 无法创建"cc6.txt": 超出磁盘限额
[alex@localhost gx2]$ ls /gx2
aquota.group  aquota.user  cc1.txt  cc2.txt  cc3.txt  cc4.txt  cc5.txt  lost+found
[alex@localhost gx2]$ dd if=/dev/zero of=/gx2/cc1.txt bs=15K count=1
sdc2: warning, group block quota exceeded.
记录了1+0 的读入
记录了1+0 的写出
15360字节(15 kB)已复制, 0.000799459 秒, 19.2 MB/秒
[alex@localhost gx2]$ dd if=/dev/zero of=/gx2/cc2.txt bs=60K count=1
sdc2: write failed, group block limit reached.
dd: 写入"/gx2/cc2.txt" 出错: 超出磁盘限额
记录了1+0 的读入
记录了0+0 的写出
12288字节(12 kB)已复制, 0.00039822 秒, 30.9 MB/秒
```

图 3-33 磁盘配额限制已生效

从图 3-33 中可以看出，创建文件时，系统出现了警告提示，这是因为创建的文件个数（5 个）超过了用户组的磁盘配额文件个数的软限制个数（3 个）。

```
[alex@localhost gx2]$ touch cc{1..5}.txt
sdc2: warning,group file quota exceeded.
```

任务 3.3 动态调整硬盘分区

3.3.1 LVM 简介

1. LVM 基本概念

LVM(Logical Volume Manager，逻辑卷管理)是 Linux 系统对硬盘分区进行管理的一种机制。它是在硬盘分区和文件系统之间添加的一个逻辑层，为文件系统屏蔽下层硬盘分区布局，并提供了一个抽象的卷组，可以把多块硬盘进行卷组合并。管理员利用 LVM 可以不必关心物理硬盘设备的底层架构和布局，就能实现对硬盘分区的动态调整。如图 3-34 所示显示了 LVM 的基本结构。

2. LVM 常用的术语

① 物理存储介质(Physical Storage Media)：系统的物理存储设备，包括物理硬盘、RAID 磁盘阵列、硬盘分区等，如：/dev/hda、/dev/sda。

② 物理卷(Physical Volume,PV)：是 LVM 的基本存储逻辑块，包含与 LVM 相关的管理参数。创建物理卷可以用硬盘分区，也可以用硬盘本身。

③ 卷组(Volume Group,VG)：类似于非 LVM 系统中的物理磁盘，一个卷组由一个或多个物理卷组成，可以在卷组上创建一个或多个逻辑卷。

④ 逻辑卷(Logical Volume,LV)：类似于非 LVM 系统中的磁盘分区，逻辑卷建立在卷组之上，可以在逻辑卷之上建立文件系统。

⑤ 物理块(Physical Extent,PE)：物理卷中可以分配的最小存储单元，它的大小可以配

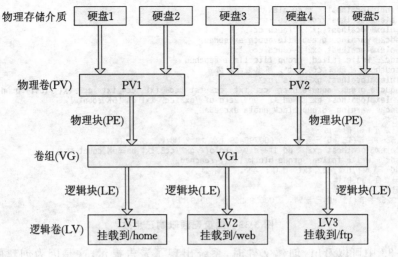

图 3-34 LVM 的基本结构

置,默认为 4MB,物理卷(PV)由大小等同的基本单元 PE 组成。

⑥ 逻辑块(Logical Extent,LE):逻辑卷中可以分配的最小存储单元,在同一个卷组中,LE 的大小和 PE 是相同的,并且一一对应。

3. LVM 常用命令

使用 LVM 进行硬盘分区的动态调整、管理时,需要依次创建、配置物理卷、卷组、逻辑卷,常用的 LVM 命令如表 3-5 所示。

表 3-5 常用的 LVM 命令

功能/命令	物理卷管理	卷组管理	逻辑卷管理
扫描	pvscan	vgscan	lvscan
建立	pvcreate	vgcreate	lvcreate
显示	pvdisplay	vgdisplay	lvdisplay
删除	pvremove	vgremove	lvremove
扩展		vgextend	lvextend
缩小		vgreduce	lvreduce

3.3.2 使用 LVM 进行硬盘分区的动态调整

1. 扩容根分区

实现步骤如下:

(1) 为系统添加一块新的硬盘/dev/sdd,并使用命令 df、lsblk,分别查看文件系统的磁盘使用情况,列出所有可用块设备的信息以及它们之间的依赖关系,如图 3-35 所示。

从图 3-35 中可以看出,当前系统的根分区大小为 17 GB,挂载到目录/dev/mapper/rhel

```
[root@localhost ~]# df -h
文件系统                容量    已用    可用   已用%  挂载点
/dev/mapper/rhel-root   17G    3.2G    14G    19%   /
devtmpfs                897M    0      897M   0%    /dev
tmpfs                   912M    0      912M   0%    /dev/shm
tmpfs                   912M   9.0M    903M   1%    /run
tmpfs                   912M    0      912M   0%    /sys/fs/cgroup
/dev/sdc2               5.8G    25M    5.5G   1%    /gx2
/dev/sdc1               4.0G    33M    4.0G   1%    /gx1
/dev/sda1               1014M  179M    836M   18%   /boot
tmpfs                   183M    20K    183M   1%    /run/user/0
[root@localhost ~]# lsblk
NAME          MAJ:MIN RM  SIZE RO TYPE MOUNTPOINT
sda             8:0    0   20G  0 disk
├─sda1          8:1    0    1G  0 part /boot
└─sda2          8:2    0   19G  0 part
  ├─rhel-root 253:0    0   17G  0 lvm  /
  └─rhel-swap 253:1    0    2G  0 lvm  [SWAP]
sdb             8:16   0   20G  0 disk
sdc             8:32   0   20G  0 disk
├─sdc1          8:33   0    4G  0 part /gx1
└─sdc2          8:34   0    6G  0 part /gx2
sdd             8:48   0   20G  0 disk
sr0            11:0    1 1024M  0 rom
```

图 3-35　查看文件系统的磁盘、可用块设备的信息

-root 上。同时,根分区在硬盘 sda 的第二个分区 sda2 上,sda2 划分为逻辑卷 rhel-root,然后以逻辑卷的形式挂载到了"/"下,根分区大小为 17 GB。

（2）在硬盘/dev/sdd 上创建分区 dev/sdd1,容量大小为 10 GB,分区结果如图 3-36 所示。

```
命令(输入 m 获取帮助): P

磁盘 /dev/sdd: 21.5 GB, 21474836480 字节, 41943040 个扇区
Units = 扇区 of 1 * 512 = 512 bytes
扇区大小(逻辑/物理): 512 字节 / 512 字节
I/O 大小(最小/最佳): 512 字节 / 512 字节
磁盘标签类型: dos
磁盘标识符: 0x7fc63d0a

   设备 Boot      Start         End      Blocks   Id  System
/dev/sdd1         2048    20973567    10485760   83   Linux

命令(输入 m 获取帮助): q

[root@localhost ~]# lsblk
NAME          MAJ:MIN RM  SIZE RO TYPE MOUNTPOINT
sda             8:0    0   20G  0 disk
├─sda1          8:1    0    1G  0 part /boot
└─sda2          8:2    0   19G  0 part
  ├─rhel-root 253:0    0   17G  0 lvm  /
  └─rhel-swap 253:1    0    2G  0 lvm  [SWAP]
sdb             8:16   0   20G  0 disk
sdc             8:32   0   20G  0 disk
├─sdc1          8:33   0    4G  0 part /gx1
└─sdc2          8:34   0    6G  0 part /gx2
sdd             8:48   0   20G  0 disk
└─sdd1          8:49   0   10G  0 part
sr0            11:0    1 1024M  0 rom
```

图 3-36　创建分区并进行查看

（3）使用分区 dev/sdd1 创建物理卷后，查看系统当前的物理卷信息，可以看到已经创建了名为/dev/sdd1 的物理卷，容量为 10 GB，如图 3-37 所示。

```
[root@localhost ~]# pvcreate /dev/sdd1
```

```
[root@localhost ~]# pvcreate /dev/sdd1
  Physical volume "/dev/sdd1" successfully created.
[root@localhost ~]# pvdisplay
  --- Physical volume ---
  PV Name               /dev/sda2
  VG Name               rhel
  PV Size               ⊲19.00 GiB / not usable 3.00 MiB
  Allocatable           yes (but full)
  PE Size               4.00 MiB
  Total PE              4863
  Free PE               0
  Allocated PE          4863
  PV UUID               MCLI8p-w32z-KOGo-Wu8I-hyCR-Cckb-3hzpr8

  "/dev/sdd1" is a new physical volume of "10.00 GiB"
  --- NEW Physical volume ---
  PV Name               /dev/sdd1
  VG Name
  PV Size               10.00 GiB
  Allocatable           NO
  PE Size               0
  Total PE              0
  Free PE               0
  Allocated PE          0
  PV UUID               D4TDFX-Vkhd-Wnii-MJPY-KqKn-pKcI-ZZnix1
```

图 3-37 创建物理卷

这时查看系统的卷组信息，卷组名称为 rhel，容量为 19 GB，将新创建的物理卷（/dev/sdd1）添加到卷组 rhel 中，如图 3-38 所示。

```
[root@localhost ~]# vgextend rhel /dev/sdd1
```

```
[root@localhost ~]# vgdisplay
  --- Volume group ---
  VG Name               rhel
  System ID
  Format                lvm2
  Metadata Areas        1
  Metadata Sequence No  3
  VG Access             read/write
  VG Status             resizable
  MAX LV                0
  Cur LV                2
  Open LV               2
  Max PV                0
  Cur PV                1
  Act PV                1
  VG Size               ⊲19.00 GiB
  PE Size               4.00 MiB
  Total PE              4863
  Alloc PE / Size       4863 / ⊲19.00 GiB
  Free  PE / Size       0 / 0
  VG UUID               CyR9bO-U0xo-oF2U-taBK-N3Bd-EbyP-kOQwWl

[root@localhost ~]# vgextend rhel /dev/sdd1
  Volume group "rhel" successfully extended
```

图 3-38 将物理卷添加到卷组

再次查看系统的卷组信息,卷组 rhel 的容量已经扩容到 28.99 GB,也就是增加了物理卷 /dev/sdd1 的容量,卷组已分配的空间为 19 GB,未分配的空间为 10 GB,如图 3-39 所示。

```
[root@localhost ~]# vgdisplay
  --- Volume group ---
  VG Name               rhel
  System ID
  Format                lvm2
  Metadata Areas        2
  Metadata Sequence No  4
  VG Access             read/write
  VG Status             resizable
  MAX LV                0
  Cur LV                2
  Open LV               2
  Max PV                0
  Cur PV                2
  Act PV                2
  VG Size               28.99 GiB
  PE Size               4.00 MiB
  Total PE              7422
  Alloc PE / Size       4863 / <19.00 GiB
  Free  PE / Size       2559 / <10.00 GiB
  VG UUID               CyR9bO-U0x0-oF2U-taBK-N3Bd-EbyP-k0QwWl
```

图 3-39　查看系统的卷组信息

(4) 查看系统逻辑卷信息,系统根分区的逻辑卷名称为 root,路径为 /dev/rhel/root,如图 3-40 所示。

```
[root@localhost ~]# lvdisplay
  --- Logical volume ---
  LV Path                /dev/rhel/swap
  LV Name                swap
  VG Name                rhel
  LV UUID                iekKvU-86fj-z2hx-LxuR-ol1i-hhLQ-Ag01L3
  LV Write Access        read/write
  LV Creation host, time localhost, 2020-03-03 00:32:37 +0800
  LV Status              available
  # open                 2
  LV Size                2.00 GiB
  Current LE             512
  Segments               1
  Allocation             inherit
  Read ahead sectors     auto
  - currently set to     8192
  Block device           253:1

  --- Logical volume ---
  LV Path                /dev/rhel/root
  LV Name                root
  VG Name                rhel
  LV UUID                9QjhBI-v8yc-G9qM-3UE9-Vx3e-VE3k-3YMyAb
  LV Write Access        read/write
  LV Creation host, time localhost, 2020-03-03 00:32:38 +0800
  LV Status              available
  # open                 1
  LV Size                <17.00 GiB
  Current LE             4351
  Segments               1
  Allocation             inherit
  Read ahead sectors     auto
  - currently set to     8192
  Block device           253:0
```

图 3-40　查看系统逻辑卷信息

将卷组 rhel 未分配的空间(10 GB)扩展到根分区逻辑卷中,使用命令 xfs_growfs(针对文件系统 xfs)或者 resize2fs(针对文件系统 ext2、ext3、ext4)进行在线扩容,如图 3-41 所示。

```
[root@localhost ~]# lvextend -l +100% FREE /dev/rhel/root
[root@localhost ~]# xfs_growfs /dev/rhel/root
```

```
[root@localhost ~]# lvextend -l +100%FREE /dev/rhel/root
  Size of logical volume rhel/root changed from <17.00 GiB (4351 extents) to 26.99 GiB (6910 extents).
  Logical volume rhel/root successfully resized.
[root@localhost ~]# xfs_growfs /dev/rhel/root
meta-data=/dev/mapper/rhel-root  isize=512    agcount=4, agsize=1113856 blks
         =                       sectsz=512   attr=2, projid32bit=1
         =                       crc=1        finobt=0 spinodes=0
data     =                       bsize=4096   blocks=4455424, imaxpct=25
         =                       sunit=0      swidth=0 blks
naming   =version 2              bsize=4096   ascii-ci=0 ftype=1
log      =internal               bsize=4096   blocks=2560, version=2
         =                       sectsz=512   sunit=0 blks, lazy-count=1
realtime =none                   extsz=4096   blocks=0, rtextents=0
data blocks changed from 4455424 to 7075840
```

图 3-41 在线扩容

再次查看系统逻辑卷信息,系统根分区的逻辑卷的容量已经扩容到 26.99 GB,如图 3-42 所示。

```
[root@localhost ~]# lvdisplay
  --- Logical volume ---
  LV Path                /dev/rhel/swap
  LV Name                swap
  VG Name                rhel
  LV UUID                iekKvU-86fj-z2hx-LxuR-ol1i-hhLQ-Ag01L3
  LV Write Access        read/write
  LV Creation host, time localhost, 2020-03-03 00:32:37 +0800
  LV Status              available
  # open                 2
  LV Size                2.00 GiB
  Current LE             512
  Segments               1
  Allocation             inherit
  Read ahead sectors     auto
  - currently set to     8192
  Block device           253:1

  --- Logical volume ---
  LV Path                /dev/rhel/root
  LV Name                root
  VG Name                rhel
  LV UUID                9QjhBI-v8yc-G9qM-3UE9-Vx3e-VE3k-3YMyAb
  LV Write Access        read/write
  LV Creation host, time localhost, 2020-03-03 00:32:38 +0800
  LV Status              available
  # open                 1
  LV Size                26.99 GiB
  Current LE             6910
  Segments               2
  Allocation             inherit
  Read ahead sectors     auto
  - currently set to     8192
  Block device           253:0
```

图 3-42 扩容成功

再次查看文件系统的磁盘使用情况以及所有可用块设备的信息,根分区已经从 17 GB 扩容到 27 GB,如图 3-43 所示。

2. 创建使用 LVM

实现步骤如下:

(1) 为系统添加两块新的硬盘/dev/sde、/dev/sdf(大小均为 10 GB),并将两块硬盘创建成物理卷,从而支持 LVM,如图 3-44 所示。

```
[root@localhost ~]# pvcreate /dev/sd{e,f}
```

可以使用 pvremoce 删除物理卷。

```
[root@localhost ~]# df -h
文件系统                   容量    已用   可用  已用%  挂载点
/dev/mapper/rhel-root      27G    3.2G   24G   12%   /
devtmpfs                   897M   0      897M  0%    /dev
tmpfs                      912M   0      912M  0%    /dev/shm
tmpfs                      912M   9.0M   903M  1%    /run
tmpfs                      912M   0      912M  0%    /sys/fs/cgroup
/dev/sdc2                  5.8G   25M    5.5G  1%    /gx2
/dev/sdc1                  4.0G   33M    4.0G  1%    /gx1
/dev/sda1                  1014M  179M   836M  18%   /boot
tmpfs                      183M   20K    183M  1%    /run/user/0
[root@localhost ~]# lsblk
NAME         MAJ:MIN RM   SIZE RO TYPE MOUNTPOINT
sda            8:0    0    20G  0 disk
├─sda1         8:1    0     1G  0 part /boot
└─sda2         8:2    0    19G  0 part
  ├─rhel-root 253:0   0    27G  0 lvm  /
  └─rhel-swap 253:1   0     2G  0 lvm  [SWAP]
sdb            8:16   0    20G  0 disk
sdc            8:32   0    20G  0 disk
├─sdc1         8:33   0     4G  0 part /gx1
└─sdc2         8:34   0     6G  0 part /gx2
sdd            8:48   0    20G  0 disk
└─sdd1         8:49   0    10G  0 part
  └─rhel-root 253:0   0    27G  0 lvm  /
sr0           11:0    1  1024M  0 rom
```

图 3-43　查看文件系统的磁盘、可用块设备的信息

```
[root@localhost ~]# pvcreate /dev/sd{e,f}
  Physical volume "/dev/sde" successfully created.
  Physical volume "/dev/sdf" successfully created.
```

图 3-44　创建物理卷

```
[root@localhost ~]# pvremove /dev/sd{e,f}
```

（2）创建卷组 vg1，如图 3-45 所示。

```
[root@localhost ~]# vgcreate vg1 /dev/sd{e,f}
```

```
[root@localhost ~]# vgcreate vg1 /dev/sd{e,f}
  Volume group "vg1" successfully created
[root@localhost ~]# vgdisplay vg1
  --- Volume group ---
  VG Name               vg1
  System ID
  Format                lvm2
  Metadata Areas        2
  Metadata Sequence No  1
  VG Access             read/write
  VG Status             resizable
  MAX LV                0
  Cur LV                0
  Open LV               0
  Max PV                0
  Cur PV                2
  Act PV                2
  VG Size               19.99 GiB
  PE Size               4.00 MiB
  Total PE              5118
  Alloc PE / Size       0 / 0
  Free  PE / Size       5118 / 19.99 GiB
  VG UUID               GCc2Ta-1Jn3-0ecx-2FXk-ld7H-nTDG-6GEeVg
```

图 3-45　创建卷组

从图 3-45 中可以看到卷组 vg1 的容量为 19.99 GB。

可以使用 vgremoce 删除卷组。

```
[root@localhost ~]# vgremove vg1
```

（3）创建逻辑卷 lv1，其容量为 3 GB，如图 3-46 所示。

```
[root@localhost~]# lvcreate -L 3G -n lv1 vg1
```

```
[root@localhost ~]# lvcreate -L 3G -n lv1 vg1
  Logical volume "lv1" created.
[root@localhost ~]# lvdisplay /dev/vg1/lv1
  --- Logical volume ---
  LV Path                /dev/vg1/lv1
  LV Name                lv1
  VG Name                vg1
  LV UUID                QoCsc6-OgwU-4AIU-avGN-Ndkm-PwYc-nrmR6X
  LV Write Access        read/write
  LV Creation host, time localhost.localdomain, 2020-03-15 23:35:36 +0800
  LV Status              available
  # open                 0
  LV Size                3.00 GiB
  Current LE             768
  Segments               1
  Allocation             inherit
  Read ahead sectors     auto
  - currently set to     8192
  Block device           253:2
```

图 3-46　创建逻辑卷

可以使用 lvremoce 删除逻辑卷。

```
[root@localhost~]# lvremove /dev/vg1/lv1
```

（4）格式化逻辑卷，并挂载到目录/data 下，然后使用，如图 3-47 所示。

```
[root@localhost~]# mkfs.ext4 /dev/vg1/lv1
```

```
[root@localhost ~]# mkdir /data
[root@localhost ~]# mount /dev/vg1/lv1 /data
[root@localhost ~]# df -h
文件系统                容量   已用  可用  已用% 挂载点
/dev/mapper/rhel-root   27G   3.2G  24G   12%   /
devtmpfs                897M  0     897M  0%    /dev
tmpfs                   912M  0     912M  0%    /dev/shm
tmpfs                   912M  9.0M  903M  1%    /run
tmpfs                   912M  0     912M  0%    /sys/fs/cgroup
/dev/sdc1               4.0G  33M   4.0G  1%    /gx1
/dev/sda1               1014M 179M  836M  18%   /boot
/dev/sdc2               5.8G  25M   5.5G  1%    /gx2
tmpfs                   183M  16K   183M  1%    /run/user/0
/dev/mapper/vg1-lv1     2.9G  9.0M  2.8G  1%    /data
```

图 3-47　格式化逻辑卷并进行挂载

（5）扩容逻辑卷，对逻辑卷 lv1 进行扩展，将其在线扩容到 5 GB，如图 3-48 所示。

```
[root@localhost~]# umount /data
[root@localhost~]# lvextend -L +2G /dev/vg1/lv1
```

```
[root@localhost ~]# lvextend -L +2G /dev/vg1/lv1
  Size of logical volume vg1/lv1 changed from 3.00 GiB (768 extents) to 5.00 GiB (1280 extents).
  Logical volume vg1/lv1 successfully resized.
```

图 3-48　扩容逻辑卷

检查硬盘完整性，使用命令 resize2fs 进行在线扩容，如图 3-49 所示。重新将逻辑卷挂载到目录/data 下，如图 3-50 所示。

```
[root@localhost~]# e2fsck -f /dev/vg1/lv1
[root@localhost~]# resize2fs /dev/vg1/lv1
```

（6）缩小逻辑卷，将逻辑卷 lv1 的容量减小至 4 GB，如图 3-51 所示。

```
[root@localhost~]# umount /data
[root@localhost~]# lvreduce -L -1G /dev/vg1/lv1
```

```
[root@localhost ~]# e2fsck -f /dev/vg1/lv1
e2fsck 1.42.9 (28-Dec-2013)
第一步：检查inode,块,和大小
第二步：检查目录结构
第3步：检查目录连接性
Pass 4: Checking reference counts
第5步：检查簇概要信息
/dev/vg1/lv1: 11/196608 files (0.0% non-contiguous), 31036/786432 blocks
[root@localhost ~]# resize2fs /dev/vg1/lv1
resize2fs 1.42.9 (28-Dec-2013)
Resizing the filesystem on /dev/vg1/lv1 to 1310720 (4k) blocks.
The filesystem on /dev/vg1/lv1 is now 1310720 blocks long.
```

图 3-49 在线扩容

```
[root@localhost ~]# mount /dev/vg1/lv1 /data/
[root@localhost ~]# df -h
文件系统              容量    已用   可用   已用%  挂载点
/dev/mapper/rhel-root  27G   3.2G   24G    12%   /
devtmpfs              897M    0    897M    0%   /dev
tmpfs                 912M    0    912M    0%   /dev/shm
tmpfs                 912M  9.0M   903M    1%   /run
tmpfs                 912M    0    912M    0%   /sys/fs/cgroup
/dev/sdc1             4.0G   33M   4.0G    1%   /gx1
/dev/sda1            1014M  179M   836M   18%   /boot
/dev/sdc2             5.8G   25M   5.5G    1%   /gx2
tmpfs                 183M   16K   183M    1%   /run/user/0
/dev/mapper/vg1-lv1   4.9G   12M   4.6G    1%   /data
```

图 3-50 将逻辑卷挂载到目录/data下

```
[root@localhost ~]# lvreduce -L -1G /dev/vg1/lv1
  WARNING: Reducing active and open logical volume to 4.00 GiB.
  THIS MAY DESTROY YOUR DATA (filesystem etc.)
Do you really want to reduce vg1/lv1? [y/n]: y
  Size of logical volume vg1/lv1 changed from 5.00 GiB (1280 extents) to 4.00 GiB (1024 extents).
  Logical volume vg1/lv1 successfully resized.
```

图 3-51 缩小逻辑卷

使用命令 resize2fs 进行在线扩容调整，重新将逻辑卷挂载到目录/data 下，如图 3-52 所示。

```
[root@localhost ~]# mkfs.ext4 /dev/vg1/lv1
[root@localhost ~]# e2fsck -f /dev/vg1/lv1
[root@localhost ~]# resize2fs /dev/vg1/lv1
```

```
[root@localhost ~]# df -h
文件系统              容量    已用   可用   已用%  挂载点
/dev/mapper/rhel-root  27G   3.2G   24G    12%   /
devtmpfs              897M    0    897M    0%   /dev
tmpfs                 912M    0    912M    0%   /dev/shm
tmpfs                 912M  9.0M   903M    1%   /run
tmpfs                 912M    0    912M    0%   /sys/fs/cgroup
/dev/sdc1             4.0G   33M   4.0G    1%   /gx1
/dev/sda1            1014M  179M   836M   18%   /boot
/dev/sdc2             5.8G   25M   5.5G    1%   /gx2
tmpfs                 183M   16K   183M    1%   /run/user/0
/dev/mapper/vg1-lv1   3.9G   16M   3.6G    1%   /data
```

图 3-52 查看文件系统的磁盘信息

3. 逻辑卷快照

LVM 快照卷的功能，类似 Windows 中的系统还原点功能、VMware Workstation 的快照功能。在创建快照时，仅拷贝原始卷里数据的元数据，并不会有数据的物理拷贝，因此快照的创建几乎是瞬间完成的,当在原始卷上执行写操作时,快照跟踪原始卷块的改变,这个时候原始卷上将要改变的数据在改变之前被拷贝到快照预留的空间,其特点有:

① 快照卷的容量必须与逻辑卷的容量相同。

② 快照卷仅一次有效,在执行还原后就会自动删除。

实现步骤如下:

(1) 在目录/data 下,创建文件 1.txt。

```
[root@localhost~]# touch /data/1.txt
[root@localhost~]# ls /data
1.txt   lost+found
```

(2) 为逻辑卷 lv1 创建快照,如图 3-53、图 3-54 所示。

```
[root@localhost ~]# lvdisplay /dev/vg1/lv1
  --- Logical volume ---
  LV Path                /dev/vg1/lv1
  LV Name                lv1
  VG Name                vg1
  LV UUID                pDCQIa-WWxt-AlrR-NVHG-9hFi-QUH9-KeQnDq
  LV Write Access        read/write
  LV Creation host, time localhost.localdomain, 2020-03-16 00:00:50 +0800
  LV snapshot status     source of
                         kuaizhao [active]
  LV Status              available
  # open                 1
  LV Size                4.00 GiB
  Current LE             1024
  Segments               1
  Allocation             inherit
  Read ahead sectors     auto
  - currently set to     8192
  Block device           253:2
```

图 3-53 查看逻辑卷信息

```
[root@localhost ~]# lvdisplay /dev/vg1/kuaizhao
  --- Logical volume ---
  LV Path                /dev/vg1/kuaizhao
  LV Name                kuaizhao
  VG Name                vg1
  LV UUID                AoDyb8-dbOk-HCmo-4t6C-D8P2-fy13-D5LNOV
  LV Write Access        read/write
  LV Creation host, time localhost.localdomain, 2020-03-16 00:43:35 +0800
  LV snapshot status     active destination for lv1
  LV Status              available
  # open                 0
  LV Size                4.00 GiB
  Current LE             1024
  COW-table size         4.00 GiB
  COW-table LE           1024
  Allocated to snapshot  0.00%
  Snapshot chunk size    4.00 KiB
  Segments               1
  Allocation             inherit
  Read ahead sectors     auto
  - currently set to     8192
  Block device           253:5
```

图 3-54 查看快照卷信息

```
[root@localhost~]# lvcreate -L 4G -s -n kuaizhao /dev/vg1/lv1
```

其中,-L 是快照卷的大小,-s 是创建快照逻辑卷,-n 是快照卷名称。

可以使用以下命令,删除快照。

```
[root@localhost~]# lvremove /dev/vg1/kuaizhao
```

(3) 创建快照后,在目录/data下,创建文件2.txt。

```
[root@localhost~]# touch /data/2.txt
[root@localhost~]# ls /data/
1.txt  2.txt  lost+found
```

(4) 对逻辑卷进行快照还原操作(图3-55)。

```
[root@localhost~]# umount /data
[root@localhost~]# lvconvert --merge /dev/vg1/kuaizhao
```

```
[root@localhost ~]# umount /data
[root@localhost ~]# lvconvert --merge /dev/vg1/kuaizhao
  Merging of volume vg1/kuaizhao started.
  lv1: Merged: 100.00%
[root@localhost ~]# mount /dev/vg1/lv1 /data
[root@localhost ~]# ls /data
1.txt  lost+found
```

图3-55 快照还原

对逻辑卷进行快照还原操作、创建快照后,新建的文件2.txt被删除了,同时快照卷也被删除,如图3-56所示。

```
[root@localhost ~]# lvdisplay /dev/vg1/kuaizhao
  Failed to find logical volume "vg1/kuaizhao"
[root@localhost ~]# lvdisplay /dev/vg1/lv1
  --- Logical volume ---
  LV Path                /dev/vg1/lv1
  LV Name                lv1
  VG Name                vg1
  LV UUID                pDCQIa-WWxt-AlrR-NVHG-9hFi-QUH9-KeQnDq
  LV Write Access        read/write
  LV Creation host, time localhost.localdomain, 2020-03-16 00:00:50 +0800
  LV Status              available
  # open                 1
  LV Size                4.00 GiB
  Current LE             1024
  Segments               1
  Allocation             inherit
  Read ahead sectors     auto
  - currently set to     8192
  Block device           253:2
```

图3-56 快照卷已被自动删除

任务3.4 提升数据的安全性

3.4.1 RAID简介

1. RAID 基本概念

独立冗余磁盘阵列(Redundant Array of Independent Disks,RAID),简称磁盘阵列。

1988 年,美国加利福尼亚大学伯克利分校在发表的文章 *A Case for Redundant Arrays of Inexpensive Disks* 首次谈到了 RAID 这个词汇,并定义了 RAID 的 5 层级。

RAID 技术通过把多个硬盘设备组合成一个容量更大的、安全性更好的磁盘阵列,把数据切割成许多区段后分别放在不同的物理磁盘上,然后利用分散读写技术来提升磁盘阵列整体的性能,同时把多个重要数据的副本同步到不同的物理设备上,从而起到了非常好的数据冗余备份效果。

2. RAID 的分类

(1) RAID 0

RAID 0 是一种简单的、无数据校验的数据条带化技术,由相同容量的两块或两块以上的硬盘组成。这种模式下会先把硬盘分隔出大小相等的区块,当有数据需要写入硬盘时,会把数据也切割成相同大小的区块,然后依次将数据写入各个硬盘。RAID 0 具有低成本、高读写性能、硬盘利用率 100% 等优点,但是它不提供数据冗余保护,一旦数据损坏,将无法恢复。RAID 0 如图 3-57 所示。

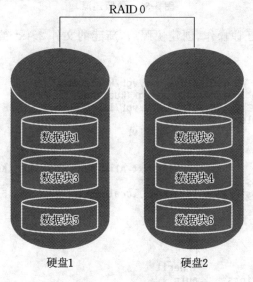

图 3-57 RAID 0

(2) RAID 1

RAID 1 也称为镜像卷,在 Windows 系统中,就可以实现镜像卷的配置。RAID 1 最少要有两个硬盘,并且硬盘数是偶数。以两个硬盘为例,当数据在写入一块硬盘的同时,会在另一块硬盘上生成数据的镜像(备份)文件,硬盘利用率为 50%。RAID 1 就具备了数据冗余功能,一旦某一块硬盘发生故障,系统自动从另外一块硬盘中读取数据,从而保证数据恢复正常。RAID 1 如图 3-58 所示。

(3) RAID 5

RAID 5(分布式奇偶校验的独立磁盘结构)是将奇偶校验信息(parity)存储于所有硬盘上,并且奇偶校验信息与相对应的数据分别存储于不同的硬盘上。RAID 5 至少需要三块硬盘,并且每个硬盘必须提供相同的硬盘空间。当有一块硬盘损坏时,采用这个奇偶校验值进行数据恢复。RAID 5 兼顾存储性能、数据安全和存储成本等各方面,可以说它是

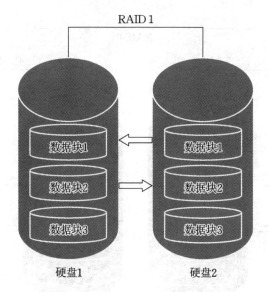

图 3-58　RAID 1

RAID 0、RAID 1 的综合体。RAID 5 如图 3-59 所示。

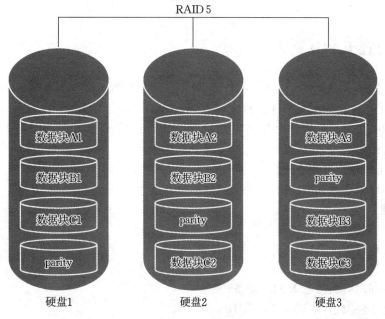

图 3-59　RAID 5

（4）RAID 10

RAID 10 是 RAID 1＋RAID 0 的结合体，它使用了镜像卷和数据条带化技术。RAID 10 至少需要四块硬盘。在使用时，两块硬盘为一组，每一组硬盘先创建 RAID 1 磁盘阵列，再将这两个 RAID 1 磁盘阵列组成 RAID 0，这样的组成方式，既具有 RAID 0 的性能优点，也具有 RAID 1 的数据冗余优点。RAID 10 如图 3-60 所示。

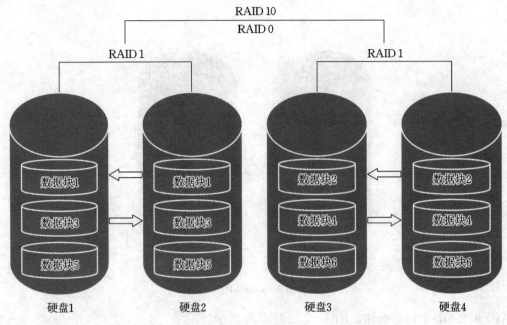

图 3-60 RAID 10

3.4.2 配置 RAID

在服务器上配置 RAID,有硬 RAID、软 RAID 两种配置方式。其中硬 RAID 通过使用 RAID 卡和主板上集成的 RAID 芯片来实现;软 RAID 通过软件实现。在 Linux 系统中可以使用 mdadm 命令来配置软 RAID,其用法为:mdadm[模式]＜RAID 设备名称＞[选项][成员设备名称]。

常见的模式与选项如下:
① Create(创建模式)
-C:创建;
-n:指定磁盘数量;
-x:热备磁盘的个数;
-l:指定 RAID 的级别;
-a:决定创建后面接的 RAID 设备;
-c:指定分块大小,默认为 512 KB;
-v:显示过程。
② Manage(管理模式)
-a:将后面的设备加入 RAID;
-r:将后面的设备从 RAID 中移除;
-f:模拟设备损坏;
-D:查看磁盘阵列详细信息;
-Q:查看磁盘阵列摘要信息;
-S:停止 RAID。

1. 创建 RAID 0

实现步骤如下：

（1）为系统添加两块新的硬盘/dev/sdb、dev/sdc，如图 3-61 所示。

```
[root@localhost ~]# fdisk -l
磁盘 /dev/sda：21.5 GB，21474836480 字节，41943040 个扇区
Units = 扇区 of 1 * 512 = 512 bytes
扇区大小(逻辑/物理)：512 字节 / 512 字节
I/O 大小(最小/最佳)：512 字节 / 512 字节
磁盘标签类型：dos
磁盘标识符：0x0000c070

   设备 Boot      Start         End      Blocks   Id  System
/dev/sda1   *        2048     2099199     1048576   83  Linux
/dev/sda2         2099200    41943039    19921920   8e  Linux LVM

磁盘 /dev/sdb：21.5 GB，21474836480 字节，41943040 个扇区
Units = 扇区 of 1 * 512 = 512 bytes
扇区大小(逻辑/物理)：512 字节 / 512 字节
I/O 大小(最小/最佳)：512 字节 / 512 字节


磁盘 /dev/sdc：21.5 GB，21474836480 字节，41943040 个扇区
Units = 扇区 of 1 * 512 = 512 bytes
扇区大小(逻辑/物理)：512 字节 / 512 字节
I/O 大小(最小/最佳)：512 字节 / 512 字节
```

图 3-61　查看系统所有分区类型

（2）使用 mdadm 命令创建 RAID 0，通过文件/proc/mdstat 查看 RAID 级别以及状态等信息，如图 3-62 所示，并对 RAID 0 进行格式化后，挂载使用，如图 3-63、图 3-64 所示。

```
[root@localhost ~]# mdadm -Cv /dev/md0 -a yes -n 2 -l 0 /dev/sd{b,c}
```

```
[root@localhost ~]# mdadm -Cv /dev/md0 -a yes -n 2 -l 0 /dev/sd{b,c}
mdadm: chunk size defaults to 512K
mdadm: Defaulting to version 1.2 metadata
mdadm: array /dev/md0 started.
[root@localhost ~]# cat /proc/mdstat
Personalities : [raid0]
md0 : active raid0 sdc[1] sdb[0]
      41910272 blocks super 1.2 512k chunks

unused devices: <none>
```

图 3-62　创建 RAID 0

```
[root@localhost ~]# mkfs.ext4 /dev/md0
```

```
[root@localhost ~]# mkdir /raid0
[root@localhost ~]# mount /dev/md0 /raid0/
[root@localhost ~]# mdadm -D /dev/md0
/dev/md0:
           Version : 1.2
     Creation Time : Tue Mar 17 18:29:53 2020
        Raid Level : raid0
        Array Size : 41910272 (39.97 GiB 42.92 GB)
      Raid Devices : 2
     Total Devices : 2
       Persistence : Superblock is persistent

       Update Time : Tue Mar 17 18:29:53 2020
             State : clean
    Active Devices : 2
   Working Devices : 2
    Failed Devices : 0
     Spare Devices : 0

        Chunk Size : 512K

Consistency Policy : none

              Name : localhost.localdomain:0  (local to host localhost.localdomain)
              UUID : e0ed8073:42638a9e:d311835d:13d3758f
            Events : 0

    Number   Major   Minor   RaidDevice State
       0       8       16        0      active sync   /dev/sdb
       1       8       32        1      active sync   /dev/sdc
```

图 3-63　查看 RAID 0 信息

```
[root@localhost ~]# df -h
文件系统              容量    已用    可用   已用%  挂载点
/dev/mapper/rhel-root  17G   3.2G    14G    19%   /
devtmpfs              897M     0    897M    0%   /dev
tmpfs                 912M     0    912M    0%   /dev/shm
tmpfs                 912M   9.1M   903M    1%   /run
tmpfs                 912M     0    912M    0%   /sys/fs/cgroup
/dev/sda1            1014M   179M   836M   18%   /boot
tmpfs                 183M   4.0K   183M    1%   /run/user/42
tmpfs                 183M    24K   183M    1%   /run/user/0
/dev/md0               40G    49M    38G    1%   /radi0
```

图 3-64　查看文件系统的磁盘信息

2. 创建 RAID 1

实现步骤如下：

（1）取消 RAID 0 的挂载，并停止之前 RAID 0。

```
[root@localhost~]# umount /raid0
[root@localhost~]# mdadm -S /dev/md0
```

（2）创建 RAID 1，并查看磁盘阵列信息，如图 3-65 所示。

```
[root@localhost~]# mdadm -Cv /dev/md0 -a yes -n 2 -l 1 /dev/sd{b,c}
```

```
[root@localhost ~]# mdadm -D /dev/md0
/dev/md0:
           Version : 1.2
     Creation Time : Tue Mar 17 19:17:36 2020
        Raid Level : raid1
        Array Size : 20955136 (19.98 GiB 21.46 GB)
     Used Dev Size : 20955136 (19.98 GiB 21.46 GB)
      Raid Devices : 2
     Total Devices : 2
       Persistence : Superblock is persistent

       Update Time : Tue Mar 17 19:19:21 2020
             State : clean
    Active Devices : 2
   Working Devices : 2
    Failed Devices : 0
     Spare Devices : 0

Consistency Policy : resync

              Name : localhost.localdomain:0  (local to host localhost.localdomain)
              UUID : 0915dce0:2245d062:f561d104:a7458c5a
            Events : 17

    Number   Major   Minor   RaidDevice State
       0       8       16        0      active sync   /dev/sdb
       1       8       32        1      active sync   /dev/sdc
```

图 3-65　查看 RAID 1 信息

3. 创建 RAID 5

（1）为系统添加三块新的硬盘/dev/sdd、/dev/sde、dev/sdf，如图 3-66 所示。

```
磁盘 /dev/sdd：21.5 GB，21474836480 字节，41943040 个扇区
Units = 扇区 of 1 * 512 = 512 bytes
扇区大小(逻辑/物理)：512 字节 / 512 字节
I/O 大小(最小/最佳)：512 字节 / 512 字节

磁盘 /dev/sde：21.5 GB，21474836480 字节，41943040 个扇区
Units = 扇区 of 1 * 512 = 512 bytes
扇区大小(逻辑/物理)：512 字节 / 512 字节
I/O 大小(最小/最佳)：512 字节 / 512 字节

磁盘 /dev/sdf：21.5 GB，21474836480 字节，41943040 个扇区
Units = 扇区 of 1 * 512 = 512 bytes
扇区大小(逻辑/物理)：512 字节 / 512 字节
I/O 大小(最小/最佳)：512 字节 / 512 字节
```

图 3-66 添加三块新的硬盘

（2）创建 RAID 5，并查看磁盘阵列信息，如图 3-67 所示。

```
[root@localhost ~]# mdadm -Cv /dev/md1 -a yes -n 3 -l 5 /dev/sd{d,e,f}

[root@localhost ~]# mdadm -D /dev/md1
/dev/md1:
           Version : 1.2
     Creation Time : Tue Mar 17 19:41:12 2020
        Raid Level : raid5
        Array Size : 41910272 (39.97 GiB 42.92 GB)
     Used Dev Size : 20955136 (19.98 GiB 21.46 GB)
      Raid Devices : 3
     Total Devices : 3
       Persistence : Superblock is persistent

       Update Time : Tue Mar 17 19:42:58 2020
             State : clean
    Active Devices : 3
   Working Devices : 3
    Failed Devices : 0
     Spare Devices : 0

            Layout : left-symmetric
        Chunk Size : 512K

Consistency Policy : resync

              Name : localhost.localdomain:1  (local to host localhost.localdomain)
              UUID : 005edff0:d84354c5:d25e45c2:263f0232
            Events : 18

    Number   Major   Minor   RaidDevice State
       0       8       48        0      active sync   /dev/sdd
       1       8       64        1      active sync   /dev/sde
       3       8       80        2      active sync   /dev/sdf
```

图 3-67 查看 RAID 5 信息

4. 创建 RAID 10

为系统添加五块新的硬盘/dev/sdg、/dev/sdh、dev/sdi、dev/sdj、dev/sdk，并使用前四块硬盘创建 RAID 10，并查看磁盘阵列信息，如图 3-68 所示。

```
[root@localhost ~]# mdadm -Cv /dev/md2 -a yes -n 4 -l 10 /dev/sd{g,h,i,j}
```

```
[root@localhost ~]# mdadm -Cv /dev/md2 -a yes -n 4 -l 10 /dev/sd{g,h,i,j}
mdadm: layout defaults to n2
mdadm: layout defaults to n2
mdadm: chunk size defaults to 512K
mdadm: size set to 20955136K
mdadm: Defaulting to version 1.2 metadata
mdadm: array /dev/md2 started.
[root@localhost ~]# mdadm -D /dev/md2
/dev/md2:
           Version : 1.2
     Creation Time : Tue Mar 17 19:55:17 2020
        Raid Level : raid10
        Array Size : 41910272 (39.97 GiB 42.92 GB)
     Used Dev Size : 20955136 (19.98 GiB 21.46 GB)
      Raid Devices : 4
     Total Devices : 4
       Persistence : Superblock is persistent

       Update Time : Tue Mar 17 19:55:29 2020
             State : clean, resyncing
    Active Devices : 4
   Working Devices : 4
    Failed Devices : 0
     Spare Devices : 0

            Layout : near=2
        Chunk Size : 512K

Consistency Policy : resync

     Resync Status : 7% complete

              Name : localhost.localdomain:2  (local to host localhost.localdomain)
              UUID : bc8c9ddb:3c8a6b56:07f84531:3ee46eb8
            Events : 1

    Number   Major   Minor   RaidDevice State
       0       8       96        0      active sync set-A   /dev/sdg
       1       8      112        1      active sync set-B   /dev/sdh
       2       8      128        2      active sync set-A   /dev/sdi
       3       8      144        3      active sync set-B   /dev/sdj
```

图 3-68　查看 RAID 10 信息

5. 磁盘阵列的修复

如在上面创建的 RAID 10 中，使用 mdadm 的 -f 选项，模拟硬盘 dev/sdj 出现损坏的情况，如图 3-69 所示。

```
[root@localhost~]# mdadm /dev/md2 -f /dev/sdj
```

```
[root@localhost ~]# mdadm /dev/md2 -f /dev/sdj
mdadm: set /dev/sdj faulty in /dev/md2
[root@localhost ~]# mdadm -D /dev/md2
/dev/md2:
           Version : 1.2
     Creation Time : Tue Mar 17 19:55:17 2020
        Raid Level : raid10
        Array Size : 41910272 (39.97 GiB 42.92 GB)
     Used Dev Size : 20955136 (19.98 GiB 21.46 GB)
      Raid Devices : 4
     Total Devices : 4
       Persistence : Superblock is persistent

       Update Time : Tue Mar 17 20:05:42 2020
             State : clean, degraded
    Active Devices : 3
   Working Devices : 3
    Failed Devices : 1
     Spare Devices : 0

            Layout : near=2
        Chunk Size : 512K

Consistency Policy : resync

              Name : localhost.localdomain:2  (local to host localhost.localdomain)
              UUID : bc8c9ddb:3c8a6b56:07f84531:3ee46eb8
            Events : 19

    Number   Major   Minor   RaidDevice State
       0       8       96        0      active sync set-A   /dev/sdg
       1       8      112        1      active sync set-B   /dev/sdh
       2       8      128        2      active sync set-A   /dev/sdi
       -       0        0        3      removed
       3       8      144        -      faulty   /dev/sdj
```

图 3-69　硬盘出现损坏

从图 3-69 中可以看出，硬盘/dev/sdj 的状态为"faulty"，RAID 10 目前只有三块硬盘。

接下来，将硬盘 dev/sdk 添加到 RAID 10 里，以修复磁盘阵列，修复完成后，如图 3-70 所示，RAID 10 磁盘阵列硬盘数又变成了四个。

```
[root@localhost ~]# mdadm /dev/md2 -a /dev/sdk
[root@localhost ~]# mdadm -D /dev/md2
/dev/md2:
           Version : 1.2
     Creation Time : Tue Mar 17 19:55:17 2020
        Raid Level : raid10
        Array Size : 41910272 (39.97 GiB 42.92 GB)
     Used Dev Size : 20955136 (19.98 GiB 21.46 GB)
      Raid Devices : 4
     Total Devices : 5
       Persistence : Superblock is persistent

       Update Time : Tue Mar 17 20:10:54 2020
             State : clean
    Active Devices : 4
   Working Devices : 4
    Failed Devices : 1
     Spare Devices : 0

            Layout : near=2
        Chunk Size : 512K

Consistency Policy : resync

              Name : localhost.localdomain:2  (local to host localhost.localdomain)
              UUID : bc8c9ddb:3c8a6b56:07f84531:3ee46eb8
            Events : 38

    Number   Major   Minor   RaidDevice State
       0       8       96        0      active sync set-A   /dev/sdg
       1       8      112        1      active sync set-B   /dev/sdh
       2       8      128        2      active sync set-A   /dev/sdi
       4       8      160        3      active sync set-B   /dev/sdk

       3       8      144        -      faulty   /dev/sdj
```

图 3-70　修复 RAID 10

这时，可以将故障硬盘从磁盘阵列中删除，如图 3-71 所示。

```
[root@localhost ~]# mdadm /dev/md2 -r /dev/sdj
[root@localhost ~]# mdadm /dev/md2 -r /dev/sdj
mdadm: hot removed /dev/sdj from /dev/md2
[root@localhost ~]# mdadm -D /dev/md2
/dev/md2:
           Version : 1.2
     Creation Time : Tue Mar 17 19:55:17 2020
        Raid Level : raid10
        Array Size : 41910272 (39.97 GiB 42.92 GB)
     Used Dev Size : 20955136 (19.98 GiB 21.46 GB)
      Raid Devices : 4
     Total Devices : 4
       Persistence : Superblock is persistent

       Update Time : Tue Mar 17 20:13:57 2020
             State : clean
    Active Devices : 4
   Working Devices : 4
    Failed Devices : 0
     Spare Devices : 0

            Layout : near=2
        Chunk Size : 512K

Consistency Policy : resync

              Name : localhost.localdomain:2  (local to host localhost.localdomain)
              UUID : bc8c9ddb:3c8a6b56:07f84531:3ee46eb8
            Events : 39

    Number   Major   Minor   RaidDevice State
       0       8       96        0      active sync set-A   /dev/sdg
       1       8      112        1      active sync set-B   /dev/sdh
       2       8      128        2      active sync set-A   /dev/sdi
       4       8      160        3      active sync set-B   /dev/sdk
```

图 3-71　将故障硬盘从磁盘阵列中删除

6. 添加磁盘阵列的备份盘

使用四个硬盘创建的 RAID 10 中,通过 mdadm 的 -x 选项,为 RAID 10 添加备份盘/dev/sdl(首先添加该硬盘到系统中),如图 3-72 所示。

```
[root@localhost ~]# mdadm -Cv /dev/md3 -a yes -n 4 -x 1 -l 10 /dev/sd{g,h,i,j,l}
```

```
[root@localhost ~]# mdadm -D /dev/md3
/dev/md3:
           Version : 1.2
     Creation Time : Tue Mar 17 20:40:50 2020
        Raid Level : raid10
        Array Size : 41910272 (39.97 GiB 42.92 GB)
     Used Dev Size : 20955136 (19.98 GiB 21.46 GB)
      Raid Devices : 4
     Total Devices : 5
       Persistence : Superblock is persistent

       Update Time : Tue Mar 17 20:44:20 2020
             State : clean
    Active Devices : 4
   Working Devices : 5
    Failed Devices : 0
     Spare Devices : 1

            Layout : near=2
        Chunk Size : 512K

Consistency Policy : resync

              Name : localhost.localdomain:3 (local to host localhost.localdomain)
              UUID : e437512d:5b7cae85:42e4ffd5:cec45e60
            Events : 17

    Number   Major   Minor   RaidDevice State
       0       8       96        0      active sync set-A   /dev/sdg
       1       8      112        1      active sync set-B   /dev/sdh
       2       8      128        2      active sync set-A   /dev/sdi
       3       8      144        3      active sync set-B   /dev/sdj

       4       8      176        -      spare   /dev/sdl
```

图 3-72 添加备份盘

如果硬盘/dev/sdg 出现损坏,再次查看 RAID 10 的信息,会发现,备份盘/dev/sdl 会自动替换掉硬盘/dev/sdg,并开始同步数据(sync 状态),如图 3-73 所示。

```
[root@localhost ~]# mdadm /dev/md3 -f /dev/sdg
```

```
[root@localhost ~]# mdadm -D /dev/md3
/dev/md3:
           Version : 1.2
     Creation Time : Tue Mar 17 20:40:50 2020
        Raid Level : raid10
        Array Size : 41910272 (39.97 GiB 42.92 GB)
     Used Dev Size : 20955136 (19.98 GiB 21.46 GB)
      Raid Devices : 4
     Total Devices : 5
       Persistence : Superblock is persistent

       Update Time : Tue Mar 17 20:52:20 2020
             State : clean
    Active Devices : 4
   Working Devices : 4
    Failed Devices : 1
     Spare Devices : 0

            Layout : near=2
        Chunk Size : 512K

Consistency Policy : resync

              Name : localhost.localdomain:3 (local to host localhost.localdomain)
              UUID : e437512d:5b7cae85:42e4ffd5:cec45e60
            Events : 36

    Number   Major   Minor   RaidDevice State
       4       8      176        0      active sync set-A   /dev/sdl
       1       8      112        1      active sync set-B   /dev/sdh
       2       8      128        2      active sync set-A   /dev/sdi
       3       8      144        3      active sync set-B   /dev/sdj

       0       8       96        -      faulty   /dev/sdg
```

图 3-73 备份盘同步数据

在实际运用中,可以配置/etc/fstab 文件,实现磁盘阵列开机自动挂载,如图 3-74 所示。

```
[root@localhost~]# mkfs.ext4 /dev/md3

# /etc/fstab
# Created by anaconda on Tue Mar  3 00:32:39 2020
#
# Accessible filesystems, by reference, are maintained under '/dev/disk'
# See man pages fstab(5), findfs(8), mount(8) and/or blkid(8) for more info
#
/dev/mapper/rhel-root   /        xfs     defaults    0 0
UUID=93abc0c5-f930-430c-ab0a-e2db81e2b643 /boot    xfs    defaults   0 0
/dev/mapper/rhel-swap   swap     swap    defaults    0 0

/dev/md3    /raid0   ext4  defaults  0 0
```

图 3-74 磁盘阵列开机自动挂载

自我测试

一、填空题

1. 在 Linux 系统中,使用_____命令来创建文件系统。
2. 在 Linux 系统中使用_____命令进行磁盘配额的配置与管理。
3. LVM(Logical Volume Manager,逻辑卷管理)是 Linux 系统对_____进行管理的一种机制。
4. 独立冗余磁盘阵列,简称_____。
5. 在服务器上配置 RAID,有_____、_____两种配置方式。

二、简答题

1. 简述 Linux 系统中两种磁盘配额的限制方法。
2. 简述 LVM 常用的术语。
3. 简述 RAID 的分类。

三、实训题

1. 某公司在 Linux 服务器上新增加了一块硬盘/dev/sdb,需要使用 fdisk 命令新建/dev/sdb1 主分区和/dev/sdb2 扩展分区,在扩展分区中新建逻辑分区/dev/sdb5,并使用 mkfs 命令创建 ext4 文件系统,并把这两个文件系统分别挂载到/web 和/ftp 目录下。

2. 某公司在 Linux 服务器上新增加了一块硬盘/dev/sdb,要求 Linux 系统的分区能自动调整磁盘容量。需要使用 fdisk 命令新建/dev/sdb1、/dev/sdb2、/dev/sdb3、/dev/sdb4 为 LVM 类型,并在这 4 个分区上创建物理卷、卷组(vg1)和逻辑卷(lv1),并将逻辑卷挂载到/volume 目录下。

3. 某公司为了提供数据的可靠性、安全性,在服务上增加了四块硬盘,要求在这四块硬盘上创建 RAID 5 卷,从而实现数据的冗余备份。

项目 4　网络配置与远程访问

📚 项目综述

最近公司的网络管理员，需要经常进行服务器的配置、调试，但是每次都需要到公司机房进行调试，如果服务器在工作时间之外出现故障，网络管理员只能赶回公司处理，非常耽误时间，而且降低了工作效率。为了解决这个问题，网络管理员准备在服务器上配置、启用远程访问功能，并配置、启用防火墙。

📚 项目目标

- 了解网络环境相关知识；
- 了解网卡链路聚合；
- 了解防火墙的分类与工作原理；
- 了解三种远程访问方式的异同；
- 掌握常见配置网络环境的方法；
- 掌握网卡链路聚合的配置方法；
- 掌握 iptables、firewalld 防火墙的配置方法；
- 掌握三种远程访问方式的配置方法。

◀ 任务 4.1　配置网络环境 ▶

不论是安装了什么类型操作系统的主机，在网络中主机之间的通信，必须要正确配置网络相关参数（主机名、IP 地址、子网掩码、默认网关、DNS 服务器等）。

4.1.1　配置主机名

主机名是主机在网络中的唯一标识，在 RHEL7 系统中，有三种定义主机名的方式：

① 静态的（static）主机名：也称为内核主机名，是系统在启动时从 /etc/hostname 自动初始化的主机名。

② 瞬态的（transient）主机名：是在系统运行时临时分配的主机名，例如，通过 DHCP 或 DNS 服务器分配的主机名。

③ 灵活的（pretty）主机名：允许使用自由形式（必须是 UTF-8 编码）的主机名，以展示给终端用户。

配置主机名有三种方法：

1. 使用 hostnamectl 命令

（1）显示当前主机名称的配置信息，如图 4-1 所示。

[root@localhost~]# hostnamectl status

```
[root@localhost ~]# hostnamectl status
     Static hostname: localhost.localdomain
           Icon name: computer-vm
             Chassis: vm
          Machine ID: 68ca77e9e75d40fb8292c38a08292762
             Boot ID: 72892a57be4742f88f6cec416ece6dd4
      Virtualization: vmware
    Operating System: Red Hat Enterprise Linux Server 7.4 (Maipo)
         CPE OS Name: cpe:/o:redhat:enterprise_linux:7.4:GA:server
              Kernel: Linux 3.10.0-693.el7.x86_64
        Architecture: x86-64
```

图 4-1　显示当前主机名称的配置信息

（2）修改主机名称，如图 4-2 所示。

[root@localhost~]# hostnamectl set-hostname jack.com

```
[root@localhost ~]# hostnamectl set-hostname jack.com
[root@localhost ~]# hostnamectl status
     Static hostname: jack.com
           Icon name: computer-vm
             Chassis: vm
          Machine ID: 68ca77e9e75d40fb8292c38a08292762
             Boot ID: 72892a57be4742f88f6cec416ece6dd4
      Virtualization: vmware
    Operating System: Red Hat Enterprise Linux Server 7.4 (Maipo)
         CPE OS Name: cpe:/o:redhat:enterprise_linux:7.4:GA:server
              Kernel: Linux 3.10.0-693.el7.x86_64
        Architecture: x86-64
```

图 4-2　修改主机名称

修改后，使用命令 bash，使配置生效，如图 4-3 所示。

```
[root@localhost ~]# bash
[root@jack ~]#
```

图 4-3　使用命令 bash 使配置生效

2. 使用 nmtui 命令

[root@localhost~]# nmtui

选择"设置系统主机名"如图 4-4 所示，配置主机名如图 4-5 所示。

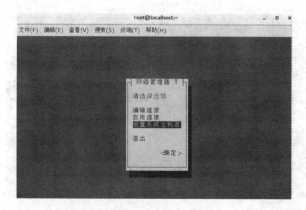

图 4-4　选择"设置系统主机名"

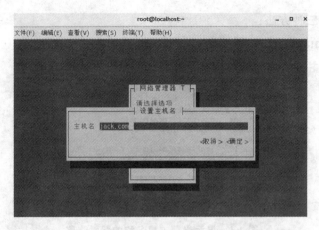

图 4-5 配置主机名

配置完成后使用命令 systemctl restart systemd-hostnamed，重启 hostnamed 服务，使配置生效，让 hostnamectl 知道静态主机名已经修改。

```
[root@localhost~]# systemctl restart systemd-hostnamed
```

3. 使用 NetworkManager 的命令行工具

在 RHEL7 中，NetworkManager 是动态网络管理和配置的守护进程，其中 nmcli 是 NetworkManager 的命令行工具，nmtui 是 NetworkManager 的文本用户界面工具。

```
[root@localhost~]# nmcli general hostname lucy.com    //设置主机名为 lucy.com
[root@localhost~]# nmcli general hostname             //查看当前主机名
lucy.com
[root@localhost~]# systemctl restart systemd-hostnamed //重启 hostnamectl 服务
```

4.1.2 配置网络参数

1. 使用 nmtui 命令图形化界面

```
[root@localhost~]# nmtui
```

图形化界面如图 4-6～图 4-11 所示。

图 4-6 选择"编辑连接"

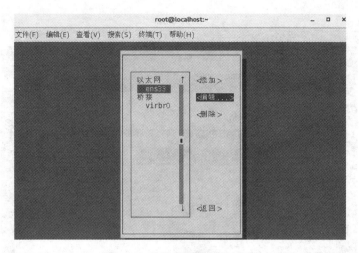

图 4-7　选择需要编辑的网卡

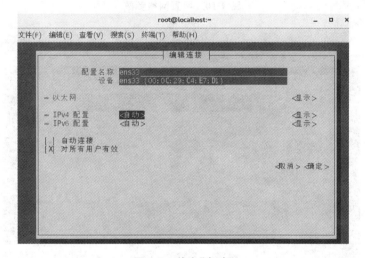

图 4-8　单击"自动"

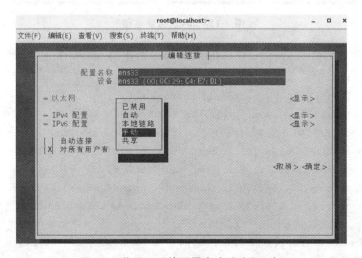

图 4-9　将"IPv4"的配置方式改为"手动"

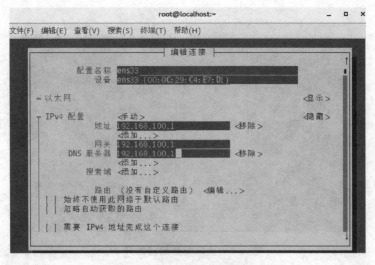

图 4-10 配置网络参数

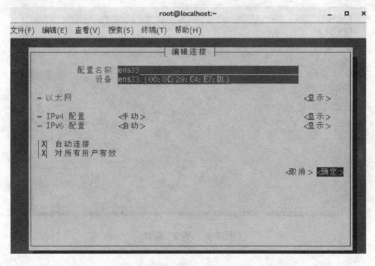

图 4-11 勾选"自动连接"

配置完成后使用命令 systemctl restart network，重启 network 服务，使配置生效，并使用 ping 命令进行测试。

```
[root@localhost~]# systemctl restart network
[root@localhost~]# ping 192.168.100.1 -c 4
PING 192.168.100.1 (192.168.100.1) 56(84) bytes of data.
64 bytes from 192.168.100.1: icmp_seq=1 ttl=64 time=0.037 ms
64 bytes from 192.168.100.1: icmp_seq=2 ttl=64 time=0.064 ms
64 bytes from 192.168.100.1: icmp_seq=3 ttl=64 time=0.057 ms
64 bytes from 192.168.100.1: icmp_seq=4 ttl=64 time=0.038 ms
--- 192.168.100.1 ping statistics ---
4 packets transmitted, 4 received, 0% packet loss, time 3002ms
rtt min/avg/max/mdev=0.037/0.049/0.064/0.011 ms
```

2. 使用系统菜单

（1）单击系统桌面右上角的"⊥"图标，打开如图 4-12 所示界面，单击"有线设置"按钮。

图 4-12　单击"有线设置"

（2）在图 4-13 所示界面中，单击"✦"图标，在图 4-14 所示界面中，选择"IPv4"配置网络参数。

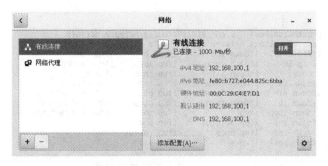

图 4-13　单击"齿轮"按钮

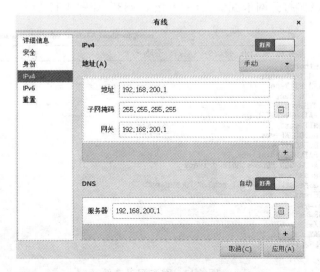

图 4-14　配置 IPv4 相关信息

完成配置后，如果配置没有生效，单击图 4-13 中的"打开"按钮来回切换网络连接的状态，以使配置生效，如图 4-15 所示。

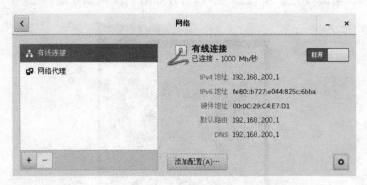

图 4-15　网络配置界面

3. 使用网卡配置文件

在 RHEL7 系统中网卡配置文件的前缀为"eth"，第一块网卡为 eth0，第二块网卡为 eth1，以此类推；在 RHEL7 系统中网卡配置文件的前缀为"ifcfg"，网卡配置文件由"ifcfg"＋网卡名称组成，如"ifcfg-ens33"。网卡配置文件在目录/etc/sysconfig/network-scripts/下，如图 4-16 所示。

```
[root@localhost ~]# cd /etc/sysconfig/network-scripts
[root@localhost network-scripts]# ls
ifcfg-ens33    ifdown-ppp      ifup-ib       ifup-Team
ifcfg-lo       ifdown-routes   ifup-ippp     ifup-TeamPort
ifdown         ifdown-sit      ifup-ipv6     ifup-tunnel
ifdown-bnep    ifdown-Team     ifup-isdn     ifup-wireless
ifdown-eth     ifdown-TeamPort ifup-plip     init.ipv6-global
ifdown-ib      ifdown-tunnel   ifup-plusb    network-functions
ifdown-ippp    ifup            ifup-post     network-functions-ipv6
ifdown-ipv6    ifup-aliases    ifup-ppp
ifdown-isdn    ifup-bnep       ifup-routes
ifdown-post    ifup-eth        ifup-sit
```

图 4-16　网卡配置文件目录

使用 VIM 编辑器打开网卡配置文件"ifcfg-ens33"，如图 4-17 所示。

```
TYPE=Ethernet
PROXY_METHOD=none
BROWSER_ONLY=no
BOOTPROTO=none
DEFROUTE=yes
IPV4_FAILURE_FATAL=no
IPV6INIT=yes
IPV6_AUTOCONF=yes
IPV6_DEFROUTE=yes
IPV6_FAILURE_FATAL=no
IPV6_ADDR_GEN_MODE=stable-privacy
NAME=ens33
UUID=a5ae4311-39fe-4c4f-aca8-43f6619d4ea6
DEVICE=ens33
ONBOOT=yes
IPADDR=192.168.200.1
PREFIX=32
GATEWAY=192.168.200.1
DNS1=192.168.200.1
```

图 4-17　网卡配置文件

网卡配置文件的参数说明如下：
- 设备类型：TYPE＝Ethernet；
- 代理方式：PROXY_METHOD＝none；
- 只是浏览器：BROWSER_ONLY＝no；
- 网卡的引导协议：BOOTPROTO＝static；
- 默认路由：DEFROUTE＝yes；
- 是否开启 IPv4 致命错误检测：IPv4_FAILURE_FATAL＝no；
- IPv6 是否自动初始化：IPv6INIT＝yes；
- IPv6 是否自动配置：IPv6_AUTOCONF＝yes；
- IPv6 是否可以为默认路由：IPv6_DEFROUTE＝yes；
- 是否开启 IPv6 致命错误检测：IPv6_FAILURE_FATAL＝no；
- IPv6 地址生成模型：IPv6_ADDR_GEN_MODE＝stable-privacy；
- 网卡物理设备名称：NAME＝ens33；
- 通用唯一识别码：UUID＝ a5ae4311-39fe-4c4f-aca8-43f6619d4ea6；
- 网卡名称：DEVICE＝ens33；
- 是否开机启动：ONBOOT＝yes；
- IP 地址：IPADDR＝192.168.200.1；
- 子网掩码：NETMASK＝255.255.255.0；
- 默认网关：GATEWAY＝192.168.200.1；
- DNS 地址：DNS1＝192.168.200.1。

配置完成后使用命令 systemctl restart network，重启 network 服务，使配置生效，可以通过 ifconfig 命令查看网卡的配置信息。

```
[root@localhost network-scripts]# systemctl restart network
[root@localhost network-scripts]# ifconfig ens33
ens33: flags=4163<UP,BROADCAST,RUNNING,MULTICAST>  mtu 1500
        inet 192.168.200.1  netmask 255.255.255.0  broadcast 192.168.200.255
        inet6 fe80::b727:e044:825c:6bba  prefixlen 64  scopeid 0x20<link>
        ether 00:0c:29:c4:e7:d1  txqueuelen 1000  (Ethernet)
        RX packets 20253  bytes 7450433 (7.1 MiB)
        RX errors 0  dropped 0  overruns 0  frame 0
        TX packets 267  bytes 38315 (37.4 KiB)
        TX errors 0  dropped 0 overruns 0  carrier 0  collisions 0
```

4. 使用 nmcli 命令

nmcli 常用命令及功能如表 4-1 所示。

表 4-1　nmcli 常用命令及功能

命令	功能
nmcli connection show	显示所有网络连接
nmcli connection show -active	显示所有活动的网络连接
nmcli connection showens33	显示指定网卡的详细信息

续表 4-1

命令	功能
nmcli device status	显示设备的连接状态
nmcli device show	显示所有设备网络设备详细信息
nmcli device show ens33	显示指定网络设备的详细信息
nmcli connection up xx	启用配置文件 xx
nmcli connection down xx	禁用配置文件 xx
nmcli device disconnect ens33	禁用激活网卡设置
nmcli connection reload	重新加载配置文件
nmcli connection add con-name xx type ethernet ifname ens33 autoconnect yes ipv4.addresses 172.16.1.1/16 gw4 172.16.1.1	创建名为 xx 的网络接口,接口类型为以太网,绑定网卡为 ens33,开机自动启动,并设置 IP 地址、默认网关
nmcli connection modify ens33 connection.autoconnect yes	设置自动启动网卡
nmcli connection modify ens33 ipv4.method manual ipv4.addresses 192.168.1.1/24 nmcli connection modify ens33 ipv4.method auto	设置 IP 地址获取方式是手动或者 DHCP
nmcli connection modify ens33 ipv4.addresses 192.168.1.1/24	修改 IP 地址
nmcli connection modify ens33 ipv4.gateway 192.168.1.1	修改网关
nmcli connection modify ens33 +ipv4.addresses 192.168.2.1/24	添加第二个 IP 地址
nmcli connection modify ens33 ipv4.dns 114.114.114.114	添加 DNS
nmcli connection modify ens33 +ipv4.dns 8.8.8.8	添加第二个 DNS

如图 4-18、图 4-19 所示,查看系统当前的网络连接、网卡的相关信息。

```
[root@localhost ~]# nmcli connection show
名称    UUID                                  类型           设备
ens33   a5ae4311-39fe-4c4f-aca8-43f6619d4ea6  802-3-ethernet ens33
virbr0  ee2e7c34-c819-4ea0-9f47-45b26370a315  bridge         virbr0
[root@localhost ~]# nmcli connection show --active
名称    UUID                                  类型           设备
ens33   a5ae4311-39fe-4c4f-aca8-43f6619d4ea6  802-3-ethernet ens33
virbr0  ee2e7c34-c819-4ea0-9f47-45b26370a315  bridge         virbr0
[root@localhost ~]# nmcli connection show ens33
connection.id:                          ens33
connection.uuid:                        a5ae4311-39fe-4c4f-aca8-43f6619d4ea6
connection.stable-id:                   --
connection.interface-name:              ens33
connection.type:                        802-3-ethernet
connection.autoconnect:                 yes
connection.autoconnect-priority:        0
connection.autoconnect-retries:         -1 (默认)
connection.timestamp:                   1584588720
connection.read-only:                   no
connection.permissions:                 --
connection.zone:                        --
connection.master:                      --
connection.slave-type:                  --
connection.autoconnect-slaves:          -1 (默认)
connection.secondaries:                 --
connection.gateway-ping-timeout:        0
connection.metered:                     未知
connection.lldp:                        -1 (default)
```

图 4-18 查看网络连接信息

```
[root@localhost ~]# nmcli device show ens33
GENERAL.设备：                          ens33
GENERAL.类型：                          ethernet
GENERAL.硬盘：                          00:0C:29:C4:E7:D1
GENERAL.MTU：                           1500
GENERAL.状态：                          100 (连接的)
GENERAL.连接：                          ens33
GENERAL.连接路径：                      /org/freedesktop/NetworkManager/ActiveConnection/15
WIRED-PROPERTIES.载波：                 开
IP4.地址[1]：                           192.168.200.1/24
IP4.网关：                              192.168.200.1
IP4.DNS[1]：                            192.168.200.1
IP6.地址[1]：                           fe80::b727:e044:825c:6bba/64
IP6.网关：                              --
[root@localhost ~]# nmcli connection down ens33
成功取消激活连接 'ens33' (D-Bus 活动路径：/org/freedesktop/NetworkManager/ActiveConnection/15)
[root@localhost ~]# nmcli connection up ens33
连接已成功激活 (D-Bus 活动路径：/org/freedesktop/NetworkManager/ActiveConnection/16)
[root@localhost ~]# nmcli device disconnect ens33
成功断开设备 'ens33'。
[root@localhost ~]# nmcli device connect ens33
成功用 'ens33' 激活了设备 'a5ae4311-39fe-4c4f-aca8-43f6619d4ea6'。
```

图 4-19　查看网卡信息、禁用（启用）网卡

创建名为 xx 的网络接口，接口类型为以太网，绑定网卡 ens33，开机自动启动，并设置 IP 地址、默认网关。

```
[root@localhost ~]# nmcli connection add con-name xx type ethernet ifname ens33 autoconnect yes ipv4.addresses 172.16.1.1/16 gw4 172.16.1.1
连接"xx"(a59b592f-dcb8-4f35-b5a5-4bf6c385fc65) 已成功添加。
[root@localhost ~]# ls /etc/sysconfig/network-scripts/
ifcfg-ens33   ifcfg-xx    ifdown-eth    ifdown-ipv6    ifdown-ppp
[root@localhost ~]# nmcli connection up xx
连接已成功激活 (D-Bus 活动路径：/org/freedesktop/NetworkManager/ActiveConnection/20)
[root@localhost ~]# nmcli device show ens33
GENERAL.设备：                          ens33
GENERAL.类型：                          ethernet
GENERAL.硬盘：                          00:0C:29:C4:E7:D1
GENERAL.MTU：                           1500
GENERAL.状态：                          100 (连接的)
GENERAL.连接：                          xx
GENERAL.连接路径：                      /org/freedesktop/NetworkManager/ActiveConnection/20
WIRED-PROPERTIES.载波：                 开
IP4.地址[1]：                           172.16.1.1/16
IP4.地址[2]：                           192.168.150.8/24
IP4.网关：                              172.16.1.1
IP4.DNS[1]：                            192.168.0.1
IP6.地址[1]：                           fe80::8665:6a33:57e9:b553/64
IP6.网关：                              --
```

修改网卡 IP 地址、默认网关，并添加 DNS 地址，这里需要注意的是，当前网卡使用的配置文件"xx"，需启用配置文件"ens33"，配置才生效（一个网卡可以有多个配置文件）。

```
[root@localhost~]# nmcli connection modify ens33 ipv4.addresses 192.168.1.1/24
[root@localhost~]# nmcli connection modify ens33 ipv4.gateway 192.168.1.1
[root@localhost~]# nmcli connection modify ens33 +ipv4.addresses 192.168.2.1/24
[root@localhost~]# nmcli connection modify ens33 ipv4.dns 114.114.114.114
[root@localhost~]# nmcli connection modify ens33 +ipv4.dns 8.8.8.8
[root@localhost~]# nmcli connection up ens33
连接已成功激活(D-Bus 活动路径:/org/freedesktop/NetworkManager/ActiveConnection/23)
[root@localhost~]# nmcli device show ens33
GENERAL.设备:              ens33
GENERAL.类型:              ethernet
GENERAL.硬盘:              00:0C:29:C4:E7:D1
GENERAL.MTU:              1500
GENERAL.状态:              100 (连接的)
GENERAL.连接:              ens33
GENERAL.连接路径:          /org/freedesktop/NetworkManager/ActiveConnection/23
WIRED-PROPERTIES.载波:     开
IP4.地址[1]:               192.168.1.1/24
IP4.地址[2]:               192.168.2.1/24
IP4.网关:                  192.168.1.1
IP4.DNS[1]:               114.114.114.114
IP4.DNS[2]:               8.8.8.8
IP6.地址[1]:               fe80::b727:e044:825c:6bba/64
IP6.网关:                  --
```

在实际应用中,可以根据不同的应用环境,配置不同的网卡文件,然后依据应用环境使用。

4.1.3 网卡链路聚合

1. 基本概念

在交换机中,可以使用端口汇聚技术,将两个交换机之间的多个物理端口捆绑在一起,形成一个更大带宽的逻辑链路,从而提高网络速度,并提供链路的备份功能。在操作系统中,可以使用网卡链路聚合来提高网速,并提供链路的备份功能。网卡链路聚合就是将多块网卡连接起来,一般常用的有"bond""team"两种模式,"bond"模式最多可以添加两块网卡,"team"模式最多可以添加八块网卡。如 Windows server 2012 可以通过"NIC 组合"来实现网卡的链路聚合。

让 Linux 内核支持网卡绑定驱动。常见的网卡绑定驱动有三种模式:mode0、mode1、mode6。

mode0(平衡负载模式"balance-rr",平衡轮询策略):平时两块网卡均工作,且自动备援,但需要在与服务器本地网卡相连的交换机设备上进行端口聚合来支持绑定技术。

mode1（平衡备援模式"active-backup"，主-备份策略）：平时只有一块网卡工作，在它故障后自动替换为另外的网卡。

mode6（平衡负载模式"balance-alb"，适配器适应性负载均衡）：平时两块网卡工作，且自动备援，无需交换机设备提供辅助支持。

2. 配置 bond 模式

bond 聚合有两种方式：

① 轮询聚合（balance-rr）：两块网卡同时工作。

② 主备聚合（active-backup）：平时只有一块网卡工作，在它故障后自动替换为另外的网卡。

实现步骤如下：

（1）在虚拟机里，为系统添加一块网卡，并确保两块网卡的"网络连接"类型相同，如图 4-20 所示。

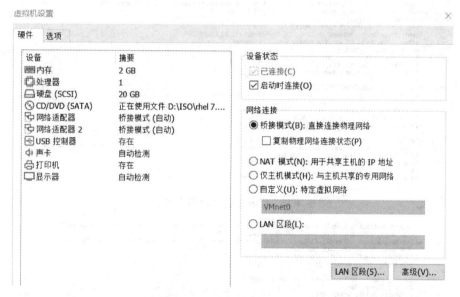

图 4-20　在虚拟机系统中添加网卡

系统重启后，可以使用命令 ifconfig 查看两块网卡的相关信息，需要注意的是，要保证网卡没有被系统使用，如图 4-21 所示。

```
[root@localhost ~]# ifconfig
ens33: flags=4163<UP,BROADCAST,RUNNING,MULTICAST>  mtu 1500
        ether 00:0c:29:c4:e7:d1  txqueuelen 1000  (Ethernet)
        RX packets 1458  bytes 357325 (348.9 KiB)
        RX errors 0  dropped 0  overruns 0  frame 0
        TX packets 0  bytes 0 (0.0 B)
        TX errors 0  dropped 0 overruns 0  carrier 0  collisions 0

ens38: flags=4163<UP,BROADCAST,RUNNING,MULTICAST>  mtu 1500
        ether 00:0c:29:c4:e7:db  txqueuelen 1000  (Ethernet)
        RX packets 0  bytes 0 (0.0 B)
        RX errors 0  dropped 0  overruns 0  frame 0
        TX packets 0  bytes 0 (0.0 B)
        TX errors 0  dropped 0 overruns 0  carrier 0  collisions 0
```

图 4-21　查看网卡信息

可以看到新添加的网卡名称为"ens38"。

(2) 配置 bond 主备聚合。

[root@localhost ~]# nmcli connection add con-name bond0 ifname bond0 type bond mode active-backup ip4 172.16.16.1/24   //创建名为 bond0 的网络接口,绑定接口 bond0,接口类型为 bond,工作模式为主备,并设置 IP 地址

[root@localhost ~]# nmcli connection add con-name ens33 ifname ens33 type bond-slave master bond0   //创建名为 ens33 的网络接口,绑定网卡 ens33,接口类型为 bond,bond-slave 表示 ens33 为 bond 的从属设备,master bond0 表示 ens33 为 bond0 服务

[root@localhost ~]# nmcli connection add con-name ens38 ifname ens38 type bond-slave master bond0

如图 4-22、图 4-23 所示,在 bond0 中,目前处于工作状态的网卡是 ens33。

```
[root@localhost ~]# nmcli connection add con-name bond0 ifname bond0 type bond mode active-backup ip4 172.16.16.1/24
连接 bond0"(4954a43c-8155-4683-b8e4-543e9088fe8f) 已成功添加。
[root@localhost ~]# ifconfig
bond0: flags=5123<UP,BROADCAST,MASTER,MULTICAST>  mtu 1500
        inet 172.16.16.1  netmask 255.255.255.0  broadcast 172.16.16.255
        ether 0a:ce:00:90:65:ab  txqueuelen 1000  (Ethernet)
        RX packets 0  bytes 0 (0.0 B)
        RX errors 0  dropped 0  overruns 0  frame 0
        TX packets 0  bytes 0 (0.0 B)
        TX errors 0  dropped 0 overruns 0  carrier 0  collisions 0

ens33: flags=4099<UP,BROADCAST,MULTICAST>  mtu 1500
        ether 00:0c:29:c4:e7:d1  txqueuelen 1000  (Ethernet)
        RX packets 1743  bytes 494063 (482.4 KiB)
        RX errors 0  dropped 0  overruns 0  frame 0
        TX packets 42  bytes 3774 (3.6 KiB)
        TX errors 0  dropped 0 overruns 0  carrier 0  collisions 0

ens38: flags=4163<UP,BROADCAST,RUNNING,MULTICAST>  mtu 1500
        ether 00:0c:29:c4:e7:db  txqueuelen 1000  (Ethernet)
        RX packets 437  bytes 243674 (237.9 KiB)
        RX errors 0  dropped 0  overruns 0  frame 0
        TX packets 15  bytes 1185 (1.1 KiB)
        TX errors 0  dropped 0 overruns 0  carrier 0  collisions 0
```

图 4-22   查看主备聚合网卡信息

```
[root@localhost ~]# nmcli connection add con-name ens33 ifname ens33 type bond-slave master bond0
连接 ens33"(29f7edea-a21f-4396-8316-eea1a53fcb0d) 已成功添加。
[root@localhost ~]# nmcli connection add con-name ens38 ifname ens38 type bond-slave master bond0
连接 ens38"(892f17fa-2ece-4800-b2c2-cd9e717893cb) 已成功添加。
[root@localhost ~]# cat /proc/net/bonding/bond0
Ethernet Channel Bonding Driver: v3.7.1 (April 27, 2011)

Bonding Mode: fault-tolerance (active-backup)
Primary Slave: None
Currently Active Slave: ens33
MII Status: up
MII Polling Interval (ms): 100
Up Delay (ms): 0
Down Delay (ms): 0

Slave Interface: ens33
MII Status: up
Speed: 1000 Mbps
Duplex: full
Link Failure Count: 0
Permanent HW addr: 00:0c:29:c4:e7:d1
Slave queue ID: 0

Slave Interface: ens38
MII Status: up
Speed: 1000 Mbps
Duplex: full
Link Failure Count: 0
Permanent HW addr: 00:0c:29:c4:e7:db
Slave queue ID: 0
[root@localhost ~]# nmcli connection show
名称    UUID                                    类型          设备
bond0   4954a43c-8155-4683-b8e4-543e9088fe8f    bond          bond0
ens33   29f7edea-a21f-4396-8316-eea1a53fcb0d    802-3-ethernet  ens33
ens38   892f17fa-2ece-4800-b2c2-cd9e717893cb    802-3-ethernet  ens38
virbr0  e2578389-03cb-4c51-bd8a-8f4cfe06f537    bridge        virbr0
```

图 4-23   查看主备聚合网卡的工作状态(一)

(3)当其中一个网卡,如使用命令 ifconfig ens33 down 的模拟网卡 ens33 出现故障(也

可以在虚拟机中将该网卡断开连接),如图 4-24 所示,网络 ens38 将自动替换网卡 ens33,从而确保网络不中断,如图 4-25 所示,网络中断到恢复,时间间隔非常短,用户基本是毫无感觉的。

```
[root@localhost ~]# ifconfig ens33 down
[root@localhost ~]# cat /proc/net/bonding/bond0
Ethernet Channel Bonding Driver: v3.7.1 (April 27, 2011)

Bonding Mode: fault-tolerance (active-backup)
Primary Slave: None
Currently Active Slave: ens38
MII Status: up
MII Polling Interval (ms): 100
Up Delay (ms): 0
Down Delay (ms): 0

Slave Interface: ens38
MII Status: up
Speed: 1000 Mbps
Duplex: full
Link Failure Count: 4
Permanent HW addr: 00:0c:29:c4:e7:db
Slave queue ID: 0

Slave Interface: ens33
MII Status: down
Speed: 1000 Mbps
Duplex: full
Link Failure Count: 1
Permanent HW addr: 00:0c:29:c4:e7:d1
Slave queue ID: 0
```

图 4-24　查看主备聚合网卡的工作状态(二)

```
Reply from 172.16.16.1: bytes=32 time<1ms TTL=64
Reply from 172.16.16.1: bytes=32 time<1ms TTL=64
Reply from 172.16.16.1: bytes=32 time<1ms TTL=64
Reply from 172.16.16.1: bytes=32 time<1ms TTL=64
Request timed out.
Reply from 172.16.16.1: bytes=32 time<1ms TTL=64
Reply from 172.16.16.1: bytes=32 time<1ms TTL=64
Reply from 172.16.16.1: bytes=32 time<1ms TTL=64
Reply from 172.16.16.1: bytes=32 time=10ms TTL=64
Reply from 172.16.16.1: bytes=32 time<1ms TTL=64
Reply from 172.16.16.1: bytes=32 time<1ms TTL=64
```

图 4-25　网络连通性测试

可以使用下面的命令删除 bond0、ens33、ens38。

```
[root@localhost ~]# nmcli connection delete bond0
[root@localhost ~]# nmcli connection delete ens33
[root@localhost ~]# nmcli connection delete ens38
```

3. 配置 team 模式

team 聚合有四种方式:
① broadcast:广播容错;
② roundrobin:轮询;

③ activebackup：主备；

④ loadbalance：负载均衡。

实现步骤如下：

(1) 配置 team 模式，使用之前配置 bond 模式的系统环境，如图 4-26 所示。

```
[root@localhost ~]# ifconfig
ens33: flags=4163<UP,BROADCAST,RUNNING,MULTICAST>  mtu 1500
        ether 00:0c:29:c4:e7:d1  txqueuelen 1000  (Ethernet)
        RX packets 2415  bytes 419039 (409.2 KiB)
        RX errors 0  dropped 0  overruns 0  frame 0
        TX packets 17  bytes 2033 (1.9 KiB)
        TX errors 0  dropped 0 overruns 0  carrier 0  collisions 0

ens38: flags=4163<UP,BROADCAST,RUNNING,MULTICAST>  mtu 1500
        ether 00:0c:29:c4:e7:db  txqueuelen 1000  (Ethernet)
        RX packets 950  bytes 61200 (59.7 KiB)
        RX errors 0  dropped 0  overruns 0  frame 0
        TX packets 0  bytes 0 (0.0 B)
        TX errors 0  dropped 0 overruns 0  carrier 0  collisions 0
```

图 4-26 查看网卡信息

(2) 配置 team 轮询聚合，如图 4-27 所示。

```
[root@localhost ~]# nmcli connection add con-name team0 ifname team0 type team  autoconnect yes config '{"runner": {"name": "roundrobin"}}'
//创建名为 team0 的网络接口，绑定接口 team0，接口类型为 team，开机启动，运行模式为轮询
[root@localhost ~]# nmcli connection add con-name ens33 ifname ens33 type team-slave master team0  //创建名为 ens33 的网络接口，绑定网卡 ens33，接口类型为 team，team-slave 表示 ens33 为 team 的从属设备，master team0 表示 ens33 为 team0 服务
[root@localhost ~]# nmcli connection add con-name ens38 ifname ens38 type team-slave master team0
[root@localhost ~]# nmcli connection modify team0 ipv4.addresses  172.16.1.1/24  //配置 team0 的 IP 地址
[root@localhost ~]# nmcli connection modify team0 ipv4.method manual
//配置 team0 的 IP 地址获取方式为手动配置
[root@localhost ~]# nmcli connection up team0
[root@localhost ~]# nmcli connection up ens33
[root@localhost ~]# nmcli connection up ens38
```

```
[root@localhost ~]# nmcli connection add con-name team0 ifname team0 type team  autoconnect yes config '{"runner": {"name": "roundrobin"}}'
连接 "team0"(7c4e693d-482f-4d7f-bc4e-8b9b81dfb35f) 已成功添加。
[root@localhost ~]# nmcli connection add con-name ens33 ifname ens33 type team-slave master team0
连接 "ens33"(178efe4b-1742-400f-bf33-a46fddc6dd17) 已成功添加。
[root@localhost ~]# nmcli connection add con-name ens38 ifname ens38 type team-slave master team0
连接 "ens38"(fb7c09da-94c9-4ac4-8f7f-aa09ebe115d5) 已成功添加。
[root@localhost ~]# nmcli connection modify team0 ipv4.addresses  172.16.1.1/24
[root@localhost ~]# nmcli connection modify team0 ipv4.method manual
[root@localhost ~]# nmcli connection up team0
成功激活（主服务器等待从服务器）连接（D-Bus 激活路径：/org/freedesktop/NetworkManager/ActiveConnection/8)
[root@localhost ~]# nmcli connection up ens33
连接已成功激活（D-Bus 活动路径：/org/freedesktop/NetworkManager/ActiveConnection/11)
[root@localhost ~]# nmcli connection up ens38
连接已成功激活（D-Bus 活动路径：/org/freedesktop/NetworkManager/ActiveConnection/12)
```

图 4-27 配置 team 轮询聚合

可以使用 teamdctl team0 state 命令查看 team0 的工作状态，如图 4-28 所示。

```
[root@localhost ~]# teamdctl team0 state
```

```
[root@localhost ~]# teamdctl team0 state
setup:
  runner: roundrobin
ports:
  ens33
    link watches:
      link summary: up
      instance[link_watch_0]:
        name: ethtool
        link: up
        down count: 0
  ens38
    link watches:
      link summary: up
      instance[link_watch_0]:
        name: ethtool
        link: up
        down count: 0
```

图 4-28 查看 team0 的工作状态

◀ 任务4.2　防火墙管理 ▶

  防火墙分为硬件防火墙和软件防火墙。硬件防火墙,如思科、华为、锐捷等网络设备厂商所生产的防火墙。软件防火墙单独使用软件系统来完成防火墙功能。在 RHEL7 中默认使用 firewalld 作为防火墙,它取代了之前的 iptables 防火墙,iptables 防火墙默认不启动,但仍可以使用。iptables、firewalld 服务都只是用来定义防火墙策略功能的"防火墙管理工具"而已,iptables 服务会把配置好的防火墙策略交由内核层面的 netfilter 网络过滤器来处理,而 firewalld 服务则是把配置好的防火墙策略交由内核层面的 nftables 包过滤框架来处理。

  在 Linux 系统中,提供给用户多个防火墙管理工具进行选择,系统管理员只需要使用其中一种管理工具就可以了,这些管理工具都是为了方便系统管理员配置防火墙策略。防火墙策略规则是按照从上到下的顺序来匹配的,因此允许的规则一定要放到拒绝规则的前面,否则所有的数据包都会被拒绝,从而导致任何主机都无法访问有关的服务。

4.2.1　iptables 防火墙

  netfilter/iptables IP 组成 Linux 系统的包过滤防火墙。netfilter 集成在内核里,它主要是定义、保存相应的规则,而 iptables 则是工具,用来修改信息的过滤规则以及其他配置。用户可以通过 iptables 来设置适合当前环境需求的规则,这些规则会保存在内核空间里。

  netfilter 是 Linux 核心中一个通用架构,它提供了一系列的"表"(tables),每个表由若干"链"(chains)组成,每条链中可以由一条或数条"规则"(rules)组成。实际上,netfilter 是表的容器,表是链的容器,链是规则的容器。

1. iptables 基本知识

(1) filter 表

filter 表有三个链:INPUT、FORWARD、OUTPUT。

INPUT 链:过滤所有目标地址是本机的数据包(对进入本机数据包的过滤)。

FORWARD 链:过滤所有经过本机的数据包(源地址和目标地址都不是本机的数据包)。

OUTPUT 链:过滤所有本机产生的数据包(对源地址的数据包过滤,也就是从本机发出的数据包)。

filter 表有三个动作:ACCEPT、REJECT、DROP。

ACCEPT:接受所有数据包。

REJECT:拒绝所有数据包。

DROP:丢弃所有数据包。

(2) NAT 表

NAT 表主要用于源 IP 地址与目的 IP 地址的转换,与本机无关,主要与主机后的局域网内的计算机相关。

NAT 表有三个链:PREROUTING、OUTPUT、POSTROUTING。

PREROUTING 链:数据包做路由选择前要用此键中的规则。主要是改变数据包的目的地址、目的端口等。

OUTPUT 链:与主机发出去的数据包有关,主要是改变主机发出数据包的目标地址。

POSTROUTING 链:发送数据包之前修改该包,主要是改变数据包的源地址、源端口等。

NAT 表有三个动作:DNAT、SNAT、MASQUERARE。

DNAT:改变数据包的目的地址,使包能从路由到某台机器(使公网能访问内网中的某台主机)。

SNAT:改变数据包的源地址(使内网能访问公网)。

MASQUERADE:和 SNAT 的作用一样,其主要针对外部接口为动态 IP 地址来进行设置,比如 ADSL 拨号、DHCP 连接等。

(3) mangle 表

mangle 表主要用在数据包的特殊操作,如修改数据包的服务类型、TTL 等。它有 PREROUTING、OUTPUT 两个链。

2. iptables 工作流程

iptables 工作流程如图 4-29 所示。

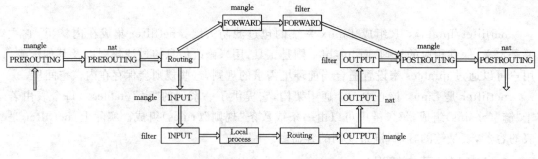

图 4-29 iptables 工作流程

(1) 数据包到达网络接口后,进入 mangle 表的 PREROUTING 链,如果有特殊的设

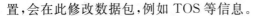

置,会在此修改数据包,例如 TOS 等信息。

(2) 数据包进入 nat 表的 PREROUTING 链,如果有规则设置,就可以进行 DNAT(改变数据包的目的地址)。

(3) 数据包经过路由,判断该包是交给本地主机还是转发给网络中的其他主机。

(4) 如果是转发,就发送给 mangle 表的 FORWARD 链,根据需要修改相应的参数,然后发送给 filter 表的 FORWARD 链,对所有转发的数据包进行过滤。然后再转给 mangle 表的 POSTROUTING 链,如有设置,可调整参数。然后发给 nat 表的 POSTROUTING 链,在这里一般都是用来做 SNAT,修改数据包的源地址,最后数据包发送给网络接口,转发给外部网络。

(5) 如果目的地址为本机,数据包就进入 mangle 表的 INPUT 链,经过处理,进入 filter 表的 INPUT 链,再经过处理,进入本机的处理进程。

(6) 处理完毕后,本机产生的数据包首先进入路由,然后分别经过 mangle、nat、filter 的 OUTPUT 链进行相应操作,再进入 mangle、nat 的 POSTROUTING 链,发给出去的网络接口。

3. iptables 基本语法

iptables 语法格式为:iptables[-t 表名]-命令 -匹配 -j 动作/目标。

参数说明如下:

① 表名

iptables 内置了 filter、nat 和 mangle 三张表,使用参数 -t 来设置对哪张表生效,也可以省略参数 -t,默认对 filter 表进行操作。

② 命令

-A:向规则链末尾中添加一条规则(--append)。

-D:从规则链中删除规则(--delete)。

-I:向规则链的开头(或者指定序号)中插入规则,未指定规则序号时,默认作为第一规则(--insert)。

-L:显示规则链中的所有规则(--list)。

-F:清除规则链中所有规则,若没有指定链,则清空指定表中所有链的规则(--flush)。

-P:定义默认的策略,所有不符合规则的包都被强制使用该策略(--policy)。

③ 匹配

-p:匹配指定的协议(--protocol)。

-sport:基于 TCP 包的源端口来匹配包,若没有指定,则匹配所有的端口(--source-port)。

-dport:基于 TCP 包的目的端口来匹配包,若没有指定,则匹配所有的端口(--destination-port)。

-s:以 IP 源地址匹配包(--source)。

-d:以 IP 目的地址匹配包(--destination)。

-i:以数据包进入本机的网络接口来匹配(--in-interface)。

-o:以数据包要离开本机所使用的网络接口来匹配(--out-interface)。

④ 动作/目标

ACCEPT：允许符合条件的数据包通过。

DROP：拒绝符合条件的数据包通过，直接丢弃数据包，不给出任何回应信息。

REJECT：拒绝符合条件的数据包通过，并给数据发送者一个响应错误信息。

LOG：用来记录与数据包相关的信息，这些信息可以用来帮助用户排除错误。

REDIRECT：端口重定向，将数据包重定向到本机或另外一台主机的某个端口。

DNAT：目标地址转换。

SNAT：源地址转换。

MASQUERADE：和 SNAT 的作用一样，其主要针对外部接口为动态的 IP 地址来进行设置。

4. 配置 iptables 规则

首先要安装 iptables 相关软件包，然后关闭 firewalld 服务，启动 iptables 服务。

```
[root@localhost~]# yum -y install iptables*
[root@localhost~]# systemctl stop firewalld
[root@localhost~]# systemctl start iptables
```

(1) 查看 iptables 规则

如图 4-30 所示，查看 filter 表中的所有链的规则。

```
[root@localhost~]# iptables -t filter -L
```

```
[root@localhost ~]# iptables -t filter -L
Chain INPUT (policy ACCEPT)
target     prot opt source               destination
ACCEPT     all  --  anywhere             anywhere             state RELATED,ESTABLISHED
ACCEPT     icmp --  anywhere             anywhere
ACCEPT     all  --  anywhere             anywhere
ACCEPT     tcp  --  anywhere             anywhere             state NEW tcp dpt:ssh
REJECT     all  --  anywhere             anywhere             reject-with icmp-host-prohibited

Chain FORWARD (policy ACCEPT)
target     prot opt source               destination
REJECT     all  --  anywhere             anywhere             reject-with icmp-host-prohibited

Chain OUTPUT (policy ACCEPT)
target     prot opt source               destination
```

图 4-30 查看 filter 表中的所有链的规则

(2) 添加、删除、修改规则

如在 fliter 表的 INPUT 链添加一条规则，规则为拒绝所有使用 ICMP 的数据包（也就是拒绝网络中的其他主机，ping 通该主机）。

```
[root@localhost~]# iptables -F INPUT        //清除 fliter 表的 INPUT 链的所有规则
[root@localhost~]# iptables -t filter -A INPUT -p icmp -j DROP
```

如在 fliter 表的 INPUT 链添加一条规则，规则为允许访问 TCP 的 8080 端口的数据包通过。

```
[root@localhost~]# iptables -t filter -A INPUT -p tcp --dport 8080 -j ACCEPT
```

如在 fliter 表的 INPUT 链的第二条规则前插入一条规则，规则为允许访问 TCP 的 9090 端口的数据包通过。

```
[root@localhost~]# iptables -t filter -I INPUT 2 -p tcp --dport 9090 -j ACCEPT
```

查看当前 fliter 表的 INPUT 链规则。

```
[root@localhost~]# iptables -L INPUT --line -n
Chain INPUT (policy ACCEPT)
num   target     prot opt source              destination
1     DROP       icmp --  0.0.0.0/0           0.0.0.0/0
2     ACCEPT     tcp  --  0.0.0.0/0           0.0.0.0/0           tcp dpt:9090
3     ACCEPT     tcp  --  0.0.0.0/0           0.0.0.0/0           tcp dpt:8080
```

如删除 fliter 表的 INPUT 链的第二条规则。

```
[root@localhost~]# iptables -D INPUT 2
[root@localhost~]# iptables -L INPUT --line -n
Chain INPUT (policy ACCEPT)
num   target     prot opt source              destination
1     DROP       icmp --  0.0.0.0/0           0.0.0.0/0
2     ACCEPT     tcp  --  0.0.0.0/0           0.0.0.0/0           tcp dpt:8080
```

如在 FORWARD 链中添加规则,允许后台主机查询 DNS 和浏览网页。

```
[root@localhost~]# iptables -A FORWARD -p tcp --dport 80 -j ACCEPT
[root@localhost~]# iptables -A FORWARD -p tcp --dport 53 -j ACCEPT
[root@localhost~]# iptables -A FORWARD -p udp --dport 53 -j ACCEPT
[root@localhost~]# iptables -A FORWARD -p tcp -m state --state RELATED,ESTABLISHED -j ACCEPT
[root@localhost~]# iptables -A FORWARD -p udp -m state --state RELATED,ESTABLISHED -j ACCEPT
```

如在 FORWARD 链中添加规则,允许使用电子邮件服务(-m multiport,可以匹配多个端口)。

```
[root@localhost~]# iptables -A FORWARD -p tcp -m multiport --dport 25,110,143,993,995 -j ACCEPT
```

如在 fliter 表的 INPUT 链添加一条规则,规则为除了 192.168.1.100 不能 ping 本机,其他主机都可以。

```
[root@localhost~]# iptables -A INPUT -s 192.168.1.100 -p icmp -j REJECT
```

如在 fliter 表的 INPUT 链添加一条规则,规则为拒绝 192.168.100.5 的主机访问本机的 Web 服务。

```
[root@localhost~]# iptables -A INPUT -p tcp -s 192.168.100.5 --dport 80 -j REJECT
```

保存 iptables 规则。

```
[root@localhost~]# service iptables save
```

(3) nat 配置

如使用 SNAT,将 192.168.1.0 网段的内网 Web 服务器主机的 IP 地址转换成外网的 IP 地址 12.12.12.1。

```
[root@localhost~]# iptables -t nat -A POSTROUTING -s 192.168.1.0/24 -o eth0 -j SNAT --to-source 12.12.12.1
```

如外网的主机 12.12.12.1 要访问内网的 Web 服务器主机 192.168.1.200，使用 DNAT 将外网的 IP 地址转换为内网地址的 192.168.1.5。

```
[root@localhost~]# iptables -t nat -A PREROUTING -d 12.12.12.1 -p tcp --dport 80 -j DNAT --to-destination 192.168.1.5:80
```

如在拨号上网时，将 192.168.1.0/24 的内部地址转换为 ppp0 的公网地址。

```
[root@localhost~]# iptables -t nat -A POSTROUTING -o ppp0 -s 192.168.1.0/24 -j MASQUERADE
```

4.2.2 firewalld 防火墙

1. firewalld 基本知识

firewalld(Dynamic Firewall Manager of Linux systems，Linux 系统的动态防火墙管理器)是 RHEL 7 下默认的防火墙配置管理工具，它拥有基于 CLI(命令行界面 firewalld-cmd)和基于 GUI(图形用户界面 firewalld-config)的两种管理方式。相较于传统的防火墙管理配置工具，firewalld 支持动态更新技术并加入了区域(zone)的概念。简单来说，区域就是 firewalld 预先准备了几套防火墙策略集合(策略模板)，用户可以根据生产场景的不同选择合适的策略集合，从而实现防火墙策略之间的快速切换。表 4-2 列出了 firewalld 中常用的区域名称(默认为 public)以及相应的策略规则。

表 4-2 firewalld 中常用的区域名称以及相应的策略规则

区域	默认策略规则
trusted	允许所有的数据包
home	拒绝流入的数据包，除非与输出流量数据包相关或是 ssh、mdns、ipp-client 与 dhcpv6-client 服务，则允许
internal	等同于 home 区域
work	拒绝流入的数据包，除非与输出流量数据包相关或是 ssh、ipp-client 与 dhcpv6-client 服务，则允许
public	拒绝流入的数据包，除非与输出流量数据包相关或是 ssh、dhcpv6-client 服务，则允许
external	拒绝流入的数据包，除非与输出流量数据包相关或是与 ssh 服务相关，则允许
dmz	拒绝流入的数据包，除非与输出流量数据包相关或是与 ssh 服务相关，则允许
block	拒绝流入的数据包，除非与输出流量相关
drop	拒绝流入的数据包，除非与输出流量相关

在 Linux 系统中，使用命令行终端，可以极大地提高工作效率。当配置 firewalld 防火墙时，可以 firewalld-cmd 命令行界面，表 4-3 列出 firewalld-cmd 命令中常用的参数以及功能。

表 4-3　firewalld-cmd 命令常用的参数以及功能

参数	功能
--get-default-zone	查询默认的区域名称，如 firewall-cmd --get-default-zone
--set-default-zone=＜区域名称＞	设置默认的区域，如 firewall-cmd --set-default-zone=public
--get-zones	显示可用区域，如 firewall-cmd --get-zones
--get-services	显示预定的服务，如 firewall-cmd --get-services
--get-active-zones	显示当前正在使用的区域与网卡名称，如 firewall-cmd --get-active-zones
--add-source=ip 地址	将来源与此 ip 或子网的流量导向指定的区域，如 firewall-cmd --add-source=8.8.8.8 --zone=dmz
--remove-source=ip 地址	不再将此 ip 地址或子网的流量导向指定区域
--add-interface=网卡名称	将来自该网卡的所有流量导向指定区域
--change-interface=网卡名称	将该网卡与区域进行关联
--list-all	显示当前区域的网卡配置参数、资源、端口及服务等信息
--list-all-zones	显示所有区域的网卡配置参数、资源、端口及服务等信息
add-service=服务名称	设置默认区域允许该服务的流量
--add-port=端口号/协议	设置默认区域允许该端口的流量
--remove-service=服务名	设置默认区域不再允许服务的流量
--remove-port=端口号/协议	设置默认区域不再允许该端口的流量
--reload	让"永久生效"的规则立即生效，同时覆盖当前配置
--panic-on	开启应急状况模式
--panic-off	关闭应急状况模式

与大多数 Linux 服务一样，在使用 firewalld 配置好防火墙策略后，这些配置只在当前生效（Runtime 模式），也就是重启系统后失效。可以在 firewalld-cmd 命令添加 --permanent 参数，从而将配置永久失效（Permanent 模式），需要注意是的，要想这些配置立即生效，必须执行 firewalld-cmd --reload 命令；否则，这些配置只有在系统重启后才生效。

2. 使用 firewalld-cmd 配置防火墙

如查询当前 firewalld 服务使用的区域。

```
[root@localhost ~]# firewall-cmd --get-default-zone
public
```

如查询网卡 ens33 在 firewalld 服务中的区域。

```
[root@localhost ~]# firewall-cmd --get-zone-of-interface=ens33
public
```

如设置 firewalld 服务的当前默认区域为 dmz。

```
[root@localhost ~]# firewall-cmd --set-default-zone=dmz
success
[root@localhost ~]# firewall-cmd --get-default-zone
dmz
```

如把 firewalld 服务中的网卡 ens33 的默认区域改为 home，查看当前模式下的区域名称。

```
[root@localhost ~]# firewall-cmd --permanent --zone=home --change-interface=ens33
success
[root@localhost ~]# firewall-cmd --reload
[root@localhost ~]# firewall-cmd --get-zone-of-interface=ens33
home
```

系统重启后生效，再次查看当前模式下的区域名称，会发现，配置已经永久失效。

```
[root@localhost ~]# firewall-cmd --get-zone-of-interface=ens33
home
```

如查询 public 区域是否允许请求 SSH、HTTP、DNS、FTP 协议的流量。

```
[root@localhost ~]# firewall-cmd --zone=public --query-service=ssh
yes
[root@localhost ~]# firewall-cmd --zone=public --query-service=http
no
[root@localhost ~]# firewall-cmd --zone=public --query-service=dns
no
[root@localhost ~]# firewall-cmd --zone=public --query-service=ftp
no
```

如设置 firewalld 服务中请求 HTTP 协议的流量为永久允许，同时立即生效。

```
[root@localhost ~]# firewall-cmd --permanent --zone=public --add-service=http
success
[root@localhost ~]# firewall-cmd --reload
success
[root@localhost ~]# firewall-cmd --zone=public --query-service=http
yes
```

如设置 firewalld 服务中请求 HTTP 协议的流量为永久拒绝，同时立即生效。

```
[root@localhost ~]# firewall-cmd --permanent --zone=public --remove-service=http
success
[root@localhost ~]# firewall-cmd --reload
success
[root@localhost ~]# firewall-cmd --zone=public --query-service=http
no
```

如设置 firewalld 服务中访问 9090 和 9091 端口的流量策略设置为允许，仅当前生效。

```
[root@localhost ~]# firewall-cmd --zone=public --add-port=9090-9091/tcp
success
[root@localhost ~]# firewall-cmd --zone=public --list-ports
9090-9091/tcp
```

端口转发功能可以将原本到某端口的数据包转发到其他端口，流量转发的命令语法格式为：firewall-cmd --permanent --zone=<区域> --add-forward-port=port=<源端口号>:proto=<协议>:toport=<目标端口号>:toaddr=<目标 IP 地址>。

如将访问 192.168.1.1 主机 8080 端口的流量请求转发至 80 端口。

```
[root@localhost~]# firewall-cmd --permanent --zone=public
--add-forward-port= port=8080:proto=tcp:toport=80:toaddr=192.168.1.1
success
```

3. 使用 firewalld-cmd 配置防火墙

在 RHEL7 中，在终端输入命令 firewalld-config 或者是依次单击"应用程序"→"杂项"→"防火墙"，打开如图 4-31 所示界面。

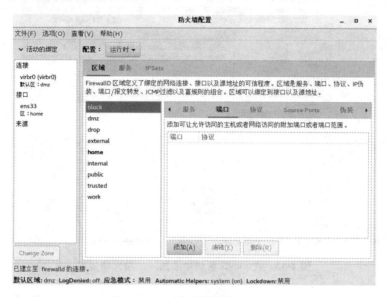

图 4-31　防火墙配置界面

使用 firewalld-cmd 配置防火墙，操作很简单，并且配置完成后，修改过的配置内容会自动保存，不需要进行二次配置。

如将 public 区域中请求 http 流量设置为允许，并当前生效，如图 4-32 所示。

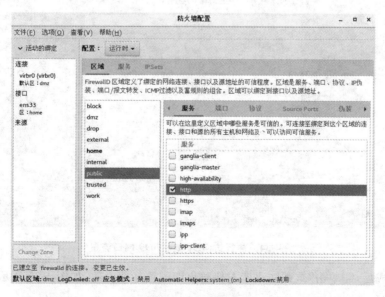

图 4-32　放行 http 流量

如将 public 区域中，添加一条防火墙规则，使其放行访问 9090～9099 端口（TCP）的流量策略，并设置为永久生效，如图 4-33、图 4-34 所示。

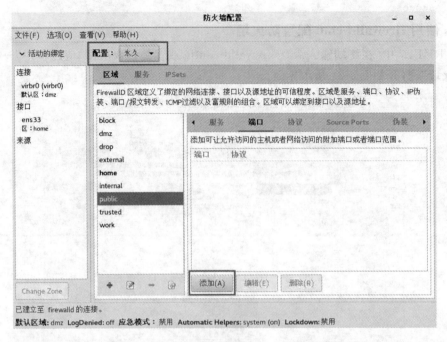

图 4-33　在 public 区域中添加一条防火墙规则

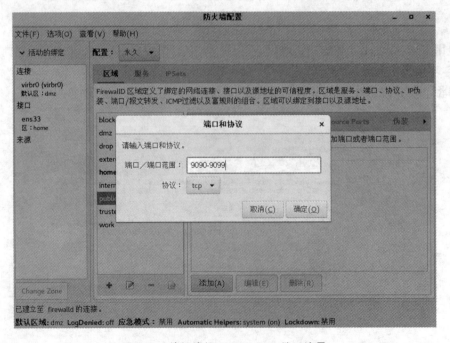

图 4-34　放行访问 9090～9099 端口流量

设置完成后，需单击"重载防火墙"，使防火墙配置立即生效，如图 4-35 所示。

图 4-35 让配置的防火墙策略规则立即生效

任务 4.3 远程访问与连接

4.3.1 远程访问方式简介

在 Linux 系统中,常用的远程访问方式主要有 Telnet、SSH、远程桌面三种方式。

1. Telnet

Telnet 协议是 TCP/IP 协议族中的一员,是 Internet 远程登录服务的标准协议和主要方式,其默认使用的端口是 23 端口。可以使用它远程连接管理网络设备(交换机、路由器、防火墙)、远程主机等。终端使用者在电脑上使用 Telnet 程序(如 Windows 系统自带的 Telnet 客户端),远程登录服务器。由于使用 Telnet 协议传输的数据未加密,容易被窃听,现在常用 SSH 协议取代。

2. SSH

SSH(Secure Shell,安全外壳协议)可以把所有传输的数据进行加密,从而有效地防止在远程访问过程中,数据被窃听、泄露,其默认使用的端口是 22 端口。

SSH 提供两种级别的安全验证。

① 基于口令的安全验证:使用账号和密码进行验证登录,所有传输的数据都会被加密。但是可能其他服务器会伪装成真正的服务器,进行数据的窃取。

② 基于密匙的安全验证:在本地生成一对密匙,并将密匙中的公匙上传到被访问的服务器。当使用 SSH 连接到服务器时,客户端软件就会向服务器发出请求,请求用密匙进行安全验证。服务器收到请求之后,先在该服务器主目录下寻找公匙,然后把它和发送过来的

公用密匙进行比较。如果两个密钥一致，服务器就用公钥加密"质询"（challenge）并把它发送给客户端软件。客户端软件收到"质询"之后就可以用私钥在本地解密再把它发送给服务器完成登录。该方式相对而言，更加安全。

3. 远程桌面

远程桌面是从 Telnet 发展而来的，可以把它比喻为图形化的 Telnet。当一台计算机开启了远程桌面连接功能后，用户就可以通过网络使用自己的计算机连接、控制、管理这台计算机，如现在人们使用 QQ 的"远程协助"功能，就是基于远程桌面来实现的。目前常用的远程桌面工具有：TeamViewer、Splashtop、PC Anywhere、UltraVNC、向日葵等。

4.3.2 Telnet

实现步骤如下：

(1) 安装 Telnet 服务、Telnet 守护进程 xinetd，RHEL7 默认没有安装这些服务。

```
[root@localhost ~]# rpm-qa |grep telnet
[root@localhost ~]# rpm-qa |grep xinetd
[root@localhost ~]# yum -y install telnet*
[root@localhost ~]# yum -y install xinetd
```

(2) 启动 Telnet 服务、xinetd 服务，设置这两个服务开机自动启动，关闭防火墙。

```
[root@localhost etc]# systemctl start telnet.socket
[root@localhost etc]# systemctl start xinetd
[root@localhost etc]# systemctl enable telnet.socket
[root@localhost etc]# systemctl enable xinetd
[root@localhost etc]# systemctl stop firewalld
```

也可以设置防火墙策略，放行 Telnet 服务使用的 23 端口。

```
[root@localhost etc]# systemctl start firewalld
[root@localhost ~]# firewall-cmd --query-port=23/tcp    //查询防火墙是否放行 23 端口
no
[root@localhost ~]# firewall-cmd --zone=public --add-port=23/tcp --permanent
                                                        //设置防火墙放行 23 端口
success
[root@localhost ~]# firewall-cmd --complete-reload      //重新加载防火墙
success
[root@localhost ~]# firewall-cmd --query-port=23/tcp    //再次查询，防火墙已放行 23 端口
yes
```

(3) 使用客户端（Windows 系统）进行测试，RHEL7 的 IP 地址为 192.168.1.1，如图 4-36、图 4-37 所示。

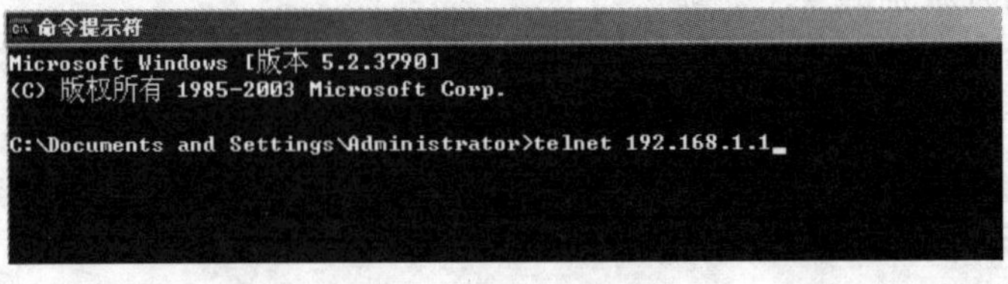

图 4-36 远程访问主机

图 4-37 普通用户 jack 登录系统

从图 4-37 中可以看出,使用普通用户 jack 可以登录。

此时,使用 root 用户无法登录系统,如图 4-38 所示。需要将/etc/securetty 文件改名后,root 用户才能登录,如图 4-39 所示。

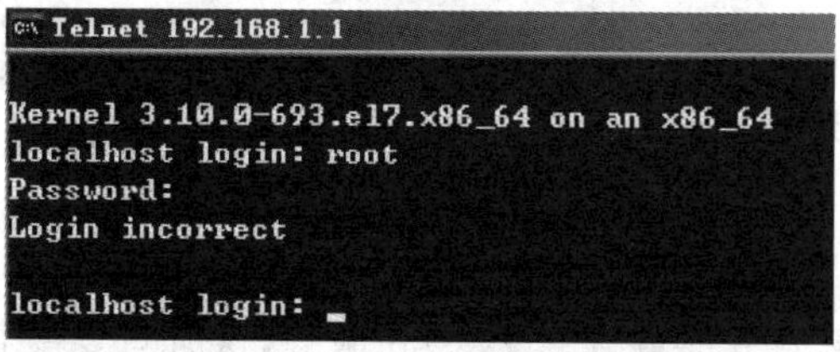

图 4-38 root 用户无法登录系统

[root@localhost~]# mv /etc/securetty /etc/securetty.bak

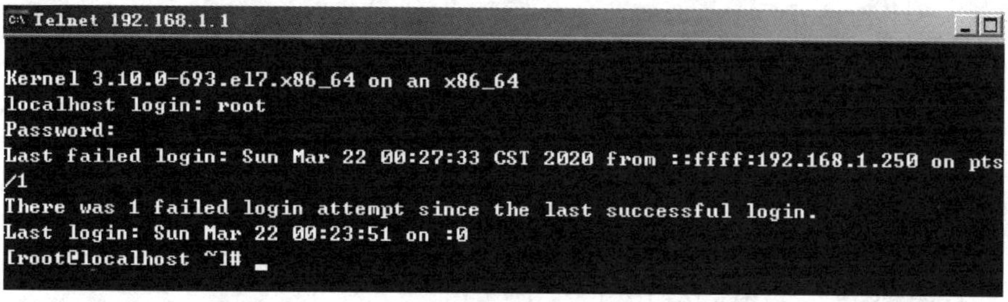

图 4-39 root 用户登录系统

4.3.3 SSH

配置 sshd 的文件为/etc/ssh/sshd_config,该文件相关参数及功能如表 4-4 所示。

表 4-4 sshd 服务配置文件的参数及功能

参数	功能
Port 22	默认 ssh 端口
ListenAddress 0.0.0.0	监听本机所有 IPv4 地址
Protocol 1/2	设置协议版本为 SSH1 或 SSH2
HostKey /etc/ssh/ssh_host_key	设置 SSHversion 1 使用的私钥
HostKey /etc/ssh/ssh_host_rsa_key	设置 SSHversion 2 使用的 RSA 私钥
HostKey /etc/ssh/ssh_host_dsa_key	设置 SSHversion 2 使用的 DSA 私钥

续表 4-4

参数	功能
Authentication	限制用户必须在指定的时限内认证成功，0 表示无限制。默认值是 120 秒
PermitRootLogin yes	设置是否允许 root 管理员直接登录
StrictModes yes	当远程用户的私钥改变时，直接拒绝连接
MaxAuthTries 6	设置最大失败尝试登录次数为 6
MaxSessions 10	最大支持 10 个 ssh 会话
PasswordAuthentication yes	是否开启密码验证机制
PermitEmptyPasswords no	是否允许用口令为空的账号登录系统

1. 基本验证

实现步骤如下：

（1）在 RHEL7 中，默认安装并开启了 sshd 服务，这时只要配置了 sshd 服务的服务器端、客户端的 IP 地址，客户端就可以使用命令"ssh [参数] 服务器 IP 地址"来远程连接服务器。如服务器 RHEL7-1，IP 地址为 192.168.1.1，客户端 RHEL7-2，IP 地址为 192.168.1.10，如图 4-40 所示。

```
[root@RHEL7-2 ~]# ifconfig ens33
ens33: flags=4163<UP,BROADCAST,RUNNING,MULTICAST>  mtu 1500
        inet 192.168.1.10  netmask 255.255.255.0  broadcast 192.168.1.255
        inet6 fe80::5e7:91b8:e26b:ee64  prefixlen 64  scopeid 0x20<link>
        ether 00:0c:29:a8:29:ee  txqueuelen 1000  (Ethernet)
        RX packets 7  bytes 554 (554.0 B)
        RX errors 0  dropped 0  overruns 0  frame 0
        TX packets 40  bytes 5237 (5.1 KiB)
        TX errors 0  dropped 0  overruns 0  carrier 0  collisions 0

[root@RHEL7-2 ~]# ssh 192.168.1.1
The authenticity of host '192.168.1.1 (192.168.1.1)' can't be established.
ECDSA key fingerprint is SHA256:ZzlzEA0jOu4CS5OQtTmIHqbwAy7LCObLiOh77WXejvM.
ECDSA key fingerprint is MD5:bb:28:00:f5:71:2e:40:f7:62:b7:8e:85:65:34:05:15
Are you sure you want to continue connecting (yes/no)? yes
Warning: Permanently added '192.168.1.1' (ECDSA) to the list of known hosts.
root@192.168.1.1's password:
Last login: Tue Mar  3 11:45:48 2020
[root@RHEL7-1 ~]# ifconfig ens33
ens33: flags=4163<UP,BROADCAST,RUNNING,MULTICAST>  mtu 1500
        inet 192.168.1.1  netmask 255.255.255.0  broadcast 192.168.1.255
        inet6 fe80::7f0f:c16b:15c9:fa5f  prefixlen 64  scopeid 0x20<link>
        ether 00:0c:29:c4:e7:d1  txqueuelen 1000  (Ethernet)
        RX packets 1558  bytes 371788 (363.0 KiB)
        RX errors 0  dropped 0  overruns 0  frame 0
        TX packets 84  bytes 12689 (12.3 KiB)
        TX errors 0  dropped 0  overruns 0  carrier 0  collisions 0
```

图 4-40 远程连接服务器

图 4-40 中，首先查看客户端的 IP 地址，然后连接服务器，此时，使用用户 root 远程登录服务器，在"root@192.168.1.1's password:"处，需要输入服务器用户 root 的密码。登录完成后，再次查看系统当前的 IP 地址，IP 地址已经变为服务器的 IP 地址 192.168.1.1，说明客户端连接到服务器。

（2）如果要禁止使用用户 root 远程登录服务器，需要将配置文件 sshd_config 的 38 行

"#PermitRootLogin yes"参数前面的"#"取消,修改参数"yes"为"no",修改后保存文件,重启 sshd 服务,从而禁止用户 root 直接登录。

```
[root@RHEL7-1~]# vim /etc/ssh/sshd_config
……
37 # LoginGraceTime 2m
38 PermitRootLogin no
39 # StrictModes yes
40 # MaxAuthTries 6
41 # MaxSessions 10
……
[root@RHEL7-1~]# systemctl restart sshd
```

此时,客户端再次使用用户 root 远程登录服务器,系统提示已经不能登录,如图 4-41 所示。

```
[root@RHEL7-2 ~]# ssh 192.168.1.1
root@192.168.1.1's password:
Permission denied, please try again.
```

图 4-41 禁止用户 root 远程登录服务器

2. 安全密钥验证

实现步骤如下:

(1) 在服务器 RHEL7-1 上创建用户 tom,并设置密码。

```
[root@RHEL7-1~]# useradd tom
[root@RHEL7-1~]# echo "123456" | passwd --stdin tom
```

(2) 在客户端 RHEL7-2 上生成"密钥对"并查看公钥。

```
[root@RHEL7-2~]# ssh-keygen
Generating public/private rsa key pair.
Enter file in which to save the key (/root/.ssh/id_rsa):    //按回车键或设置密钥的储存路径
Enter passphrase (empty for no passphrase):    //直接按回车键或设置密钥的密码
Enter same passphrase again:    //再次按回车键或设置密钥的密码
Your identification has been saved in /root/.ssh/id_rsa.
Your public key has been saved in /root/.ssh/id_rsa.pub.
The key fingerprint is:
SHA256:gDyqDnaLwfRsuaWBfojjLGKzDdnSc4e+USmZWgc5OJM root@RHEL7-2
The key's randomart image is:
+---[RSA 2048]----+
|                 |
| .o..            |
| E++.            |
| .o.=..          |
|.. =+S           |
|o.B+=            |
|+O.% =.          |
|@ + O.X o        |
|=O+=o.           |
```

```
       +----[SHA256]-----+
       [root@RHEL7-2~]# cat /root/.ssh/id_rsa.pub
       ssh-rsa
       AAAAB3NzaC1yc2EAAAADAQABAAABAQDWRSLzNzfkw6wgGy3YyqeN84nDUeeZ34bby        +
5HqjbIy95Ry7DoaUhj7slMbySj5n62ZT0nnM5lFDDH99Yr50MUjpx3u3tVntJzXajW3oj99Xm5c
1iJ5vVo+0tRQCVtfRFTM3LbE0+lf751wJHxeeEKHsQpmbPKvjracMYyeZuzyeq1sBTFGfs5QH38p
avChPF3ltDrGFLhQRCrjNPF6iI7d4r5XdkLcs1HBEkzkSq9bwWszf8x1Q2CKG7jR7EzNLh/9Zmp8
wREKqGtOvyyBiIIy6rCuh5yJnaQEyfVMQWMrNn81xG5+3gZa2MKf6/orm0EVB9EmfGDHMXSI8ImY
R4v root@RHEL7-2
       [root@RHEL7-2~]# cat /root/.ssh/id_rsa
       -----BEGIN RSA PRIVATE KEY-----
       MIIEowIBAAKCAQEA1kUi8zc35MOsIBst2MqnjfOJw1Hnmd+G28vuR6o2yMveUcuw
       6GlIY+7JTG8ko+Z+tmU9J5zOZRQwx/fWK+dDFI6cd7t7VZ7Sc12o1t6I/fV5uXNY
       ……
```

（3）把客户端 RHEL7-2 生成的公钥文件转送给服务器 RHEL7-1。

```
       [root@RHEL7-2~]# ssh-copy-id tom@192.168.1.1
       /usr/bin/ssh-copy-id: INFO: attempting to log in with the new key(s),to filter
out any that are already installed
        /usr/bin/ssh-copy-id: INFO: 1 key(s) remain to be installed--if you are prompted
now it is to install the new keys
       tom@192.168.1.1's password:          //输入用户 tom 的密码
       Number of key(s) added: 1
       Now try logging into the machine,with: "ssh 'tom@192.168.1.1'"
       and check to make sure that only the key(s) you wanted were added.
```

（4）在服务器 RHEL7-1 将配置文件 sshd_config 的 65 行"PasswordAuthentication yes"改为"PasswordAuthentication no"，修改后保存文件，开启密钥验证机制，重启 sshd 服务。

```
       [root@RHEL7-1~]# vim /etc/ssh/sshd_config
       ……
       62 # To disable tunneled clear text passwords,change to no here!
       63 # PasswordAuthentication yes
       64 # PermitEmptyPasswords no
       65 PasswordAuthentication no
       ……
       [root@RHEL7-1~]# systemctl restart sshd
```

（5）在客户端 RHEL7-2 上使用用户 tom 远程登录服务器，此时无须输入密码，即可登录。

```
       [root@RHEL7-2~]# ssh tom@192.168.1.1
       Last login: Thu Mar 26 01:58:12 2020
       [tom@RHEL7-1~]$
```

3. 远程传输命令

scp（secure copy）是 Linux 系统进行远程拷贝文件的命令，和它类似的命令有 cp，不过 cp 只在本机进行拷贝，不能进行网络传输。同时，它是基于 SSH 协议在网络中进行安全传输的命令，其用法为：scp [参数] 本地文件 远程账户@远程 IP 地址:远程目录。

参数说明如下：

① -v:显示详细的连接进度。
② -P:指定远程主机的 sshd 端口号。
③ -r:传送文件夹。
④ -6:使用 ipv6 协议。

如将服务器 RHEL7-1 的文件/web/1.txt 发送到客户端 RHEL7-2 的目录/home。

```
[root@RHEL7-1~]# mkdir /web
[root@RHEL7-1~]# echo "nihao"> /web/1.txt
[root@RHEL7-1~]# scp /web/1.txt 192.168.1.10:/home
The authenticity of host '192.168.1.10 (192.168.1.10)' can't be established.
ECDSA key fingerprint is SHA256:BbVqfioMQqfx0xv2HFteuSk0tFQsg7CFx9jg1luwKPs.
ECDSA key fingerprint is MD5:54:bd:50:bb:4b:fc:82:9d:96:9b:da:b6:51:7c:a2:ea.
Are you sure you want to continue connecting (yes/no)? yes
Warning: Permanently added '192.168.1.10' (ECDSA) to the list of known hosts.
root@192.168.1.10's password:    //输入 RHEL7-2 用户 root 的密码
1.txt                               100%    6    3.3KB/s    00:00
[root@RHEL7-2~]# ls /home/
1.txt   gg   tom
```

如将客户端 RHEL7-2 的目录/home 下的 333.txt 文件,下载到服务器 RHEL7-1 的目录/home 下。

```
[root@RHEL7-2~]# touch /home/333.txt
[root@RHEL7-1~]# scp 192.168.1.10:/home/333.txt /home/
root@192.168.1.10's password:    //输入 RHEL7-1 用户 root 的密码
333.txt                             100%    0    0.0KB/s    00:00
[root@RHEL7-1~]# ls /home/
333.txt   jack   tom
```

4.3.4 远程桌面

1. Windows 客户端远程访问 Linux 服务器的桌面

在 Linux 系统安装 VNC 相关软件包,然后在 Windows 系统中,安装软件 VNC Viewer,此时,就可以在 Windows 系统中使用 VNC Viewer 访问 Linux 服务器的桌面。

实现步骤如下:

(1) 在服务器 RHEL7-1 上安装 VNC 相关软件包,创建一个新的 VNC 配置文件 vncserver@:3.service,修改该文件的[Service]部分为黑色加粗的内容,这里使用用户 root 连接服务器,并启动 VNC 服务。

```
[root@RHEL7-1~]# yum -y install tigervnc*
[root@RHEL7-1~]# cp /lib/systemd/system/vncserver@.service /lib/systemd/system/vncserver@:3.service    //复制 vncserver@.service 文件并重命名为 vncserver@:3.service,其中 3 代表开启 3 号窗口作为远程桌面
[root@RHEL7-1~]# vim /lib/systemd/system/vncserver@:3.service
……
[Service]
```

```
Type=forking
# Clean any existing files in /tmp/.X11-unix environment
ExecStartPre=/bin/sh -c '/usr/bin/vncserver -kill %i > /dev/null 2>&1 || :'
ExecStart=/sbin/runuser -l root -c "/usr/bin/vncserver %i"
PIDFile=/root/.vnc/%H%i.pid
ExecStop=/bin/sh -c '/usr/bin/vncserver -kill %i > /dev/null 2>&1 || :'
……
[root@RHEL7-1~]# vncpasswd      //设置用户 root 的 VNC 密码
Password:
Verify:
[root@RHEL7-1~]# systemctl start vncserver@:3.service     //启动 VNC 服务
[root@RHEL7-1~]# firewall-cmd --permanent --add-service=vnc-server
                                                          //设置防火墙允许 VNC 服务
success
[root@RHEL7-1~]# firewall-cmd --reload
```

注意：如不使用用户 root，使用 tom 用户连接服务器，操作如下：

```
[root@RHEL7-1~]# vim /lib/systemd/system/vncserver@:3.service
……
[Service]
Type=forking
User=root
# Clean any existing files in /tmp/.X11-unix environment
ExecStartPre=/bin/sh -c '/usr/bin/vncserver -kill %i > /dev/null 2> &1 || :'
ExecStart=/sbin/runuser -l tom -c "/usr/bin/vncserver %i"
PIDFile=/home/tom/.vnc/%H%i.pid
ExecStop=/bin/sh -c '/usr/bin/vncserver -kill %i > /dev/null 2> &1 || :'
……
[root@localhost~]# systemctl daemon-reload
[root@localhost~]# systemctl start vncserver@:3.service
```

（2）在客户端 Windows 系统中使用 VNC Viewer 连接服务器，如图 4-42～图 4-44 所示。

图 4-42　输入服务器的地址

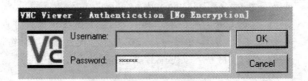

图 4-43　输入登录密码

图 4-44 Windows 客户端远程访问 Linux 服务器桌面

2. Linux 客户端远程访问 Linux 服务器的桌面

实现步骤如下：

（1）在客户端 RHEL7-2 上安装 VNC 相关软件包。

[root@RHEL7-2~]# yum -y install tigervnc*

（2）在客户端 RHEL7-2 上运行 vncvierer，连接服务器，如图 4-45～图 4-47 所示。

[root@RHEL7-2~]# vncviewer

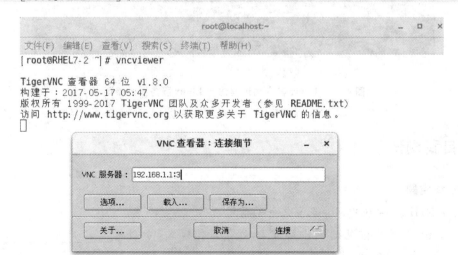

图 4-45 输入服务器的地址

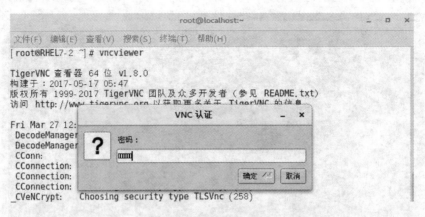

图 4-46 输入登录密码

图 4-47 Linux 客户端远程访问 Linux 服务器的桌面

自我测试

一、填空题

1. 在 RHEL7 系统中网卡配置文件的前缀为"_____"。
2. Linux 系统中常见的网卡绑定驱动有三种模式，分别是_____、_____、_____。
3. netfilter 是_____的容器，表是_____的容器，链是_____的容器。
4. NAT 表主要用作_____与_____的转换。

5. firewalld 是 RHEL 7 下默认的防火墙配置管理工具,它拥有基于_____和基于_____的两种管理方式。

6. Telnet 协议默认使用的端口是_____端口。

二、简答题

1. 简述 Linux 系统中三种定义主机名的方式。
2. 简述 Linux 系统中常见的网卡绑定驱动的三种模式。
3. 简述 NAT 表的三个链的功能。

三、实训题

使用 root 管理员配置网络环境,具体工作如下:

(1) 临时设置主机名为 myhost;

(2) 设置网卡信息:IP 地址为 202.40.20.254,网关是 202.40.20.1,子网掩码是 255.255.255.0,DNS 服务器是 200.200.200.200;

(3) 对于每一步的设置,管理员都要做相应的查看以确认设置是否成功。

项目 5 Samba 和 NFS 服务器

项目综述

在日常办公中,公司的各部门之间通常会建立自己部门所属的文件目录,以存放本部门的资料,同时,将该目录共享到网络中,便于本部门员工浏览、使用。为了满足这些办公需求,网络管理员将在服务器上配置 Samba 和 NFS 服务器,实现文件共享。

项目目标

- 了解 Samba 服务相关知识;
- 掌握 Samba 服务器的配置方法;
- 掌握 Samba 客户端的配置方法;
- 了解 NFS 服务相关知识;
- 掌握 NFS 服务器的配置方法;
- 掌握 NFS 客户端的配置方法。

◀ 任务 5.1 配置和管理 Samba 服务器 ▶

5.1.1 Samba 服务器简介

Samba 是在类 UNIX、Linux 和 Windows 系统中实现 SMB 协议的一个免费软件,由服务器及客户端程序构成。SMB(Server Messages Block,信息服务块)是在局域网上共享文件和打印机的一种通信协议,它为局域网内的不同计算机之间提供文件及打印机等资源的共享服务。SMB 协议是客户端/服务器型协议,客户端通过该协议可以访问服务器上的共享文件系统、打印机及其他资源。通过设置"NetBIOS over TCP/IP"使得 Samba 不但能与局域网内的主机分享资源,还能与全世界的计算机分享资源。Samba 常用于实现 Linux 系统与 Windows 系统之间的资源共享。

1. Samba 工作原理

Samba 服务器功能十分强大,这与其通信基于 SMB 协议有关。SMB 不仅提供目录和打印机的共享,还支持认证、权限设置。在早期,SMB 运行于 NBT 协议上,使用 UDP 协议的 137、138 端口及 TCP 协议的 139 端口。后期 SMB 经过开发,可以直接运行 TCP/IP 协议，没有额外的 NBT 层，使用 TCP 协议的 445 端口。

(1) Samba 工作流程

当客户端访问服务器时，信息通过 SMB 协议进行传输，其工作过程可以分为 4 个部分：

① 协议协商：客户端在访问 Samba 服务器时，发送 negport 指令数据包，告知目标计算机其支持的 SMB 类型。Samba 服务器根据客户端的情况，选择最优的 SMB 类型，并做出回应。

② 建立连接：当 SMB 类型确认之后，客户端会发送 session setup 数据包，提交账号和密码，请求与 Samba 服务器建立连接。如果客户端通过身份验证，Samba 服务器会对 session setup 报文做出回应，并为用户分配唯一的 UID，在客户端与其通信时使用。

③ 访问共享资源：客户端访问 Samba 共享资源时，发送 tree connect 指令数据包，通知服务器需要访问的共享资源名。如果设置允许，Samba 服务器会为每个客户端与共享资源的连接分配 TID，客户端即可以访问需要的共享资源。

④ 断开连接：共享使用完毕，客户端向服务器发送 tree disconnect 指令数据包，关闭共享，与服务器断开连接。

(2) Samba 使用 nmbd、smbd 服务来控制定位主机和权限分配

① nmbd：主要用来解析 NetBIOS 名，主要利用 UDP 协议开启 137、138 端口。

② smbd：主要用来管理 Samba 服务器上的共享资源（如共享目录、打印机等），主要利用 TCP 协议传输数据，使用 139 端口。

2. Samba 基本配置流程

Samba 基本配置流程为：

① 编辑主配置文件 smb.conf(文件路径/etc/samba/smb.conf)，指定需要共享的目录，并设置共享目录的共享权限；

② 在 smb.conf 文件中指定日志文件名称和存放路径；

③ 设置共享目录的本地文件权限；

④ 重新加载配置文件或重新启动 SMB 服务，使配置生效；

⑤ 设置防火墙、SELinux 放行相关 SMB 服务。

如图 5-1 所示，当客户端访问 Samba 服务器的 myshare 共享目录时，流程如下：

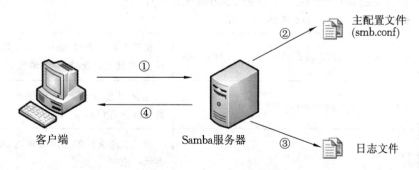

图 5-1　Samba 的工作流程

① 客户端请求访问 Samba 服务器上的 myshare 共享目录；

② Samba 服务器接收到请求后，会查询主配置文件 smb.conf，看是否共享了 myshare 目录，如果共享了这个目录，则查看客户端是否有权限访问；

③ Samba 服务器会将本次访问信息记录在日志文件中，日志文件的名称和路径都需要设置；

④ 如果客户端满足访问权限设置，则允许客户端进行访问。

5.1.2 Samba 服务参数

Samba 服务的主配置文件为 smb.conf，文件路径/etc/samba/smb.conf，表 5-1 列出了该文件中常见的配置参数及其作用。

表 5-1 Samba 服务程序配置文件中的参数及作用

选项	参数	作用
[global]	workgroup = MYGROUP	设置 Samba 服务器所在的工作组或域名
	server string = Samba Server Version %v	设置 Samba 服务器描述信息，参数%v 为显示 SMB 版本号
	netbios name = MYSERVER	设置 Samba 服务器的 NetBIOS 名称
	interfaces = lo eth0 192.168.12.2/24 192.168.13.2/24	设置 Samba 服务器使用的网络接口，可以使用网卡接口名称或 IP 地址
	hosts allow = 127. 192.168.12. 192.168.13.	设置允许访问 Samba 服务器的网络地址、主机地址以及域，多个参数用空格隔开
	♯hosts allow=192.168.0.1:允许该 IP 地址访问 Samba 服务器
	♯hosts allow=192.168.0:允许该网络段访问 Samba 服务器
	♯hosts allow=.jx.com:允许该域访问 Samba 服务器
	♯hosts allow=xx:允许计算机名为 xx 的客户端访问 Samba 服务器
	♯hosts allow=all:允许所有主机访问 Samba 服务器
	♯hosts allow=172.16.1. EXCEPT172.16.1.60:允许 172.16.1.0 网段的主机访问 Samba 服务器，但 172.16.1.60 除外
	hosts deny = 127. 192.168.12. 192.168.13.	设置不允许访问 Samba 服务器的网络地址、主机地址以及域，多个参数用空格隔开
	log file= /var/log/samba/log.%m	定义日志文件的存放位置与名称，参数%m 为客户端的主机名
	max log size = 50	定义日志文件的最大容量为 50 KB，"0"表示不限制
	security = user	设置 Samba 服务器安全验证的方式，总共有五种

续表 5-1

选项	参数	作用
[global]	#share：不需要提供用户名密码就可以访问 Samba 服务器，比较方便，但安全性很差	
	#user：需要提供用户名和密码，而且身份验证由 Samba 服务器负责，提升了安全性（系统默认方式）	
	#server：需要提供用户名和密码，使用独立的远程主机验证来访主机提供的口令（集中管理账户）	
	#domain：需要提供用户名和密码，指定 Windows 域控制器进行身份验证。Samba 服务器只能是域的成员客户端	
	#ads：需要提供用户名和密码，指定 Windows 域控制器进行身份验证。具有 domain 级别的所有功能，Samba 服务器可以成为域控制器	
	passdb backend = tdbsam	定义用户后台的类型，总共有三种
	#smbpasswd：使用 smbpasswd 命令为系统用户设置 Samba 服务程序的密码。smbpasswd 文件默认在 /etc/samba 目录下，不过有时候要手工建立该文件	
	#tdbsam：创建数据库文件并使用 pdbedit 命令建立 Samba 服务程序的用户。数据库文件为 passdb.tdb，默认在 /etc/samba 目录下。passdb.tdb 用户数据库可以使用 smbpasswd -a 来建立 Samba 用户，不过要建立的 Samba 用户必须先建立同名的系统用户，也可以使用 pdbedit 命令来建立 Samba 账户	
	#ldapsam：基于 LDAP 服务进行账户验证	
	encry passwords = yes	设置是否对 Samba 密码进行加密
	password server = 192.168.1.1	设置身份验证服务器地址，只有在设置 security 为 ads、server、domain 时，才会生效
	wins support = yes	设置 Samba 服务器是否启用 WINS 服务支持
	wins server = w.x.y.z	设置 Samba 服务器使用 WINS 服务器的 IP 地址
	wins proxy = yes	设置 Samba 服务器是否启用 WINS 代理支持
	dns proxy = yes	设置 Samba 服务器是否启用 DNS 代理支持
	load printers = yes	设置是否允许 Samba 打印机共享
	cups options = raw	打印机的选项
	printing = cups	设置 Samba 打印机的类型
	config file = /etc/samba/%U.smb.conf	引用子配置文件，%U 当前登录的用户名，%G 当前登录的用户组名
	include = /etc/samba/%U.smb.conf	引用子配置文件，%U 当前登录的用户名，%G 当前登录的用户组名。使用 include 时，如果是以用户 jack 的身份访问 Samba 服务器，除了可以看到 jack，在 smb.conf 中定义的其他共享资源也可以看到；使用 config file 时，如果是以用户 jack 的身份访问 Samba 服务器，只能看到 jack，在 smb.conf 中定义的其他共享资源都无法看到

续表 5-1

选项	参数	作用
共享定义	[ ]	设置共享目录的共享名称
	comment	设置共享目录的描述信息
	path	设置共享目录的完整路径
	browseable	设置在浏览资源时是否显示共享目录(yes/no)
	printable	设置是否允许打印(yes/no)
	public	设置是否允许匿名用户访问共享资源(yes/no)，只有当 security=share 时，该项才有作用
	guest ok	设置是否允许所有人访问(yes/no)，等效于"public"，只有当 security=share 时，该项才有作用
	guest only	设置是否只允许匿名用户访问(yes/no)
	guest account	指定访问共享目录的用户账户
	read only	设置是否允许以只读方式读取目录(yes/no)
	writable	设置是否允许以可写方式读取目录(yes/no)
	valid users	设置只有此名单内的用户和组才能访问共享资源
	invalid users	设置只有此名单内的用户和组不能访问共享资源，该参数要优于 valid users 参数的设置
	read list	设置只有此名单内的用户和组，以只读方式访问共享资源，如 read list=jack,@jack，前者表示用户 jack，后者表示用户组 jack
	write list	设置只有此名单内的用户和组，以可写方式访问共享资源，如 write list=jack,@jack，前者表示用户 jack，后者表示用户组 jack
	create mode	设置默认创建文件时的权限，类似 create mask
	directory mode	设置默认创建目录时的权限，类似 directory mask
	force group	设置默认创建文件时的用户组
	force user	设置默认创建文件时的所有者
	hosts allow	设置允许访问该共享资源的网络地址、主机地址以及域，多个参数用空格隔开
	hosts deny	设置不允许访问该共享资源的网络地址、主机地址以及域，多个参数用空格隔开

在实际的工作中，这些参数不一定能满足人们的具体要求，这时，可以使用项目 2 中关于文件权限设置的"文件访问控制列表"来进行配置，以满足特定环境的需求。

例如，某企业服务器上有个共享目录/opt/public，要求用户 lisi 能够查看和删除所有人的文件，用户 wangwu 能够查看所有人的文件，但不能删除别人的文件，这时就可以使用"文件访问控制列表"来进行配置。

```
[root@localhost~]# useradd lisi
[root@localhost~]# useradd wangwu
[root@localhost~]# mkdir/opt/public
[root@localhost~]# usermod-G lisi wangwu
[root@localhost~]# id wangwu
uid=1008(wangwu) gid=1008(wangwu)组=1008(wangwu),1007(lisi)
[root@localhost~]# chmod 777/opt/public
[root@localhost~]# chown lisi:lisi/opt /public
[root@localhost~]# setfacl-m d:u:lisi:rwx/opt/public
[root@localhost~]# setfacl-m d:u:wangwu:rx/opt/public
[root@localhost~]# getfacl/opt/public
getfacl: Removing leading '/' from absolute path names
# file: opt/public
# owner: lisi
# group: lisi
user::rwx
group::rwx
other::rwx
default:user::rwx
default:user:lisi:rwx
default:user:wangwu:r-x
default:group::rwx
default:mask::rwx
default:other::rwx
```

5.1.3 配置 Samba 服务器

公司网络管理员根据公司的经理室、财务部、销售部分别创建用户组 finance、sales、manager，并创建四个用户 user1、user2、user3、manager1，其中 user1 和 user2 属于 finance 组，user3 属于 sales 组，manager1 属于 manager 组。建立共享目录/opt/finance_share、/opt/sales_share、/opt/public_share。根据公司的部分权限设置：①finance 组的用户对目录 finance 共享读写权限；②sales 组的用户对目录 sales_share 共享读写权限；③manger 组的用户对所有目录均有读写权限；④将目录/opt/public_share 共享，共享名为 share，创建用户 lucy，此用户不具有登录系统功能，允许所有用户访问 public_share，只具有读取权限（服务器 IP 地址为 192.168.1.1）。

实现步骤如下：

(1) 安装 Samba 服务。

```
[root@localhost~]# yum-y install samba
```

(2) 添加用户组与用户。

```
[root@localhost~]# groupadd finance
[root@localhost~]# groupadd sales
```

```
[root@localhost ~]# groupadd manager
[root@localhost ~]# useradd -g finance user1
[root@localhost ~]# useradd -g finance user2
[root@localhost ~]# useradd -g sales user3
[root@localhost ~]# useradd -g manager manager1
[root@localhost ~]# useradd -s /sbin/nologin lucy
```

(3) 将本地用户转变成 Samba 用户。

```
[root@localhost ~]# pdbedit -a user1
new password:
retype new password:
Unix username:        user1
NT username:
Account Flags:        [U          ]
User SID:             S-1-5-21-113550305-1410460033-2856495339-1000
Primary Group SID:    S-1-5-21-113550305-1410460033-2856495339-513
Full Name:
Home Directory:       \\localhost\user1
HomeDir Drive:
Logon Script:
Profile Path:         \\localhost\user1\profile
Domain:               LOCALHOST
Account desc:
Workstations:
Munged dial:
Logon time:           0
Logoff time:          三,06 2月 2036 23:06:39 CST
Kickoff time:         三,06 2月 2036 23:06:39 CST
Password last set:    一,30 3月 2020 19:21:15 CST
Password can change:  一,30 3月 2020 19:21:15 CST
Password must change: never
Last bad password   : 0
Bad password count  : 0
Logon hours         : FFFFFFFFFFFFFFFFFFFFFFFFFFFFFFFFFFFFFFFFFFFF
[root@localhost ~]# pdbedit -a user2
……
[root@localhost ~]# pdbedit -a user3
……
[root@localhost ~]# pdbedit -a manager1
……
[root@localhost ~]# pdbedit -a lucy
……
[root@localhost ~]# pdbedit -L                    //查看 Samba 账号列表
user1:1001:
user2:1002:
```

```
user3:1003:
manager1:1004:
lucy:1005:
```

（4）创建共享目录，以及目录所属用户、用户组，并设置共享目录权限为 777。

```
[root@localhost~]# mkdir -p /opt/finance_share
[root@localhost~]# mkdir -p /opt/sales_share
[root@localhost~]# mkdir -p /opt/public_share
[root@localhost~]# chmod -R 777 /opt
[root@localhost~]# chown -R user1:finance /opt/finance_share/
[root@localhost~]# chown -R user3:sales /opt/sales_share/
```

（5）配置 Samba 服务器主配置文件 smb.conf。

```
[root@localhost~]# vim /etc/samba/smb.conf
```

在该文件中，添加以下配置内容：

```
[finance]                              # 财务部共享名称
comment=finance                        # 备注信息
path=/opt/finance_share                # 共享目录路径
public=no                              # 不允许匿名用户访问
browseable=yes                         # 允许浏览目录列表
writable=yes                           # 允许读写
valid users=@finance,@manager          # 允许 finance 和 manager 组用户访问
[sales]                                # 销售部共享名称
comment=sales                          # 备注信息
path=/opt/sales_share                  # 共享目录路径
public=no                              # 不允许匿名用户访问
browseable=yes                         # 允许浏览目录列表
writable=yes                           # 允许读写
valid users=@sales,@manager            # 允许 sales 和 manager 组用户访问
[share]                                # 共享目录的共享名称
comment=share                          # 备注信息
path=/opt/public_share                 # 共享目录路径
public=yes                             # 允许匿名用户登录
browseable=yes                         # 允许浏览目录列表
readonly=yes                           # 设置只有读取权限
```

（6）使用 testparm 命令检查配置文件是否有语法错误。

```
[root@localhost~]# testparm
Load smb config files from /etc/samba/smb.conf
rlimit_max: increasing rlimit_max (1024) to minimum Windows limit (16384)
Processing section "[homes]"
Processing section "[printers]"
Processing section "[print$ ]"
Processing section "[finance]"
Processing section "[sales]"
Processing section "[share]"
```

```
            Loaded services file OK.
            Server role: ROLE_STANDALONE
            Press enter to see a dump of your service definitions
```

（7）启动 Samba 服务，使用 chcon 命令进行 SELinux 安全上下文的修改（context 值），并配置防火墙允许 Samba 服务通过。

```
            [root@localhost ~]# systemctl start smb
            [root@localhost ~]# chcon -t samba_share_t /opt/finance_share/ -R
            //将 samba 目录 finance_share 共享给其他用户，若不配置，如用户 manager1 进入共享目录 finance，只能查看里面的文件，不能进行写操作，如创建文件。
            [root@localhost ~]# chcon -t samba_share_t /opt/sales_share/ -R
            [root@localhost ~]# chcon -t samba_share_t /opt/public_share/ -R
            [root@localhost ~]# firewall-cmd --permanent --add-service=samba
            success
            [root@localhost ~]# firewall-cmd --reload
            success
            [root@localhost ~]# firewall-cmd --list-all   //Samba 服务已被防火墙允许通过
            public (active)
              target: default
              icmp-block-inversion: no
              interfaces: ens33
              sources:
              services: ssh dhcpv6-client samba
              ports:
              protocols:
              masquerade: no
              forward-ports:
              source-ports:
              icmp-blocks:
              rich rules:
```

在上述配置中，如果不对 SELinux 安全上下文进行修改，也可以直接关闭 SELinux。

```
            [root@localhost ~]# getenforce
            Enforcing
            [root@localhost ~]# setenforce 0
            [root@localhost ~]# getenforce
            Permissive
```

配置到这里，基本上满足公司需求的功能都实现了，但是进行测试会发现，用户不能访问自己的"家目录"，如 user1 不能访问自己的文件夹"user1"，这是因为 SELinux 关闭了"samba_enable_home_dirs"该项功能，可以通过以下命令进行查看及设置，解决这个问题。

```
            [root@localhost ~]# getsebool -a |grep samba
            samba_create_home_dirs --> off
            samba_domain_controller --> off
            samba_enable_home_dirs --> off
            samba_export_all_ro --> off
```

```
samba_export_all_rw --> off
samba_load_libgfapi --> off
samba_portmapper --> off
samba_run_unconfined --> off
samba_share_fusefs --> off
samba_share_nfs --> off
sanlock_use_samba --> off
tmpreaper_use_samba --> off
use_samba_home_dirs --> off
virt_use_samba --> off
[root@localhost ~]# setsebool -P samba_enable_home_dirs on
[root@localhost ~]# getsebool -a |grep samba
samba_create_home_dirs --> off
samba_domain_controller --> off
samba_enable_home_dirs --> on
……
```

(8) 在共享目录中,创建测试文件。

```
root@localhost ~]# touch /opt/finance_share/fin.txt
[root@localhost ~]# touch /opt/sales_share/sal.txt
[root@localhost ~]# touch /opt/public_share/pub.txt
[root@localhost ~]# echo "nihao">/opt/finance_share/fin.txt
[root@localhost ~]# echo "nihao">/opt/sales_share/sal.txt
[root@localhost ~]# echo "nihao">/opt/public_share/pub.txt
```

(9) 测试。

① Windows 客户端测试

A. UNC 路径法:在"运行"中输入"\\服务器 IP 地址"进行访问,如之前的 Linux 服务器地址为 192.168.1.1,输入后,打开如图 5-2 所示界面,输入用户名及密码,登录服务器。

图 5-2 "Windows 安全"对话框

可以使用之前的五个用户分别进行测试,测试完成后,会发现 Samba 服务器的功能完全符合公司的需求。这里,用 manager1 用户进行测试,如图 5-3～图 5-9 所示。

图 5-3 输入映射账号及密码

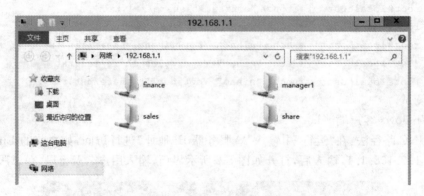

图 5-4 访问 Samba 服务器的共享资源

打开目录"finance",可以浏览、修改文件"fin.txt",并可以创建新的文件"1.txt"(图 5-5、图 5-6)。

图 5-5 创建新的文件

图 5-6 完成新文件的创建

打开目录"sales",可以浏览、修改文件"sal.txt",并可以创建新的文件"2.txt"(图 5-7)。

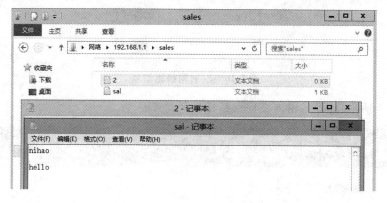

图 5-7 修改文件内容

打开目录"share",只能浏览文件"pub.txt",不能进行任何写入操作(图 5-8)。

图 5-8 浏览文件

能正常访问自己的"家目录"(图 5-9)。

图 5-9 访问"家目录"

B. 映射网络驱动器法：如在 Windows Server 2012 中，如图 5-10 所示界面中，单击"映射网络驱动器"。

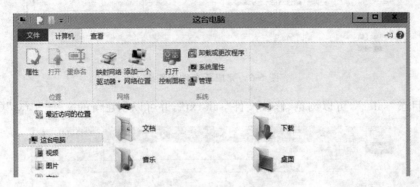

图 5-10　映射网络驱动器

在图 5-11 所示界面中，输入文件夹路径。

图 5-11　指定驱动器的名称及文件夹路径

在图 5-12 所示界面中，已经将共享目录"sales"挂载到 Z 盘。

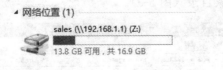

图 5-12　完成网络驱动器的映射

②Linux 客户端测试

A. 在 Linux 客户端上安装 samba -client(samba 客服端服务)、cifs -utils(支持文件共

享服务)。

```
[root@RHEL7-2~]# yum -y install samba-client
[root@RHEL7-2~]# yum -y install cifs-utils
```

B. 使用 smbclient 命令连接服务器,查看服务器的文件夹内的文件,如图 5-13 所示。smbclient 命令格式为:smbclient//目的 IP 地址/共享目录 -U 登录名。

```
[root@RHEL7-2~]# smbclient//192.168.1.1/sales -U user3
```

```
[root@RHEL7-2 ~]# smbclient //192.168.1.1/sales -U user3
Enter SAMBA\user3's password:
Domain=[LOCALHOST] OS=[Windows 6.1] Server=[Samba 4.6.2]
smb: \> ls
  .                                   D        0  Wed Apr  1 00:39:03 2020
  ..                                  D        0  Tue Mar 31 23:49:28 2020
  2.txt                               A        0  Wed Apr  1 00:27:59 2020
  sal.txt                             A       23  Wed Apr  1 00:36:35 2020

              17811456 blocks of size 1024. 14507488 blocks available
```

图 5-13 使用 smbclient 命令连接服务器

C. 创建文件夹 kk,如图 5-14 所示。

```
smb: \> mkdir kk
smb: \> ls
  .                                   D        0  Wed Apr  1 01:02:38 2020
  ..                                  D        0  Tue Mar 31 23:49:28 2020
  2.txt                               A        0  Wed Apr  1 00:27:59 2020
  sal.txt                             A       23  Wed Apr  1 00:36:35 2020
  kk                                  D        0  Wed Apr  1 01:02:38 2020

              17811456 blocks of size 1024. 14507528 blocks available
```

图 5-14 创建文件夹

D. 使用 mount 命令挂载共享目录,如图 5-15 所示。

mount 命令格式为:mount -t cifs//目的 IP 地址/共享目录 挂载点 -o username=登录名。

```
[root@RHEL7-2~]# mount -t cifs//192.168.1.1/finance/finance/ -o username=user2
```

```
[root@RHEL7-2 ~]# mkdir /finance
[root@RHEL7-2 ~]# mount -t cifs //192.168.1.1/finance /finance/ -o username=user2
Password for user2@//192.168.1.1/finance:  ******
[root@RHEL7-2 ~]# ls /finance/
1.txt  fin.txt
```

图 5-15 挂载共享目录

这时,可以将共享目录的文件拷贝到本地,或者将本地的文件拷贝到服务器。

```
[root@RHEL7-2~]# mkdir /gongxiang
[root@RHEL7-2~]# cp /finance/fin.txt /gongxiang/
[root@RHEL7-2~]# ls /gongxiang/
fin.txt
[root@RHEL7-2~]# touch /gongxiang/5.txt
[root@RHEL7-2~]# cp /gongxiang/5.txt /finance/
[root@RHEL7-2~]# ls /finance/
1.txt  5.txt  fin.txt
```

可以查看服务器的共享文件夹,验证拷贝是否成功。

```
[root@localhost ~]# ls /opt/finance_share/
1.txt  5.txt  fin.txt
```

任务 5.2 配置和管理 NFS 服务器

5.2.1 NFS 服务器简介

NFS(Network File System,网络文件系统)由 Sun 公司于 1984 年开发出来,它允许一个系统在网络上与他人共享目录和文件。主要用在类 UNIX 系统上,实现系统之间的资源共享,NFS 在文件传送或信息传送过程中依赖于 RPC 协议。

1. NFS 工作原理

用户可以将 NFS 服务器共享的目录挂载到本地计算机的文件系统中,用户使用这些资源,就像使用本地资源一样,非常便利。如图 5-16 所示,客户端将 NFS 服务器上的共享目录挂载到本地的挂载点进行使用,可以在服务器端对客户端的读写权限进行配置,以满足具体环境需求。

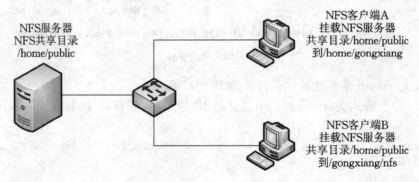

图 5-16 NFS 服务器工作原理

2. NFS 与 RPC

与绝大多数的网络服务一样,NFS 同样需要网络端口来进行客户端的连接,为客户端提供服务。但是由于 NFS 支持的功能相当多(如文件传输、身份验证等),每一个功能都会占用一个端口。为了避免 NFS 服务占用过多的固定端口,它采用动态端口的运行方式,每个功能在提供服务时,都会随机使用一个小于 1024 的端口。对于客户端而言,如此多的端口,到底目前正在使用哪个端口呢?

这时,就需要使用 RPC(Remote Procedure Call,远程进程调用)服务来统一管理、记录 NFS 每个服务对应的端口,它对外统一使用端口 111。当客户端使用 NFS 服务时,就直接访问服务器 RPC 服务的 111 端口,RPC 会将 NFS 的服务端口告诉客户端。

如图 5-17 所示为 NFS 和 RPC 共同为客户端提供服务,当客户端访问 NFS 服务器时,流程如下:

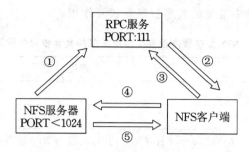

图5-17　NFS 和 RPC 共同为客户端提供服务

① NFS 服务器启动时,自动选择端口小于 1024 的端口,同时向 RPC(端口 111)发出指令,RPC 注册端口信息;
② 客户端使用 NFS 服务,向服务端的 RPC 服务请求查询 NFS 服务的端口;
③ 服务端的 RPC 回应客户端,NFS 服务使用的端口;
④ 客户端连接 NFS 服务使用的端口,并请求服务;
⑤ NFS 服务进行权限认证,允许客户端访问主机的共享资源。

5.2.2　NFS 服务的组件以及主要配置文件

1. NFS 服务的组件

NFS 服务主要由 6 个部分组成,其中前三个是必须的,后三个是可选的,具体名称及作用如表 5-2 所示。

表 5-2　NFS 服务的组件及作用

名称	作用
rpcbind	进行端口映射。当客户端访问 RPC 提供的服务时,rpcbind 会响应对 RPC 服务的请求,将所管理的与服务所对应的端口号提供给客户端,从而使客户端使用该端口向服务器请求服务
rpc.nfsd	判断、检查客户端是否具备登录服务器的权限,负责处理 NFS 请求
rpc.mountd	管理 NFS 的文件系统。当客户端通过 rpc.nfsd 登录服务器后,在使用 NFS 服务器提供的共享资源之前,它将检查客户端的权限,符合权限规则的客户端才能访问共享资源
rpc.locked	帮助 rpc.stated 守护进程,使用本进程来处理崩溃系统的锁定恢复问题的处理
rpc.stated	处理客户端与服务器之间的文件锁定问题,以确保文件的一致性
rpc.quota	提供 NFS 和配额管理程序之间的接口

2. NFS 服务的主要配置文件

NFS 服务的主要配置文件为 exports,文件路径//etc/exports,在 RHEL7 中,打开该文件后,会发现没有任何内容,此时需要安装"共享目录的路径　允许访问的 NFS 客户端(共享权限参数)"的格式,定义需要共享的资源及其对应的权限。

如:/home/pub　192.168.1.10/24(rw,sync,root_squash)

表 5-3 列出了该文件中常见的共享权限参数及作用。

表 5-3 NFS 服务程序配置文件中常见的共享权限参数及作用

参数	作用
ro	只读权限
rw	读/写权限
no_root_squash	当 NFS 客户端以 root 管理员访问时,映射为 NFS 服务器的 root 管理员,这个设置很不安全
root_squash	当 NFS 客户端以 root 管理员访问时,映射为 NFS 服务器的匿名用户,通常它的 UID 与 GID 都会变成 nobody(nfsnobody)这个系统账户的身份
all_squash	默认设置,无论 NFS 客户端使用什么账户访问,均映射为 NFS 服务器的匿名用户,通常它的 UID 与 GID 都会变成 nobody(nfsnobody)这个系统账户的身份。在多个 NFS 客户端同时读写 NFS 服务器的共享数据时,这个参数可以确保大家写入的数据的权限是一样的,但不同系统有可能匿名用户的 UID、GID 不同。因为此处需要使服务端和客户端之间的用户是一样的,例如:服务端指定匿名用户的 UID 为 1000,那么客户端也一定要存在 UID 为 1000 这个账户
sync	同时将数据写入内存与硬盘中,保证不丢失数据
async	优先将数据保存到内存,然后再写入硬盘;效率更高,但可能会丢失数据
anonuid	匿名用户的 UID,通常为 nobody,也可以自行设置,但是必须保证该 UID 保存在 /etc/passwd 文件中
anongid	匿名用户的 GID

5.2.3 配置 NFS 服务器

公司现有需求如下(服务器 IP 地址为 192.168.1.1):

① 共享目录 /nfsfile/caiwu:允许 192.168.1.0/24 和 192.168.2.0/24 网段的客户端访问,并且具有该目录的只读权限;

② 共享目录 /nfsfile/public:允许所有人具有读写权限,并且将 root 用户映射为匿名用户;

③ 共享目录 /nfsfile/jimi:允许 192.168.1.2/24 的客户端访问,并且具有该目录的读写权限。

实现步骤如下:

(1) 安装 NFS 服务。默认情况下,RHEL7 已经安装了 NFS 服务软件包。

```
[root@localhost ~]# rpm -qa |grep nfs-utils
nfs-utils-1.3.0-0.48.el7.x86_64
[root@localhost ~]# rpm -qa |grep rpcbind
rpcbind-0.2.0-42.el7.x86_64
[root@localhost ~]# yum -y install nfs-utils
[root@localhost ~]# yum -y install rpcbind
```

(2) 创建共享目录,并设置共享目录权限为 777。

```
[root@localhost~]# mkdir /nfsfile
[root@localhost~]# mkdir /nfsfile/caiwu
[root@localhost~]# mkdir /nfsfile/public
[root@localhost~]# mkdir /nfsfile/jimi
[root@localhost~]# chmod -R 777 /nfsfile/
[root@localhost~]# ls -al /nfsfile/
```

(3) 配置 NFS 服务器主配置文件 exports。

```
[root@localhost~]# vim /etc/exports
/nfsfile/caiwu    192.168.1.0/24(ro)    192.168.2.0/24(ro)
/nfsfile/public   *(rw,sync,root_squash)
/nfsfile/jimi     192.168.1.2/24(rw,sync)
```

(4) 启动 RPC、NFS 服务,并配置防火墙允许 NFS 相关服务通过。

```
[root@localhost~]# systemctl restart rpcbind
[root@localhost~]# systemctl restart nfs-server
[root@localhost~]# systemctl restart nfs
[root@localhost~]# firewall-cmd --permanent --add-service=nfs
success
[root@localhost~]# firewall-cmd --permanent --add-service=rpc-bind
success
[root@localhost~]# firewall-cmd --permanent --add-service=mountd
success
[root@localhost~]# firewall-cmd --reload
success
[root@localhost~]# firewall-cmd --list-all
public (active)
  target: default
  icmp-block-inversion: no
  interfaces: ens33
  sources:
  services: ssh dhcpv6-client samba nfs rpc-bind mountd
……
```

可以通过命令 rpcinfo -p 或者 rpcinfo -u 192.168.1.1 nfs 查看 NFS 使用的端口信息,以及 RPC 的注册状态,同时,通过 /var/lib/nfs/etab 文件可以查看 NFS 的共享目录配置信息。

```
[root@localhost~]# rpcinfo -p
   program vers proto   port  service
……
    100005    1   udp   20048  mountd
    100005    1   tcp   20048  mountd
    100005    2   udp   20048  mountd
    100005    2   tcp   20048  mountd
    100005    3   udp   20048  mountd
    100005    3   tcp   20048  mountd
```

```
        100003    3    tcp    2049    nfs
        100003    4    tcp    2049    nfs
        100227    3    tcp    2049    nfs_acl
        100003    3    udp    2049    nfs
        100003    4    udp    2049    nfs
        100227    3    udp    2049    nfs_acl
……
[root@localhost~]# rpcinfo -u 192.168.1.1 nfs
program 100003 version 3 ready and waiting
program 100003 version 4 ready and waiting
[root@localhost~]# cat /var/lib/nfs/etab
/nfsfile/jimi
192.168.1.2/24(rw,sync,wdelay,hide,nocrossmnt,secure,root_squash,no_all_
squash,no_subtree_check,secure_locks,acl,no_pnfs,anonuid=65534,anongid=65534,
sec=sys,secure,root_squash,no_all_squash)
/nfsfile/caiwu
192.168.1.0/24(ro,sync,wdelay,hide,nocrossmnt,secure,root_squash,no_all_
squash,no_subtree_check,secure_locks,acl,no_pnfs,anonuid=65534,anongid=65534,
sec=sys,secure,root_squash,no_all_squash)
/nfsfile/caiwu
192.168.2.0/24(ro,sync,wdelay,hide,nocrossmnt,secure,root_squash,no_all_
squash,no_subtree_check,secure_locks,acl,no_pnfs,anonuid=65534,anongid=65534,
sec=sys,secure,root_squash,no_all_squash)
/nfsfile/public
*(rw,sync,wdelay,hide,nocrossmnt,secure,root_squash,no_all_squash,no_
subtree_check,secure_locks,acl,no_pnfs,anonuid=65534,anongid=65534,sec=sys,
secure,root_squash,no_all_squash)
```

（5）测试。客户端（IP 地址为 192.168.1.2）使用 showmount 命令查看 NFS 服务器共享目录信息，该命令的常用参数如下：

① -e：显示 NFS 服务器的共享列表。

② -a：显示本机挂载的文件资源的情况、NFS 资源的情况。

③ -v：显示版本号。

```
[root@RHEL7-2~]# showmount -e 192.168.1.1
Export list for 192.168.1.1:
/nfsfile/public  *
/nfsfile/jimi    192.168.1.2/24
/nfsfile/caiwu   192.168.2.0/24,192.168.1.0/24
```

现在，可以将 NFS 服务器共享的目录，挂载到客户端的本地目录，并进行读写权限的验证。

```
[root@RHEL7-2~]# mkdir /public
[root@RHEL7-2~]# mkdir /jimi
[root@RHEL7-2~]# mkdir /caiwu
[root@RHEL7-2~]# mount -t nfs 192.168.1.1:/nfsfile/public /public
[root@RHEL7-2~]# mount -t nfs 192.168.1.1:/nfsfile/jimi /jimi
```

```
[root@RHEL7-2~]# mount -t nfs 192.168.1.1:/nfsfile/caiwu /caiwu
[root@RHEL7-2~]# cd /public/
[root@RHEL7-2 public]# mkdir ceshi          //具有读写权限
[root@RHEL7-2 public]# cd /jimi
[root@RHEL7-2 jimi]# mkdir ceshi
[root@RHEL7-2 jimi]# cd /caiwu              //具有读写权限
[root@RHEL7-2 caiwu]# mkdir ceshi
mkdir:无法创建目录"ceshi":只读文件系统    //具有只读权限
[root@RHEL7-2 caiwu]# ll /public/
总用量 0
drwxr-xr-x. 2 nfsnobody nfsnobody 6 4月  2 17:13 ceshi //将 root 用户映射为匿名用户
[root@RHEL7-2 caiwu]# ll /jimi/
总用量 0
drwxr-xr-x. 2 nfsnobody nfsnobody 6 4月   2 17:14 ceshi //将 root 用户映射为匿名
用户,默认设置为 all_squash
[root@RHEL7-2 caiwu]# ll /caiwu/
总用量 0
```

在服务器端可以查看当前客户端连接的情况,并可以使用命令 exportfs -rv 重新共享所有目录并输出详细信息,使用命令 exportfs -au 卸载所有共享目录。

```
[root@localhost~]# netstat -an | grep 192.168.1.1
tcp    0    0 192.168.1.1:2049     192.168.1.2:812        ESTABLISHED
ot@localhost~]# exportfs -rv
exporting 192.168.1.2/24:/nfsfile/jimi
exporting 192.168.1.0/24:/nfsfile/caiwu
exporting 192.168.2.0/24:/nfsfile/caiwu
exporting *:/nfsfile/public
[root@localhost~]# exportfs -au
```

客户端为了避免每次系统重启后反复进行挂载操作,可以配置 fstab 文件,启动自动挂载 NFS 文件系统。

```
[root@RHEL7-2 caiwu]# vim /etc/fstab
192.168.1.1:/nfsfile/public   /public  nfs  defaults 0 0
192.168.1.1:/nfsfile/jimi     /jimi    nfs  defaults 0 0
192.168.1.1:/nfsfile/caiwu    /caiwu   nfs  defaults 0 0
//重启系统后,查看系统挂载信息
[root@RHEL7-2~]# # df -h
文件系统                     容量   已用   可用   已用   挂载点
……
192.168.1.1:/nfsfile/jimi    17G   3.2G   14G   19%   /jimi
192.168.1.1:/nfsfile/public  17G   3.2G   14G   19%   /public
192.168.1.1:/nfsfile/caiwu   17G   3.2G   14G   19%   /caiwu
……
```

除了配置 fstab 文件来实现启动自动挂载外,在 Linux 系统中还可以通过 autofs 自动挂载服务来实现启动自动挂载。

首先安装 autofs 服务。

```
[root@RHEL7-2~]# yum -y install autofs
```

然后配置 autofs 服务的主配置文件 auto.mater，在该文件的第 9 行，配置子配置文件的路径为：/mynfs /etc/aa.misc（/mynfs 为自动挂载的主目录，/etc/aa.misc 为子配置文件，它用于定义挂载的动作，本案例是挂载到客户端的/mynfs/files 目录下）。

```
[root@RHEL7-2~]# vim/etc/auto.master
 1 #
 2 # Sample auto.master file
 3 # This is a 'master' automounter map and it has the following format:
 4 # mount-point[map-type[,format]:]map[options]
 5 # For details of the format look at auto.master(5).
 6 #
 7 /misc    /etc/auto.misc
 8
 9 /mynfs   /etc/aa.misc
10 # NOTE: mounts done from a hosts map will be mounted with the
11 #       "nosuid" and "nodev" options unless the "suid" and "dev"
12 #       options are explicitly given.
```

最后，对子配置文件按照"挂载目录 挂载文件类型及权限 :设备名称"的格式进行修改，然后重启 autofs 服务，并检验挂载是否成功。

```
[root@RHEL7-2~]# mkdir -p /mynfs/files
[root@RHEL7-2~]# cp -p /etc/auto.misc /etc/aa.misc
[root@RHEL7-2~]# vim /etc/aa.misc
files   -fstype=nfs      192.168.1.1:/nfsfile/public    //files 挂载的子目录
[root@RHEL7-2~]# systemctl restart autofs
[root@RHEL7-2~]# ls /mynfs/files/           //挂载成功
ceshi
[root@RHEL7-2~]# df -h
文件系统                          容量    已用   可用   已用%   挂载点
……
192.168.1.1:/nfsfile/public       17G    3.2G   14G   19%    /mynfs/files
……
```

现在验证客户端是否能在该共享目录/nfsfile/public 下进行读写操作。

```
[root@RHEL7-2~]# cd /mynfs/files/
[root@RHEL7-2 files]# # mkdir 111
[root@RHEL7-2 files]# # touch 1.txt
[root@RHEL7-2 files]# # ls
111   1.txt   ceshi
[root@RHEL7-2 files]# ]# cp -rf ceshi /home/
[root@RHEL7-2 files]# # ls /home/
ceshi   gg
```

可以看出，客户端具有该共享目录的读写权限。

自我测试

一、填空题

1. _____是一种在局域网上共享文件和打印机的一种通信协议。
2. Samba 服务的主配置文件为_____,文件路径_____。
3. RPC(Remote Procedure Call,远程进程调用)服务对外统一使用端口_____。
4. 独立冗余磁盘阵列,简称_____。
5. 在服务器上配置 RAID,有_____、_____两种配置方式。

二、简答题

1. 简述 Samba 的工作流程。
2. 简述 NFS 的工作原理。

三、实训题

1. 配置 Samba 服务器,Linux 系统的网卡地址是 192.168.20.25/24,在/share 目录下创建 1.txt,并编辑文件。在 Windows 和 Linux 系统上实现共享文件 1.txt。

2. 配置 NFS 服务器,要求如下:

(1) 共享/ftp 目录,允许所有客户端访问该目录,并仅有只读权限。

(2) 共享/nfs/public 目录,允许 192.168.20.0/24 和 192.168.30.0/24 网段的客户端访问,并对该目录仅有只读权限。

(3) 共享/nfs/web 目录,允许 192.168.20.0/24 网段的客户端访问,并对该目录仅有只读权限,同时将 root 用户映射为匿名用户。

(4) 共享/nfs/jimi 目录,允许 192.168.20.26/24 客户端访问,并对该目录有读写权限。

项目 6　DHCP 服务器

项目综述

随着公司业务的发展，公司的终端数量也日益增加，这时如果还使用静态地址分配方式来管理、使用公司内部的 IP 地址，稍不注意就会导致地址配置错误或者 IP 地址冲突，严重影响公司职员的正常工作，网络管理员将使用 DHCP 服务，彻底解决这个问题。

项目目标

- 了解 DHCP 服务相关知识；
- 了解 DHCP 服务工作原理；
- 掌握不同环境下的 DHCP 服务器的配置方法；
- 掌握 DHCP 客户端的配置、测试方法。

任务 6.1　DHCP 服务器简介

计算机之间在通信时，需要给每一台计算机分配一个唯一的 IP 地址，分配的方法有两种：一种是静态设置 IP 地址；一种是通过 DHCP 服务器，自动分配 IP 地址，通过这种方式分配 IP 地址，可以极大地减轻网络管理员的工作负担，避免局域网中 IP 地址冲突。

DHCP(Dynamic Host Configuration Protocol，动态主机配置协议)是 TCP/IP 协议簇中的一种，主要用来给网络客户端分配动态的 IP 地址，同时简化了客户端 TCP/IP 的设置。

6.1.1　DHCP 工作原理

DHCP 采用"客户端/服务器"的工作方式。以安装 DHCP 服务组件的计算机作为 DHCP 服务器，为 DHCP 客户端提供服务，作为客户端的工作站向 DHCP 器发出请求，以获得动态 IP 地址。

DHCP 客户端获取 DHCP 服务器的 IP 地址的全过程如图 6-1 所示。

(1) DHCP 客户端向网络发送 DHCPDISCOVER 广播信息，寻找网络中的 DHCP 服务器。

(2) 网络中的 DHCP 服务器收到客户端的 DHCPDISCOVER 信息后，就在它的地址池里选取一个未租用的 IP 地址作为 DHCPOFFER 广播信息，发送给客户端。在尚未与客户端完成 IP 地址租用程序之前，此 IP 地址将暂时保留，以避免重复分配给其他客户端。

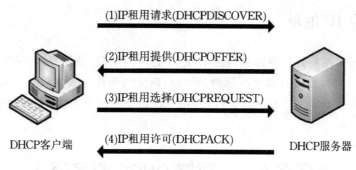

图 6-1 DHCP 工作过程

(3) DHCP 客户端收到 DHCPOFFER 广播信息后,再向网络中的 DHCP 服务器发送 DHCPREQUEST 广播信息,申请分配 IP 地址。

如果有多台 DHCP 服务器向 DHCP 客户端发送 DHCPOFFER 广播信息,则 DHCP 客户端只接受第一个收到的 DHCPOFFER 广播信息,然后它就向该 DHCP 服务器发送 DHCPREQUEST 广播信息。之所以用广播方式发送信息,是因为除了通知已被选择的 DHCP 服务器,还要通知其他未被选择的 DHCP 服务器,使它们能够及时释放原本准备分配给该 DHCP 客户端的 IP 地址,供其他客户端使用。

(4) 网络中的 DHCP 服务器收到 DHCP 客户端的 DHCPREQUEST 广播信息后,向 DHCP 客户端发送包含它所提供的 IP 地址、子网掩码、默认网关等的 DHCPACK 确认信息。

当 DHCP 客户端发生以下几种情况时,将向 DHCP 服务器申请分配 IP 地址:
① 客户端计算机第一次申请使用 DHCP 地址分配;
② DHCP 客户端释放了原先租用的 IP 地址,并要重新申请租用 IP 地址;
③ DHCP 客户端计算机租用的 IP 地址已经到期,IP 地址已经被 DHCP 服务器收回并分配出去;
④ DHCP 客户端计算机更换了物理网卡或重新接入其他局域网中。

6.1.2 DHCP 的三种 IP 地址分配方式

(1) 自动分配方式

DHCP 服务器为客户端指定一个永久性的 IP 地址,一旦 DHCP 客户端第一次成功从 DHCP 服务器端租用到 IP 地址,就可以永久性地使用该地址。

(2) 动态分配方式

DHCP 服务器为客户端分配一个具有租约期限的 IP 地址,租约到期后,DHCP 服务器收回该地址,并可提供给其他客户端使用。

(3) 手工分配方式

客户端的 IP 地址是由 DHCP 服务器配置的(保留地址),DHCP 服务器只将这些 IP 地址分配给客户端。

6.1.3 续约 IP 地址

若 DHCP 客户端要延长目前租约的 IP 地址的使用期限，就需要向 DHCP 服务器申请续约该 IP 地址。可以通过自动更新续约、手动更新续约两种方式，实现 IP 地址的续约。如在 Windows 客户端计算机 CMD 命令行，使用 ipconfig/renew 命令，就可进行手动续约 IP 地址。

◀ 任务 6.2　部署 DHCP 服务器 ▶

6.2.1 安装 DHCP 服务器

1. 查询系统是否已经安装 DHCP 服务器

```
[root@localhost~]# rpm-qa |grep dhcp
```

2. 安装 DHCP 服务器

```
[root@localhost~]# yum-y install dhcp
[root@localhost~]# rpm-qa |grep dhcp
dhcp-4.2.5-58.el7.x86_64
dhcp-common-4.2.5-58.el7.x86_64
dhcp-libs-4.2.5-58.el7.x86_64
```

6.2.2 认识 DHCP 服务器主配置文件

1. 主配置文件

DHCP 服务器默认的配置文件是 /etc/dhcp/dhcpd.conf，但是当打开该文件后，会发现文件只有三行注释内容，如图 6-2 所示，其中第二行语句告诉使用者去查看系统给出的参考配置文件 dhcpd.conf.example。在配置 DHCP 服务器时，需要将该文件复制到 /etc/dhcp/ 目录下，并重命名为 dhcpd.conf。

```
# DHCP Server Configuration file.
#   see /usr/share/doc/dhcp*/dhcpd.conf.example
#   see dhcpd.conf(5) man page
```

图 6-2　打开主配置文件

```
[root@localhost~]# vim /usr/share/doc/dhcp-4.2.5/dhcpd.conf.example
```

当打开该参考配置文件时，会发现文件的内容比较多，该文件分为全局配置和局部配置两块。其中，全局配置包含参数或选项，该部分对整个 DHCP 服务器生效；局部配置通常由声明部分表示，该部分仅对局部生效。该文件的格式如下：

```
# 全局配置
参数或选项;
# 局部配置
声明{
    参数或选项;
}
```

2. 常见参数

参数表明如何执行任务,是否需要执行任务,或者哪些网络配置选项发送给客户。表 6-1 列出了常见的参数及其作用。

表 6-1 dhcpd 服务程序配置文件常见的参数及其作用

参数	作用
ddns-update-style 类型	定义 DNS 服务动态更新的类型,类型包括:none(不支持动态更新)、interim(互动更新模式)、ad-hoc(特殊更新模式)
[allow/ignore] client-updates	允许/忽略客户端更新 DNS 记录
default-lease-time 600	默认超时时间,单位是秒
max-lease-time 7200	最大超时时间,单位是秒
hardware 00:0C:29:C4:E7:D1	指定网卡接口的类型与 MAC 地址
server-name "server1.kk.com"	向 DHCP 客户端通知 DHCP 服务器的主机名
fixed-address 192.168.1.200	将某个固定的 IP 地址分配给指定主机
time-offset 偏移差	指定客户端与格林尼治时间的偏移差

3. 常见声明

声明用来描述网络布局、提供给客户端的 IP 地址等。表 6-2 列出了常见的声明及其作用。

表 6-2 dhcpd 服务程序配置文件常见的声明及其作用

声明	作用
shared-network	用来告知是否允许一些子网分享相同的网络
subnet	描述一个 IP 地址是否属于该子网
range 192.168.1.10 192.168.1.100	提供动态分配 IP 的范围
host 主机名称	参考特别的主机
group	为一组参数提供声明
allow unknown-clients; deny unknown-client	是否动态分配 IP 给未知的使用者
allow bootp;deny bootp	是否响应激活查询
allow booting;deny booting	是否响应使用者查询

4. 常见选项

选项用来配置 DHCP 可选参数，全部用 option 关键字作为开始。表 6-3 列出了常见的选项及其作用。

表 6-3 dhcpd 服务程序配置文件常见的选项及其作用

选项	作用
subnet-mask 255.255.255.0	为客户端设定子网掩码
domain-name "domain.org"	为客户端指明 DNS 名字
domain-name-servers 192.168.1.1	为客户端指明 DNS 服务器 IP 地址
host-name	为客户端指定主机名称
routers 192.168.1.1	为客户端设定默认网关
broadcast-address 192.168.1.254	为客户端设定广播地址
ntp-server 192.168.1.1	为客户端设定网络时间服务器 IP 地址
nis-servers 192.168.1.1	为客户端设定 NIS 域服务器的地址
time—offset	为客户端设定与格林尼治时间的偏移时间，单位是秒

6.2.3 配置 DHCP 服务器

1. 情景一

公司人事部计算机的网络拓扑如图 6-3 所示，具体要求如下：

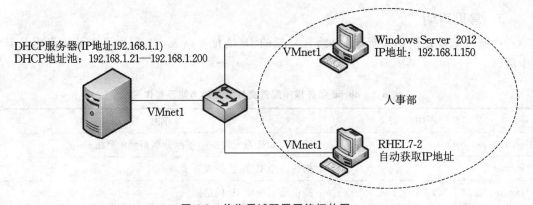

图 6-3 单作用域配置网络拓扑图

① DHCP 服务器、DNS 服务器的地址均为 192.168.1.1，该部门 IP 地址范围是 192.168.1.1—192.168.1.254，子网掩码是 255.255.255.0，默认网关是 192.168.1.254。

② 固定地址为 192.168.1.1—192.168.1.20。

③ 客户端获取的地址分别为：192.168.1.21—192.168.1.200，其中 192.168.1.150 为保留地址，分配给客户端 Windows Server 2012，其网卡的 MAC 地址为 00:0c:29:0f:9d:41。

④ RHEL7-2 通过自动获取方式，从 DHCP 服务器自动获取 IP 地址等信息。

实现步骤如下：

（1）复制 dhcpd.conf.example 文件到/etc/dhcp/目录下，并重命名为 dhcpd.conf。

[root@localhost ~]# cp -p /usr/share/doc/dhcp-4.2.5/dhcpd.conf.example /etc/dhcp/dhcpd.conf
  cp:是否覆盖"/etc/dhcp/dhcpd.conf"? yes

（2）修改主配置文件 dhcpd.conf，添加如下内容到配置文件中，并重启 dhcpd 服务。

```
[root@localhost ~]# vim /etc/dhcp/dhcpd.conf
subnet 192.168.1.0 netmask 255.255.255.0 {
  range 192.168.1.21 192.168.1.149;
  range 192.168.1.151 192.168.1.200;
  option domain-name-servers 192.168.1.1;
  option routers 192.168.1.254;
  default-lease-time 600;
  max-lease-time 7200;
}

host win2012 {
  hardware ethernet 00:0c:29:0f:9d:41;
  fixed-address 192.168.1.150;
}
[root@localhost ~]# systemctl restart dhcpd
```

（3）客户端测试

为保证在 VMware 上测试成功，避免虚拟机自带的 DHCP 服务与配置的 DHCP 服务器发生冲突，DHCP 服务器与客户端的网卡需使用"仅主机模式"（使用 VMnet1 网卡），并且需要关闭虚拟机自带的 DHCP 功能，依次在单击虚拟机菜单栏"编辑"→"虚拟网络编辑器"，在图 6-4 所示界面中，选择"VMnet1"取消勾选"使用本地 DHCP 服务将 IP 地址分配给虚拟机"选项，并应用。

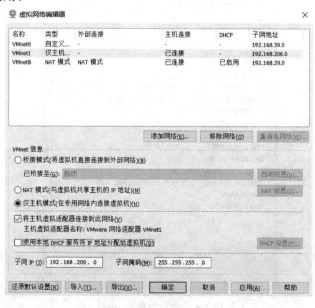

图 6-4 虚拟网络编辑器

① Windows Server 2012

将网卡的"TCP/IPv4"属性中的设置改为自动获取,如图6-5所示。

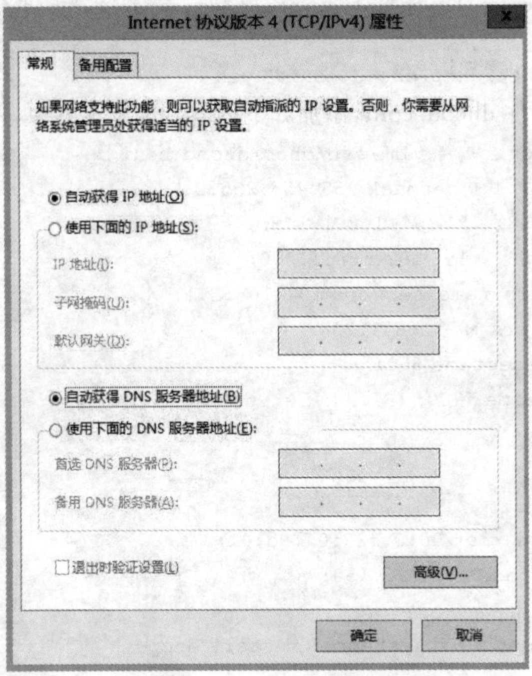

图 6-5 "TCP/IPv4"属性界面

在命令提示符下使用命令 ipconfig /release 释放 IP 地址,命令 ipconfig /renew 重新获取 IP 地址,查看是否获取到 DHCP 服务器分配的固定地址 192.168.1.150,如图 6-6 所示。

图 6-6 成功获取 IP 地址

② RHEL7-2

修改网卡配置文件，确保网卡的引导协议为 dhcp，重启网络服务，并查看网卡是否已获取到 DHCP 服务器分配的 IP 地址等信息。

```
[root@RHEL7-2~]# vim /etc/sysconfig/network-scripts/ifcfg-ens33
……
BOOTPROTO=dhcp
……
[root@RHEL7-2~]# systemctl restart network
[root@RHEL7-2~]# ifconfig ens33
ens33: flags=4163<UP,BROADCAST,RUNNING,MULTICAST> mtu 1500
inet 192.168.1.21  netmask 255.255.255.0  broadcast 192.168.1.255
        inet6 fe80::cb63:e50c:cf2c:f2d  prefixlen 64  scopeid 0x20<link>
        ether 00:0c:29:a8:29:ee  txqueuelen 1000  (Ethernet)
        RX packets 582  bytes 50541 (49.3 KiB)
        RX errors 0  dropped 0  overruns 0  frame 0
        TX packets 94  bytes 11587 (11.3 KiB)
        TX errors 0  dropped 0 overruns 0  carrier 0  collisions 0
```

这时，在服务器上，可以通过查询 DHCP 服务的租约文件，查看 DHCP 服务分配的 IP 地址信息。

```
[root@localhost~]# cat /var/lib/dhcpd/dhcpd.leases
# The format of this file is documented in thedhcpd.leases(5) manual page.
# This lease file was written by isc-dhcp-4.2.5
server-duid "\000\001\000\001&\033\013z\000\014)\304\347\321";
lease 192.168.1.21 {
  starts 6 2020/04/04 18:07:09;
  ends 6 2020/04/04 18:17:09;
cltt 6 2020/04/04 18:07:09;
  binding state active;
  next binding state free;
  rewind binding state free;
  hardware ethernet 00:0c:29:a8:29:ee;        //RHEL7-2 网卡的 MAC 地址
```

2. 情景二

公司技术部、财务部两个部门使用不同的网段，从而保证各自部门的数据安全。这时，DHCP 服务器需要使用两块网卡分别连接两个部门（网卡 1 使用 VMnet1 模式连接技术部；网卡 2 使用 VMnet8 模式"即 NAT 模式"连接财务部），并配置 DHCP 服务分别为两个部门分配 IP 地址，公司网络拓扑如图 6-7 所示。

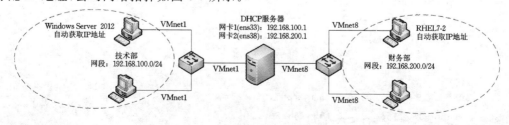

图 6-7　多作用域配置网络拓扑图

实现步骤如下：

（1）在虚拟机中为服务器添加一块网卡，其中网卡 1 使用 VMnet1 模式，网卡 2 使用 VMnet8 模式，同时，与情景一一样，需要在"虚拟网络编辑器"中取消勾选它们的"使用本地 DHCP 服务将 IP 地址分配给虚拟机"选项。并配置两块网卡的 IP 地址，如图 6-8、图 6-9 所示。

设备	摘要
内存	2 GB
处理器	1
硬盘 (SCSI)	20 GB
CD/DVD (SATA)	正在使用文件 D:\ISO\rhel 7.....
网络适配器	自定义 (VMnet1)
网络适配器 2	自定义 (VMnet8 (NAT))
USB 控制器	存在
声卡	自动检测
打印机	存在
显示器	自动检测

图 6-8　虚拟机系统添加两块网卡

```
[root@localhost ~]# ifconfig
ens33: flags=4163<UP,BROADCAST,RUNNING,MULTICAST>  mtu 1500
       inet 192.168.100.1  netmask 255.255.255.0  broadcast 192.168.100.255
       inet6 fe80::dd10:3a30:92df:363f  prefixlen 64  scopeid 0x20<link>
       ether 00:0c:29:c4:e7:d1  txqueuelen 1000  (Ethernet)
       RX packets 925  bytes 85845 (83.8 KiB)
       RX errors 0  dropped 0  overruns 0  frame 0
       TX packets 272  bytes 32612 (31.8 KiB)
       TX errors 0  dropped 0  overruns 0  carrier 0  collisions 0

ens38: flags=4163<UP,BROADCAST,RUNNING,MULTICAST>  mtu 1500
       inet 192.168.200.1  netmask 255.255.255.0  broadcast 192.168.200.255
       inet6 fe80::99dc:e104:75a0:cd93  prefixlen 64  scopeid 0x20<link>
       ether 00:0c:29:c4:e7:db  txqueuelen 1000  (Ethernet)
       RX packets 16  bytes 3043 (2.9 KiB)
       RX errors 0  dropped 0  overruns 0  frame 0
       TX packets 28  bytes 3943 (3.8 KiB)
       TX errors 0  dropped 0  overruns 0  carrier 0  collisions 0
```

图 6-9　查看网卡信息

（2）修改主配置文件 dhcpd.conf，添加如下内容到配置文件中，并重启 dhcpd 服务。

```
[root@localhost ~]# vim /etc/dhcp/dhcpd.conf
subnet 192.168.100.0 netmask 255.255.255.0 {
    range 192.168.100.21 192.168.100.200;
    option domain-name-servers 192.168.100.1;
    option routers 192.168.100.254;
    default-lease-time 600;
    max-lease-time 7200;
}

subnet 192.168.200.0 netmask 255.255.255.0 {
    range 192.168.200.21 192.168.200.200;
    option domain-name-servers 192.168.200.1;
```

```
        option routers 192.168.200.254;
        default-lease-time 600;
        max-lease-time 7200;
}
[root@localhost~]# dhcpd          //检查配置文件是否有错误
[root@localhost~]# systemctl restart dhcpd
```

(3) 客户端测试

① 在技术部的计算机 Windows Server 2012 上设置网卡模式为 VMnet1,网卡获取地址后,查看网卡信息,如图 6-10、图 6-11 所示。

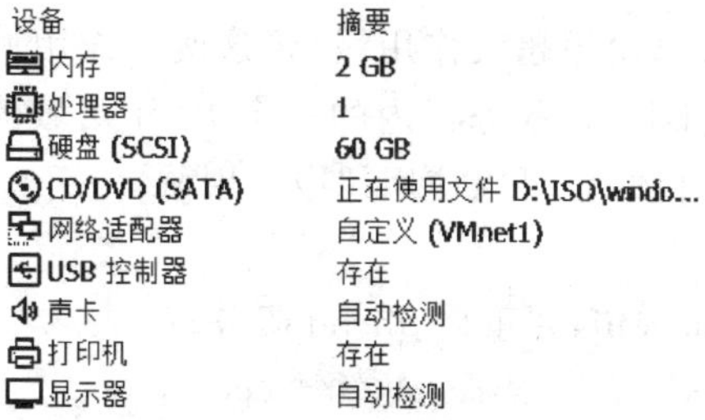

图 6-10　设置网卡模式为 VMnet1

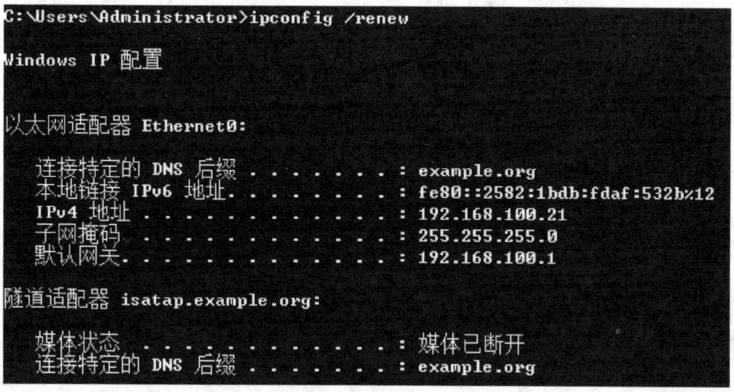

图 6-11　成功获取 IP 地址

② 在财务部的计算机 RHEL7-2 上设置网卡模式为 VMnet8,网卡获取地址后,查看网卡信息,如图 6-12 所示。

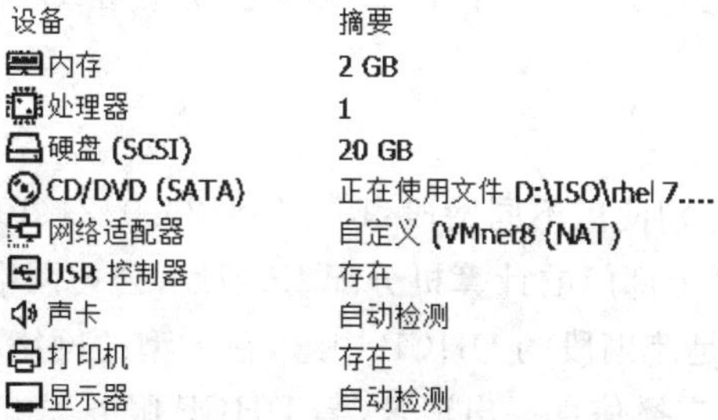

图 6-12　设置网卡模式为 VMnet8

```
[root@RHEL7-2~]# systemctl restart network
[root@RHEL7-2~]# ifconfig ens33
ens33: flags=4163<UP,BROADCAST,RUNNING,MULTICAST> mtu 1500
    inet 192.168.200.21  netmask 255.255.255.0  broadcast 192.168.200.255
        inet6 fe80::cb63:e50c:cf2c:f2d  prefixlen 64  scopeid 0x20<link>
        ether 00:0c:29:a8:29:ee  txqueuelen 1000  (Ethernet)
        RX packets 695  bytes 65900 (64.3 KiB)
        RX errors 0  dropped 0  overruns 0  frame 0
        TX packets 294  bytes 38824 (37.9 KiB)
        TX errors 0  dropped 0 overruns 0  carrier 0  collisions 0
```

在 DHCP 服务器中，可以使用超级作用域，将这两个作用域组合在一起。超级作用域由多个作用域组合而成，当 DHCP 服务器内的一个 IP 作用域的 IP 地址已经分配完，不够其他 DHCP 客户端使用的时候，就可以使用超级作用域，使用其他作用域的 IP 地址分配给客户端。

修改主配置文件 dhcpd.conf，并重启 dhcpd 服务。

```
[root@localhost~]# vim /etc/dhcp/dhcpd.conf
shared-network superscope{
  default-lease-time 600;
  max-lease-time 7200;
  subnet 192.168.100.0 netmask 255.255.255.0 {
  range 192.168.100.21 192.168.100.200;
  option domain-name-servers 192.168.100.1;
  option routers 192.168.100.254;
}

  subnet 192.168.200.0 netmask 255.255.255.0 {
  range 192.168.200.21 192.168.200.200;
  option domain-name-servers 192.168.200.1;
  option routers 192.168.200.254;
}
}
[root@localhost~]# dhcpd          //检查配置文件是否有错误
[root@localhost~]# systemctl restart dhcpd
```

当设置超级作用域后，会发现一个有趣的情况，就是两台客户端虽然网卡模式不一样，但是有可能获取的 IP 地址属于同一个网段。因为超级作用域相当于是给单作用域的网段进行扩充。

3. 情景三

公司营销部、经理室分别使用不同网段：192.168.20.0/24、192.168.30.0/24，现在需要使用 DHCP 服务器为这两个部门的计算机分配 IP 地址，公司网络拓扑如图 6-13 所示。

在图 6-13 中，其实就是跨网段的 DHCP 中继，在大型的网络中，可能会存在多个子网。DHCP 客户端是通过网络广播信息提出请求，当 DHCP 服务器响应后，才得到其分配的 IP 地址。但广播消息是不能跨越子网的。因此，如果 DHCP 客户端和 DHCP 服务器在不同

项目 6 DHCP 服务器

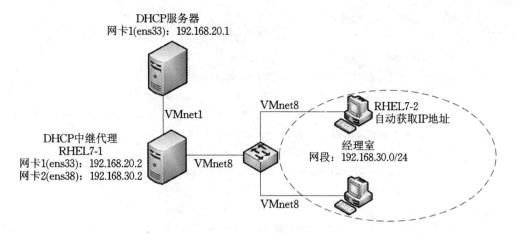

图 6-13 DHCP 中继代理网络拓扑图

的子网,就需要使用 DHCP 中继代理来为跨网段的 DHCP 客户端提供 DHCP 服务。如图 6-14 所示,DHCP 服务器与 DHCP 客户端位于不同的网络中,这时 DHCP 服务器就需要通过 DHCP 中继代理,从而为 DHCP 客户端提供服务。

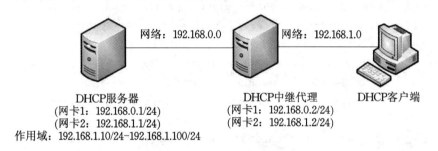

图 6-14 DHCP 中继代理

DHCP 中继代理实现步骤如下:

(1) 在 DHCP 服务器上,修改网卡 1 使用 VMnet1 模式,配置主配置文件 dhcpd.conf,添加如下内容到配置文件中,并重启 dhcpd 服务。

```
[root@localhost ~]# vim /etc/dhcp/dhcpd.conf
subnet 192.168.20.0 netmask 255.255.255.0 {
  range 192.168.20.21 192.168.20.200;
  option domain-name-servers 192.168.20.1;
  option routers 192.168.20.254;
  default-lease-time 600;
  max-lease-time 7200;
}

subnet 192.168.30.0 netmask 255.255.255.0 {
  range 192.168.30.21 192.168.30.200;
  option domain-name-servers 192.168.30.1;
  option routers 192.168.30.254;
  default-lease-time 600;
```

```
        max-lease-time 7200;
    }
    [root@localhost ~]# ifconfig ens33
    ens33: flags=4163<UP,BROADCAST,RUNNING,MULTICAST> mtu 1500
    inet 192.168.20.1  netmask 255.255.255.0  broadcast 192.168.20.255
            inet6 fe80::dd10:3a30:92df:363  fprefixlen 64  scopeid 0x20<link>
            ether 00:0c:29:c4:e7:d1  txqueuelen 1000  (Ethernet)
            RX packets 10882   bytes 985725 (962.6 KiB)
            RX errors 0  dropped 0   overruns 0   frame 0
            TX packets 1824   bytes 230845 (225.4 KiB)
            TX errors 0  dropped 0 overruns 0  carrier 0  collisions 0
```

（2）在 RHEL7-1 上，修改两块网卡的模式，其中网卡 1 使用 VMnet1 模式，网卡 2 使用 VMnet8 模式，配置 IPv4 转发功能。

```
    [root@RHEL7-1 ~]# ifconfig
    ens33: flags=4163<UP,BROADCAST,RUNNING,MULTICAST> mtu 1500
    inet 192.168.20.2  netmask 255.255.255.0  broadcast 192.168.20.255
            inet6 fe80::ed51:6df1:c6e:6be0  prefixlen 64  scopeid 0x20<link>
            ether 00:0c:29:ba:fa:43  txqueuelen 1000  (Ethernet)
            RX packets 293   bytes 28527 (27.8 KiB)
            RX errors 0  dropped 0   overruns 0   frame 0
            TX packets 64   bytes 8729 (8.5 KiB)
            TX errors 0  dropped 0 overruns 0  carrier 0  collisions 0

    ens38: flags=4163<UP,BROADCAST,RUNNING,MULTICAST> mtu 1500
    inet 192.168.30.2  netmask 255.255.255.0  broadcast 192.168.30.255
            inet6 fe80::fc53:57ab:2d5a:ef40  prefixlen 64  scopeid 0x20<link>
            ether 00:0c:29:ba:fa:4d  txqueuelen 1000  (Ethernet)
            RX packets 219   bytes 47459 (46.3 KiB)
            RX errors 0  dropped 0   overruns 0   frame 0
            TX packets 42   bytes 5457 (5.3 KiB)
            TX errors 0  dropped 0 overruns 0  carrier 0  collisions 0
    [root@RHEL7-1 ~]# vim /etc/sysctl.conf       //配置 IPv4 转发功能
    net.ipv4.ip_forward=1
    [root@RHEL7-1 ~]# sysctl -p    //载入 sysctl 配置文件，启用转发功能
    net.ipv4.ip_forward=1
```

（3）在 RHEL7-1 上，配置 DHCP 中继代理服务，并启动 DHCP 中继代理服务。

```
    [root@RHEL7-1 ~]# yum -y install dhcp
    [root@RHEL7-1 ~]# cp /lib/systemd/system/dhcrelay.service /etc/systemd/system/
    [root@RHEL7-1 ~]# vim /etc/systemd/system/dhcrelay.service
    ……
    [Service]
    Type=notify
```

```
         ExecStart=/usr/sbin/dhcrelay -d --no-pid 192.168.20.1   //指定 DHCP 服务器的 IP 地址
         ……
    [root@RHEL7-1~]# systemctl --system daemon-reload     //重载配置文件
    [root@RHEL7-1~]# systemctl restart dhcrelay       //重新启动 DHCP 中继代理服务
```

(4) 在客户端 RHEL7-2 上测试。

```
    [root@RHEL7-2~]# systemctl restart network
    [root@RHEL7-2~]# ifconfig ens33
    ens33: flags=4163<UP,BROADCAST,RUNNING,MULTICAST> mtu 1500
        inet 192.168.30.21   netmask 255.255.255.0   broadcast 192.168.30.255
            inet6 fe80::cb63:e50c:cf2c:f2d  prefixlen 64   scopeid 0x20<link>
            ether 00:0c:29:a8:29:ee   txqueuelen 1000  (Ethernet)
            RX packets 3010   bytes 316123 (308.7 KiB)
            RX errors 0   dropped 0   overruns 0   frame 0
            TX packets 1170   bytes 148883 (145.3 KiB)
            TX errors 0   dropped 0 overruns 0   carrier 0   collisions 0
```

自我测试

一、填空题

1. DHCP 采用"＿＿＿＿＿＿"的工作方式。
2. DHCP 服务器默认的配置文件是＿＿＿＿＿＿。
3. 当 DHCP 客户端要延长目前租约的 IP 地址的使用期限时,可以通过＿＿＿＿＿＿、
   ＿＿＿＿＿＿两种方式,实现 IP 地址的续约。

二、简答题

1. 简述 DHCP 协议的功能。
2. 简述 DHCP 客户端获取 DHCP 服务器的 IP 地址的全过程。
3. 简述 DHCP 的三种 IP 地址分配方式。

三、实训题

配置 DHCP 服务器,为 192.168.20.0/24 网段的用户提供动态 IP 地址分配服务。IP 地址范围在 192.168.20.50—192.168.20.100,默认网关是 192.168.20.1,该网段其余地址保留为静态地址分配,同时,为 MAC 地址 00:0C:08:06:DD:45 的主机设置静态 IP 地址 192.168.20.20。然后使用 Windows 系统获取 IP 地址,并使用 Linux 系统,网卡的 MAC 地址为 00:0C:08:06:DD:45,测试是否获取到 IP 地址 192.168.20.20。

项目 7　DNS 服务器

项目综述

公司最近正在建设公司的门户网站,为提前做好门户网站的宣传工作,已经申请了域名。现在公司网络管理员需要使用该域名,并搭建 DNS 服务器。

项目目标

- 了解 DNS 服务相关知识;
- 了解 DNS 服务工作原理;
- 掌握不同环境下 DNS 服务器的配置方法;
- 掌握 DNS 客户端的配置、测试方法。

任务 7.1　DNS 服务器简介

在网络中,计算机之间在通信时,为了相互识别,必须给每一台计算机分配一个唯一的 IP 地址,但是这些 IP 地址难以形象地描述相关的机构或企业,随着网络中的计算机越来越多,IP 地址也越来越多。为了使人们不用去记这些 IP 地址数字,还能方便地识别这些计算机,人们就采用域名来代替 IP 地址,这就叫作域名系统(DNS,Domain Name System)。域名解析,就是进行域名与 IP 地址之间的转换。DNS 服务器就是进行域名解析的服务器。

7.1.1　DNS 域名空间

域名空间是用于组织名称的域的层次结构。如图 7-1 所示,域名空间是一个树状结构,根域位于最顶部,紧接着是几个顶级域,每个顶级域下面又划分了不同的二级域,二级域下面再划分子域,子域下面可以是主机,或者再划分子域,直到最后的主机。

图 7-1 左边的主机,完整的名称是 www.jack.kk.com,这个完整的名称被称为 FQDN (Fully Qualified Domain Name,全限定域名),在命名时只能用字母"a~z"、字母"A~Z"、数字"0~9"和连字符"-",域名长度不能超过 256 个字节,并不区分大小写,"."号只能在域名标志符之间或者 FQDN 的结尾使用。

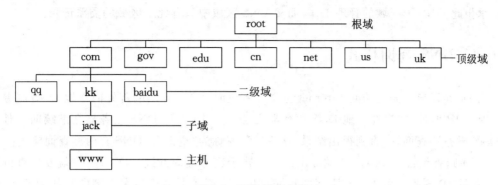

图 7-1 DNS 域名空间

7.1.2 DNS 区域

DNS 区域（ZONE）即 DNS 域名空间中树状结构的一部分，它能将域名空间按照需要划分为若干较小的管理区域，以减轻网络管理的工作负担。区域开始于一个顶级域，一直到一个子域或者其他域的开始。区域管辖的域名空间，包含了 DNS 树状结构上一个节点下的所有域名，不过不包括由其他区域管辖的域名。

一台 DNS 服务器可以存储一个或者多个区域的数据，同时，一个区域的数据也可以被存储到多个 DNS 服务器内。如图 7-2 所示，用户把域 kk.com 分为 kk.com 和 bj.kk.com 两个区域。其中，区域 kk.com 管辖 kk.com 域的子域 gz.kk.com 和 sc.kk.com。kk.com 的子域 bj.kk.com 及其下级子域 sh.bj.kk.com 和 nj.bj.kk.com，由区域 bj.kk.com 管辖。

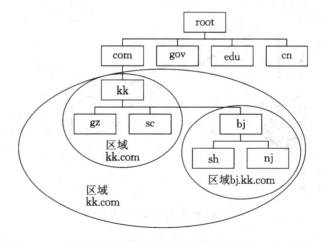

图 7-2 DNS 域名空间分层结构

7.1.3 区域委派

默认情况下，DNS 区域管理自己的子区域，并且子区域伴随 DNS 区域一起进行复制和更新。不过，用户可以将子区域委派给其他 DNS 服务器来进行管理。此时，被委派的服务器将承担此 DNS 子区域的管理工作,而父 DNS 区域中只有此子区域的委派记录。

7.1.4 DNS 查询模式

当 DNS 客户端需要访问 Internet 上某一主机时,DNS 客户端首先向本地 DNS 服务器查询对方 IP 地址,如果在本地 DNS 服务器无法查询出,本地 DNS 服务器会继续向另外一台 DNS 服务器查询,一直到得出结果,这一过程就称为"查询"。DNS 有两种查询模式:

① 递归查询:由 DNS 客户端提出的查询属于这种查询方式。DNS 客户端发出查询请求后,如果 DNS 服务器内没有需要的数据,则 DNS 服务器会代替客户端向其他的 DNS 服务器查询。

② 迭代查询:DNS 服务器与 DNS 服务器之间的查询属于这种查询方式。当第一台 DNS 服务器向第二台 DNS 服务器提出查询要求后,如果第二台 DNS 服务器内没有所需要的数据,那么它会提供第三台 DNS 服务器的 IP 地址给第 1 台 DNS 服务器。

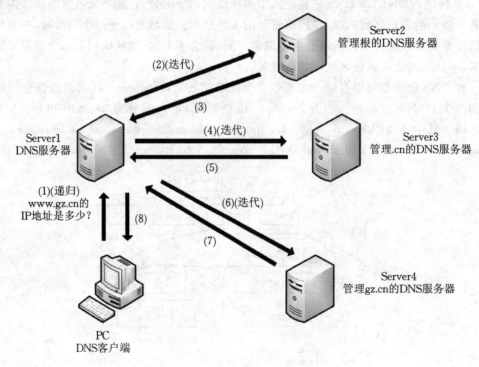

图 7-3 DNS 域名解析工作过程

如图 7-3 所示,DNS 客户端向 DNS 服务器 Server1 查询 www.gz.cn 的 IP 地址流程如下:

(1) DNS 客户端向 DNS 服务器 Server1 查询 www.gz.cn 的 IP 地址,这属于递归查询。

(2) 如果 Server1 内没有此主机记录,Server1 会将此查询要求转发到根内的 DNS 服务器 Server2,这属于迭代查询。

(3) Server2 依据主机名 www.gz.cn 得知此主机位于顶级域名 .cn 之下,所以会将管辖.cn 的 DNS 服务器 Server3 的 IP 地址告知 Server1。

（4）Server1 得到该地址后,会向 Server3 查询 www.gz.cn 的 IP 地址。

（5）Server3 依据主机名 www.gz.cn 得知此主机位于 gz.cn 域内,所以会将管辖 gz.cn 的 DNS 服务器 Server4 的 IP 地址发送给 Server1。

（6）Server1 得到 Server4 的 IP 地址后,会向 Server4 查询 www.gz.cn 的 IP 地址。

（7）Server4 将 www.gz.cn 的 IP 地址发送给 Server1。

（8）最后,Server1 将该 IP 地址发送给 DNS 客户端。

7.1.5　DNS 服务器的分类

DNS 服务器主要分为以下 4 类:

1. 主 DNS 服务器

主 DNS 服务器(Master 或 Primary)负责维护所管辖域的域名服务信息。一个域有且只有一个主域名服务器。它从域管理员构造的本地磁盘文件中加载域信息,该文件(区文件)包含该服务器具有管理权的一部分域结构的准确信息。配置主域服务器需要一整套的配置文件,在 RHEL7 系统中,包括主配置文件(/etc/named.conf)、正向域的区文件、反向域的区文件、高速缓存初始化文件(/var/named/named.ca)和回送文件(/var/named/named.local)。

2. 辅助 DNS 服务器

辅助 DNS 服务器(Slave 或 Secondary)用于分担主 DNS 服务器的查询负载任务。当主 DNS 服务器关闭、出现故障或负载过重时,辅助 DNS 服务器作为备份服务器提供域名解析服务。辅助 DNS 服务器从主 DNS 服务器获得授权,并定期向主 DNS 服务器询问是否有新数据,如果有则调入并更新域名解析数据,以达到与主 DNS 服务器同步的目的。它的区文件是从主 DNS 服务器中转移出来的,并作为本地磁盘文件被存储在辅助服务器中。配置辅助 DNS 服务器不需要生成本地区文件,该文件可以从主服务器上下载。因此,在配置时,只需要配置主配置文件、高速缓存初始化文件、回送文件。

3. 缓存 DNS 服务器

缓存 DNS 服务器(Caching-only DNS Server)可运行域名服务器软件但是没有域名数据库,供本地网络上的客户端进行域名转换。它提供查询其他 DNS 服务器服务并将获得的信息存放在它的高速缓存中,为客户端提供信息查询服务。缓存 DNS 服务器不是权威性的服务器,因为它提供的所有信息都是间接信息。

4. 转发 DNS 服务器

转发 DNS 服务器(Forwarding Name Server)可以向其他 DNS 服务器转发解析请求。转发 DNS 服务器接到查询请求时,在其缓存中查找,如没有找到,则需要向其他指定的 DNS 服务器转发解析请求;其他 DNS 服务器完成解析后,将结果返回,转发 DNS 服务器就将该解析结果缓存在自己的 DNS 缓存中,并向客户端发送解析结果。在缓冲期内,如果客户端再次请求解析相同的域名,则转发 DNS 服务器会立即回应客户端,反之,将会再次进行转发解析。

有两种解析方式：一种是正向解析，将域名转换为 IP 地址；另一个是反向解析，将 IP 地址转换为域名。

7.1.6 DNS 的区域类型

DNS 的区域类型主要有三种：主要区域、辅助区域和存根区域。

① 主要区域(Primary Zone)：该区域保存的是该区域所有主机数据记录的主副本。当在 DNS 服务器内建立主要区域后，可直接在此区域内新建、修改、删除记录。区域内的记录可以存储在文件或者活动目录数据库中。

② 辅助区域(Secondary Zone)：是主要区域的备份，从主要区域直接复制而来。辅助区域内的记录是只读的，不能进行修改。当一个区域内创建了一个辅助区域后，这个 DNS 服务器就是这个区域的辅助 DNS 服务器。

③ 存根区域(Stub Zone)：同样存储着区域的副本记录，不过该区域只包含少数的记录(SOA、NS、A 记录等)，目的是利用这些记录找到此区域的授权服务器。

7.1.7 资源记录

资源记录(Resource Records，简称 RRs)是指每个域所包含的与之相关的资源。例如，一些资源记录将域名与 IP 地址相互绑定，也有一些记录不仅包含 DNS 域中服务器的信息，而且还用于定义域，如 SOA、NS 资源记录。常见的资源记录及功能如表 7-1 所示。

表 7-1 常见的资源记录及功能

名 称	功 能
NS 资源记录	也称为名称服务器记录，用于标识区域的 DNS 服务器，也就是说负责此 DNS 区域的权威名称服务器，用哪一台 DNS 服务器来解析该区域。NS 记录说明了在这个区域里，有多少台服务器来承担解析的任务。因此，一个区域有可能有多条 NS 记录，例如 gg.com 可能有一个主服务器和多个辅助服务器
SOA 资源记录	也称为起始授权机构记录，用于一个区域的开始，SOA 记录后的所有信息均是用于控制这个区域的，每个区域数据库文件都必须包括一个 SOA 记录，并且必须是其中的第一个资源记录，用以标识 DNS 服务器管理的起始位置，SOA 说明在众多 NS 记录里，谁才是能解析这个区域的 DNS 主服务器
A 资源记录	也称为主机记录，将 FQDN 映射到 IP 地址，即正向解析
PTR 记录	也称为指针记录，将 IP 地址映射到 FQDN，即反向解析
CNAME 资源记录	也称为别名记录，用于定义 A 资源记录的别名
MX 记录	也称为邮件交换器记录，用于为 DNS 域名指定邮件交换服务器。邮件交换服务器是用于 DNS 域名处理或转发邮件的主机。数字越小，其优先级越高

任务7.2 部署 DNS 服务器

7.2.1 安装 DNS 服务器

1. 查询系统是否已经安装 bind 软件包

```
[root@localhost ~]# rpm-qa |grepbind
```

2. 安装 bind 软件包

```
[root@localhost ~]# yum-y installbind
[root@localhost ~]# rpm-qa | grep bind
rpcbind-0.2.0-42.el7.x86_64
bind-libs-lite-9.9.4-50.el7.x86_64
bind-license-9.9.4-50.el7.noarch
bind-9.9.4-50.el7.x86_64
bind-utils-9.9.4-50.el7.x86_64
keybinder3-0.3.0-1.el7.x86_64
bind-libs-9.9.4-50.el7.x86_64
```

7.2.2 认识 BIND 配置文件

安装 DNS 服务器后,为了使用 DNS 服务,需要进行 DNS 服务器的配置。在 Linux 系统中,配置 DNS 服务,大致分为 4 步:

① 配置主配置文件:设置/etc/named.conf 文件,设定监听地址、权威 DNS 转发器、转发方式、服务对象等参数。

② 配置区域配置文件:设置/etc/named.rfc1912.zones 文件,根据 named.conf 文件中指定的路径,建立区域文件。这些文件主要用来保存域名和 IP 地址对应关系所在的位置。

③ 配置正向、反向解析文件:设置/var/named 内的数据配置文件,这些文件是在区域配置文件中指定,主要用于域名和 IP 地址的正向、反向解析。

④ 重启 named 服务,使配置失效,并进行测试。

1. 主配置文件的配置

主配置文件 named.conf,用于实现 DNS 服务器的基本配置,密钥文件 named.root.key 和区域配置文件 named.rfc1912.zones 与主配置文件通常都保存在/etc 目录下。named.conf 的主要内容如下:

```
options {
listen-on port 53 {127.0.0.1;};        //named 监听端口号为 53
listen-on-v6 port 53 {::1;};           //支持 IPv6
directory    "/var/named";             //区域文件默认的存储目录
dump-file    "/var/named/data/cache_dump.db";//域名缓存文件的存储位置和文件名
statistics-file "/var/named/data/named_stats.txt";//状态统计文件的存储位置和
                                                    文件名
```

```
        memstatistics-file "/var/named/data/named_mem_stats.txt";
                                      //服务器输出的内存使用统计文件的存储位置和文件名
        allow-query     {localhost;};    //指定接受DNS查询请求的客户端
        recursion yes;    //是否允许递归查询
        dnssec-enable yes;    //是否支持dnssec开关
        dnssec-validation yes;    //是否进行dnssec确认开关。若设置为no,可以忽略SELinux
                                   的影响。注意:当dnssec-enable设置为no时,该项将不用设
                                   置,可直接设置为no。
        bindkeys-file "/etc/named.iscdlv.key";
        managed-keys-directory "/var/named/dynamic"
        pid-file "/run/named/named.pid";
        session-keyfile "/run/named/session.key";
        };
        logging {     //定义bind服务的日志
              channel default_debug {
                                //日志输出方式,只有当服务器的debug级别非0时,才产生输出
                    file "data/named.run";    //输出到的存储位置和文件名
                    severity dynamic;          //消息的严重性等级
              };
        };
        zone "." IN {    //定义根区域,一般不进行改动。该区域是互联网中所有域名的开始,只有能
                         够访问该区域后,DNS服务器才能提供正常的域名解析服务
        type hint;
        file "named.ca";
        };

        include "/etc/named.rfc1912.zones";    //指定区域配置文件
        include "/etc/named.root.key";    //指定根区域的密钥文件
```

表7-2中列出了named.conf支持的主要配置语句及其功能。

表7-2　named.conf支持的主要配置语句及其功能

名　　称	功　　能
acl	定义一个主机匹配顺序,用于访问控制等
controls	定义rndc工具与bind服务进程的通信
include	把其他文件包含到配置文件中
key	定义授权的安全密钥
logging	定义系统日志信息
opitons	定义全局配置选项和缺省值
masters	定义主域名列表
server	定义服务器的属性
trunsted-keys	定义信任的dnssec加密密钥
zone	定义区域
view	定义视图

(1) options 语句

options 语句用来定义影响整个 DNS 服务器的全局配置选项，该语句在 named.conf 文件中只能出现一次。options 语句支持的选项比较多，表 7-3 中列出了该语句常见的配置选项及功能。

表 7-3 options 语句常见的配置选项及功能

名　称	功　能
directory	设置域名服务的工作目录，默认为/var/named。如果没有设置 directory，则系统默认使用"."为工作目录
dump-file	运行 rndc dumpdb 备份缓存资料后，域名缓存文件的存储位置和文件名
statistics-file	运行 rndc stats 后，状态统计文件的存储位置和文件名
listen-on port	指定监听的 IP 地址和端口。若没有指定，则默认监听 DNS 服务器的所有 IP 地址的 53 端口
allow-query	指定接受 DNS 查询请求的客户端。localhost 表示匹配本地主机使用的所有 IP 地址。也可以使用具体 IP 地址来进行匹配，如接受 192.168.1.0/24 网段的主机的 DNS 查询请求：allow-query{192.168.1.0/24;}
blackhole	指定拒绝哪些主机的查询请求，如指定 192.168.2.0/24 网段的主机进行 DNS 查询请求：blackhole {192.168.1.0/24;}
recursion	是否允许进行递归查询
forwarders	用于定义 DNS 转发器。forward 选项仅在 forwarders 转发器列表不为空时，生效。其用法为"forwarder first | only"，默认值为 first，DNS 服务器会将用户的域名查询请求先转发给 forwarders 设置的转发器去进行域名的解析，当其他 DNS 服务器无法解析时，该服务器不再进行解析，由 DNS 服务器本身完成域名解析。若设置为"only"，则 DNS 服务器仅将用户的域名查询请求转发给转发器，若转发器无法解析，DNS 服务器本身也不尝试进行域名解析 如设置 DNS 服务器 192.168.1.1、DNS 服务器 192.168.2.1 为转发器： forwarders {192.168.1.1;192.168.2.1;}; forwarder first;
max-cache-size	设置缓存文件的最大容量

(2) zone 语句

zone 语句是 named.conf 文件的核心部分。其用来定义区域及相关选项，该语句主要的配置选项有 type、file 和 allow-update。表 7-4 中列出了该语句常见的配置选项及功能。

表 7-4 zone 语句常见的配置选项及功能

名　称	功　能
type	定义区域类型，区域类型可以是： ● hint：根 DNS 服务器集 ● maser：主 DNS 服务器，拥有区域数据文件，并对此区域提供管理数据服务 ● slave：辅助 DNS 服务器，拥有主 DNS 服务器的区域数据文件的副本，辅助 DNS 服务器会从主 DNS 服务器同步所有区域数据 ● stub：与辅助 DNS 服务器类似，但是只保存 DNS 服务器的 NS 记录 ● forward：定义转发域名服务器

续表 7-4

名 称	功 能
file	定义区域数据文件
allow-update	允许哪些主机动态更新区域数据信息

2. 正向解析文件

正向解析文件的默认内容如下：

```
$TTL 1D
@ IN SOA @ rname.invalid. (
                0;serial
                1D;refresh
                1H;retry
                1W;expire
                3H );minimum
  NS  @
  A   127.0.0.1
```

该文件的各项参数说明如表 7-5 所示。

表 7-5　正向解析文件的参数及功能

名 称	功 能
@	该域的替代符，如正向解析文件 jack.com.zone，@ 就代表 jack.com
IN	表示网络类型
SOA	表示资源记录类型
origin	表示该域的主域名服务器的 FQDN，以"."结尾
contact	表示该 DNS 服务器的管理员邮件地址，因为 @ 在 SOA 记录表示域名，所以用"."代替，如 jack.sk.com 表示 jack@sk.com
serial	表示区域文件的版本号，该数据用于主 DNS 服务器与辅助 DNS 服务器之间进行时间同步，每次更改区域文件都应该更新该数字
refresh	表示更新时间间隔。辅助 DNS 服务器根据此时间间隔周期性地检查主 DNS 服务器的 serial 是否改变。若改变，则更新自己的数据库文件
retry	表示重试时间。当辅助 DNS 服务器在主 DNS 服务器不能使用时，重试向主 DNS 服务器发出请求应等待的时间
expire	表示过期时间。在辅助 DNS 服务器无法与主 DNS 服务器通信的情况下，其区域信息保存的时间
TTL	表示最小时间间隔，单位为秒
NS	表示区域内的权威 DNS 服务器。如"@ IN dns.jack.com."代表该域的域名服务器
A	表示域名到 IP 地址的映射。如"www IN A 192.168.1.1"
PTR	表示 IP 地址到域名的映射。如"1 IN A www.jack.com."
MX	表示邮箱，MX 后面的数字表示优先级。数字后面是邮箱的服务器地址
CNAME	表示别名

3. 反向解析文件

反向解析文件的默认内容如下：

```
$ TTL 1D
@ IN SOA @ rname.invalid. (
                0;serial
                1D;refresh
                1H;retry
                1W;expire
                3H );minimum
NS @
PTR localhost.
```

该文件的参数与正向解析文件的参数相同，具体参考正向解析文件的参数。

7.2.3 配置 DNS 服务器

1. 主 DNS 服务器

公司现已申请到域名 kcorp.com，公司 DNS 服务器的 FQDN 为 dns.kcorp.com，IP 地址为 192.168.1.1，现要求配置 DNS 服务，实现以下域名的正反向解析，如表 7-6 所示。

表 7-6 域名与 IP 地址规划

域 名	IP 地址	备 注
dns.kcorp.com	192.168.1.1	DNS 服务器
www.kcorp.com	192.168.1.1	WEB 服务器
ftp.kcorp.com	192.168.1.2	FTP 服务器
mail.kcorp.com	192.168.1.3	邮件服务器
slave.corp.com	192.168.1.3	辅助 DNS 服务器

实现步骤如下：

（1）修改主配置文件 /etc/named.conf，将 named 监听端口号、指定接受 DNS 查询请求的客户端均改为 any。

```
options {
listen-on port 53 {any;};
listen-on-v6 port 53 {any;};
directory      "/var/named";
dump-file      "/var/named/data/cache_dump.db";
statistics-file "/var/named/data/named_stats.txt";
memstatistics-file "/var/named/data/named_mem_stats.txt";
allow-query     {any;};
……
```

(2) 配置区域配置文件/etc/named.rfc1912.zones，并指定正向解析文件、反向解析文件。

```
……
zone "kcorp.com" IN {
    type master;
    file "named.localhost";
    allow-update {none;};
};
zone "1.168.192.in-addr.arpa" IN {
    type master;
    file "named.loopback";
    allow-update {none;};
};
……
```

(3) 配置正向解析文件/var/named/named.localhost。

```
$ TTL 1D
@ IN SOA@root.kcorp.com. (
        0;serial
        1D;refresh
        1H;retry
        1W;expire
        3H );minimum
@  IN   NS   dns.kcorp.com.
@  IN   MX   10    mail.kcorp.com.
dns IN   A    192.168.1.1
www IN   A    192.168.1.1
ftp IN   A    192.168.1.2
mail  IN   A    192.168.1.3
slave  IN   A    192.168.1.3
```

(4) 配置反向解析文件/var/named/named.loopback。

```
$ TTL 1D
@ IN SOA@root.kcorp.com. (
        0;serial
        1D;refresh
        1H;retry
        1W;expire
        3H );minimum
@ IN   NS   dns.kcorp.com.
@ IN   MX   10    mail.kcorp.com.
1 IN   PTR  dns.kcorp.com.
1 IN   PTR  www.kcorp.com.
2 IN   PTR  ftp.kcorp.com.
3 IN   PTR  mail.kcorp.com.
3 IN   PTR  slave.kcorp.com.
```

注意：在配置正向、反向解析文件时，也可以不使用系统默认的正向、反向解析文件来配置，可以通过复制系统默认的正向、反向解析文件，重命名这两个文件进行配置，如：正向解析文件为 kcorp.com.zone，反向解析文件为 192.168.1.arpa。这时，配置区域配置文件/

etc/named.rfc1912.zones,并指定这两个文件。

```
zone "kcorp.com" IN {
    type master;
    file "kcorp.com.zone";
    allow-update {none;};
};

zone "1.168.192.in-addr.arpa" IN {
    type master;
    file "192.168.1.arpa ";
    allow-update {none;};
};
```

(5) 配置网卡文件,重启 named 服务,配置防火墙允许 dns 服务通过。

```
[root@localhost~]# vim /etc/sysconfig/network-scripts/ifcfg-ens33
……
DEVICE=ens33
ONBOOT=yes
IPADDR=192.168.1.1
NETMASK=255.255.255.0
DNS=192.168.1.1
……
[root@localhost~]# systemctl restart network
[root@localhost~]# systemctl restart named
[root@localhost~]# firewall-cmd --permanent --add-service=dns
success
[root@localhost~]# firewall-cmd --reload
success
```

(6) 配置 DNS 客户端,并进行测试。
① Windows Server 2012
在网卡的"TCP/IPv4"属性中设置首选 DNS 服务器、备用 DNS 服务器的 IP 地址,如图 7-4 所示。

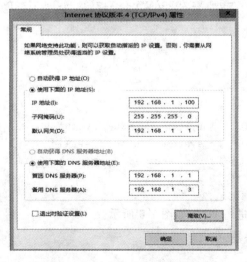

图 7-4 "TCP/IPv4"属性界面(一)

在 CMD 下使用 nslookup 命令进行测试,如图 7-5 所示。

图 7-5　nslookup 命令测试

② RHEL

配置网卡文件,修改/etc/resolv.conf 文件,指定 DNS 服务器。

```
[root@localhost~]# vi /etc/sysconfig/network-scripts/ifcfg-ens33
……
DEVICE=ens33
ONBOOT=yes
IPADDR=192.168.1.50
NETMASK=255.255.255.0
……
[root@localhost~]# systemctl restart network
[root@localhost~]# systemctl stop firewalld
[root@localhost~]# vim /etc/resolv.conf
nameserver 192.168.1.1
nameserver192.168.1.3
```

使用 nslookup 命令进行测试,如图 7-6 所示。

2. 辅助 DNS 服务器

实现步骤如下:

```
[root@localhost ~]# nslookup www.kcorp.com
Server:         192.168.1.1
Address:        192.168.1.1#53

Name:   www.kcorp.com
Address: 192.168.1.1

[root@localhost ~]# nslookup ftp.kcorp.com
Server:         192.168.1.1
Address:        192.168.1.1#53

Name:   ftp.kcorp.com
Address: 192.168.1.2

[root@localhost ~]# nslookup mail.kcorp.com
Server:         192.168.1.1
Address:        192.168.1.1#53

Name:   mail.kcorp.com
Address: 192.168.1.3

[root@localhost ~]# nslookup slave.kcorp.com
Server:         192.168.1.1
Address:        192.168.1.1#53

Name:   slave.kcorp.com
Address: 192.168.1.3
```

图 7-6 nslookup 命令测试(二)

(1) 修改主 DNS 服务器的区域配置文件/etc/named.rfc1912.zones 的 allow_update 选项的值为辅助 DNS 服务器的 IP 地址,从而允许辅助 DNS 服务器的更新请求,重启 named 服务。

```
[root@localhost~]# vim /etc/named.rfc1912.zones
……
zone "kcorp.com" IN {
    type master;
    file "named.localhost";
    allow-update {192.168.1.3;};
};
zone "1.168.192.in-addr.arpa" IN {
    type master;
    file "named.loopback";
    allow-update {192.168.1.3;};
};
……
[root@localhost~]# systemctl restart named
```

(2) 在辅助 DNS 服务器 RHEL-2 上,安装 bind 软件包。修改主配置文件/etc/named.conf,将 named 监听端口号、指定接受 DNS 查询请求的客户端均改为 any。

```
[root@RHEL-2~]# yum -y install bind
[root@RHEL-2~]# vim /etc/named.conf
options {
listen-on port 53 {any;};
listen-on-v6 port 53 {any;};
directory      "/var/named";
dump-file      "/var/named/data/cache_dump.db";/
statistics-file "/var/named/data/named_stats.txt";
memstatistics-file "/var/named/data/named_mem_stats.txt";
allow-query    {any;};
……
```

(3) 修改辅助 DNS 服务器的区域配置文件/etc/named.rfc1912.zones,设置区域类型为 slave,并指定主 DNS 服务器、正向解析文件、反向解析文件。

```
……
zone "kcorp.com" IN {
    type slave;
    masters {192.168.1.1;};
    file "slave /kcorp.com.zone";
};

zone "1.168.192.in-addr.arpa" IN {
    type slave;
    masters {192.168.1.1;};
    file "slave/192.168.1.arpa";
};
……
```

(4) 配置网卡文件,重启 named 服务,关闭防火墙。

```
[root@RHEL-2~]# vim /etc/sysconfig/network-scripts/ifcfg-ens33
DEVICE=ens33
ONBOOT=yes
IPADDR=192.168.1.3
NETMASK=255.255.255.0
[root@RHEL-2~]# systemctl restart network
……
[root@RHEL-2~]# systemctl restart network
[root@RHEL-2~]# systemctl restart named
[root@RHEL-2~]# setenforce 0
```

(5) 查看同步数据。当辅助 DNS 服务器重启后,会自动从主 DNS 上同步数据配置文件,这些文件存放在区域配置文件所指定的目录中。

```
[root@RHEL7-2~]# cd /var/named/slave/
[root@RHEL7-2 slave]# ls
192.168.1.arpa  kcorp.com.zone
```

(6) 配置 DNS 客户端,并进行测试。

在 Windows Server 2012 系统网卡的"TCP/IPv4"属性中设置首选 DNS 服务器的 IP 地址为辅助 DNS 服务器的 IP 地址,如图 7-7 所示。

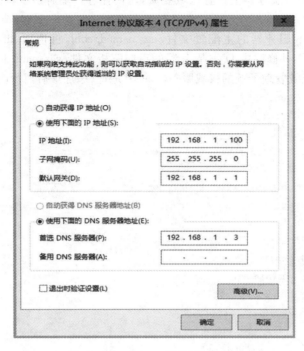

图 7-7 "TCP/IPv4"属性界面(二)

在 CMD 下使用 nslookup 命令进行测试,如图 7-8 所示。

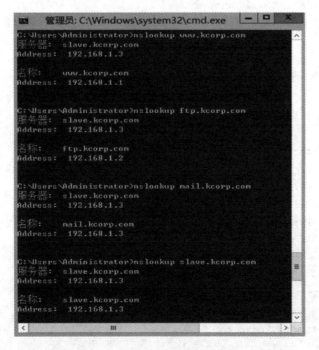

图 7-8 nslookup 命令测试(三)

3. 子域及区域委派

现在公司有一个子域 sub.kcorp.com，需要在 RHEL7-3（IP 地址 192.168.1.30）上建立该子域，并在主 DNS 服务器上设置区域委派，以便让客户端解析子域。

实现步骤如下：

（1）修改主 DNS 服务器的主配置文件/etc/named.conf，将选项 dnssec-enable、dnssec-validation 设置为"no"，并修改主 DNS 服务器的正向解析文件/var/named/named.localhost，添加子域的委派记录以及管理子域的权威服务器的 IP 地址（添加最后两行内容）。

```
[root@localhost ~]# vim /etc/named.conf
dnssec-enable no;
dnssec-validation no;
[root@localhost ~]# vim /var/named/named.localhost
$TTL 1D
@   IN SOA kcorp.com.   root.kcorp.com. (
            0;serial
            1D;refresh
            1H;retry
            1W;expire
            3H );minimum
@   IN    NS    dns.kcorp.com.
@   IN    MX    10    mail.kcorp.com.
dns  IN    A     192.168.1.1
www  IN    A     192.168.1.1
ftp  IN    A     192.168.1.2
mail IN    A     192.168.1.3
slave IN   A     192.168.1.3
sub.kcorp.com. IN NS dns1.sub.kcorp.com.    //指定委派区域的管理工作由域名服务器
                                              dns1.sub.kcorp.com 负责
dns1.sub.kcorp.com. IN A 192.168.1.30    //指定子域 sub.kcorp.com 的权威服务器
```

（2）修改主 DNS 服务器的反向解析文件/var/named/named.loopback（添加最后一行内容）。

```
[root@localhost ~]# vim /var/named/named.loopback
$TTL 1D
@   IN SOA kcorp.com.   root.kcorp.com. (
            0;serial
            1D;refresh
            1H;retry
            1W;expire
            3H );minimum
@   IN    NS    dns.kcorp.com.
@   IN    MX    10    mail.kcorp.com.
1   IN    PTR   dns.kcorp.com.
1   IN    PTR   www.kcorp.com.
2   IN    PTR   ftp.kcorp.com.
3   IN    PTR   mail.kcorp.com.
```

```
            3 IN    PTR    slave.kcorp.com.
            30 IN   PTR    dns1.sub.kcorp.com.
```

（3）修改主 DNS 服务器的主配置文件和区域文件是属组为 named，重启 named 服务。

```
[root@localhost~]# chgrp named/etc/named.conf
[root@localhost~]# systemctl restart named
```

（4）在子域 DNS 服务器上的区域配置文件/etc/named.rfc1912.zones，添加 sub.kcorp.com 区域记录。

```
zone "sub.kcorp.com" IN {
    type master;
    file "sub.kcorp.com.zone";
    allow-update {none;};
};

zone "1.168.192.in-addr.arpa" IN {
    type master;
    file "30.1.168.192.zone";
    allow-update {none;};
};
```

（5）配置子域 DNS 服务器的正向解析文件/var/named/sub.kcorp.com.zone。

```
[root@RHEL7-3~]# vim /var/named/sub.kcorp.com.zone
$TTL 1D
@ IN SOA sub.kcorp.com.   root.sub.kcorp.com. (
                0;serial
                1D;refresh
                1H;retry
                1W;expire
                3H );minimum
@  IN   NS   dns1.sub.kcorp.com.
dns1  IN  A   192.168.1.30
www2  IN  A   192.168.1.30
```

（6）配置子域 DNS 服务器的反向解析文件/var/named/30.1.168.192.zone。

```
[root@RHEL7-3~]# vim /var/named/30.1.168.192.zone
$ TTL 1D
@ IN SOA sub.kcorp.com.   root.sub.kcorp.com. (
                0;serial
                1D;refresh
                1H;retry
                1W;expire
                3H );minimum
@  IN   NS   dns1.sub.kcorp.com.
30 IN   PTR    dns1.sub.kcorp.com.
30 IN   PTR    www2.sub.kcorp.com.
```

（7）修改委派 DNS 服务器的主配置文件和区域文件是属组为 named，重启 DNS 服务，配置防火墙允许 DNS 服务通过。（为了避免启动 DNS 服务错误，这里需要将之前辅助 DNS 服务器的区域配置文件中的反向解析文件注释。）

```
[root@RHEL7-3 ~]# chgrp named /etc/named.conf
[root@RHEL7-3 ~]# systemctl restart named
[root@RHEL7-3 ~]# firewall-cmd --permanent --add-service=dns
[root@RHEL7-3 ~]# firewall-cmd --reload
```

(8) 配置 DNS 客户端，并进行测试。

在 Windows Server 2012 系统网卡的"TCP/IPv4"属性中设置首选 DNS 服务器的 IP 地址为主 DNS 服务器的 IP 地址，如图 7-9 所示。

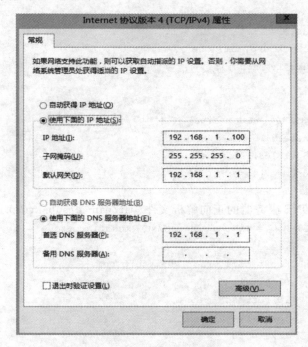

图 7-9 "TCP/IPv4"属性界面（三）

在 CMD 下使用 nslookup 命令进行测试，如图 7-10 所示，此时，由于主 DNS 服务器上没有 www2.sub.kcorp.com 的主机记录，客户端是通过主 DNS 服务器的区域委派解析该主机记录。

图 7-10 nslookup 命令测试（四）

4. 转发 DNS 服务器

转发 DNS 服务器可以向其他 DNS 服务器转发解析请求。按照转发类型,可以分为完全转发 DNS 服务器、条件转发 DNS 服务器。下面在 RHEL7-4(IP 地址 192.168.1.40)上进行完全转发 DNS 服务器的配置。

实现步骤如下:

(1) 修改 RHEL7-4 服务器的主配置文件/etc/named.conf,在 options 字段中,确保有以下内容。

```
[root@RHEL7-4~]# vim /etc/named.conf
options {
……
    directory     "/var/named";
    recursion yes;
    dnssec-enable yes;
    dnssec-validation no;
    forwarders {192.168.1.1;};   //设置转发查询请求 DNS 服务器的地址
    forward   only;              //仅执行转发操作
……
};
```

(2) 配置 DNS 客户端,并进行测试。

在 Windows Server 2012 系统网卡的"TCP/IPv4"属性中设置首选 DNS 服务器的 IP 地址为转发 DNS 服务器的 IP 地址,如图 7-11 所示。

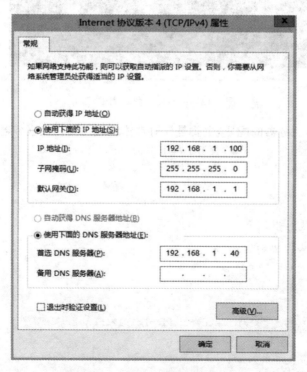

图 7-11 "TCP/IPv4"属性界面(四)

在 CMD 下使用 nslookup 命令进行测试,如图 7-12 所示。

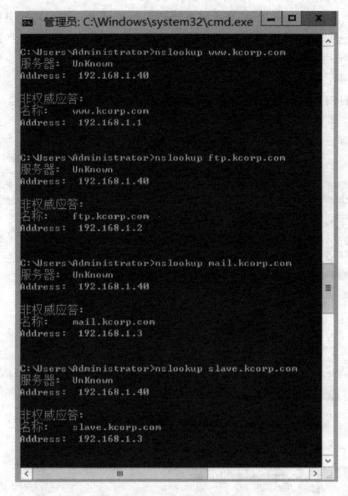

图 7-12 nslookup 命令测试(五)

条件转发 DNS 服务器的配置,主要是针对具体的域设置转发服务器。

```
[root@RHEL7-4~]# vim /etc/named.conf
options {
……
    directory     "/var/named";
recursion yes;
dnssec-enable yes;
dnssec-validation no;
……
};
[root@RHEL7-4~]# vim /etc/named.rfc1912.zones
……
zone "kcorp.com" IN {
    type forward;   //指定该区域为条件转发类型
```

```
        forwarders {192.168.1.1;};    //设置转发查询请求 DNS 服务器的地址
    };
    ......
```

配置完成后,重启 named 服务后,客户端就可以通过该条件转发 DNS 服务器进行域名解析。

自我测试

一、填空题

1. 域名解析,就是进行_____与_____地址之间的转换。
2. DNS 区域是 DNS _____中树状结构的一部分,它能将域名空间按照需要划分为若干较小的管理区域,以减轻网络管理的工作负担。
3. 主 DNS 服务器负责维护所管辖域的域名服务信息。一个域有且只有_____主域名服务器。
4. 辅助 DNS 服务器用于分担_____DNS 服务器的查询负载。
5. 转发 DNS 服务器可以向其他_____转发解析请求。
6. DNS 的区域类型主要有三种:_____、_____和_____。

二、简答题

1. 简述 DNS 的功能。
2. 简述 DNS 的两种查询模式。
3. 简述 DNS 服务器的分类。
4. 简述 DNS 的区域的三种类型。

三、实训题

配置 DNS 服务器,IP 地址为 192.168.200.20/24,域名为 dns.kq.com。要求正向、反向解析:

    dns.kq.com   192.168.200.20
    www.kq.com   192.168.200.30
    mail.kq.com   192.168.200.45

如果 DNS 不能解析时,则请求 220.11.168.55 服务器解析。并在 Windwos 客户端上使用 nslookup 命令进行测试。

项目 8　Apache 服务器

项目综述

公司已建设门户网站，搭建了 DNS 服务器。现在，公司网络管理员需要将公司网站发布到互联网上，让人们访问。

项目目标

- 了解 Web 服务器相关知识；
- 了解 Apache 服务器相关知识；
- 掌握 Apache 服务器的配置方法；
- 掌握基于虚拟主机的 Apache 服务器的配置方法；
- 掌握 https 安全认证网站的配置方法；
- 掌握用户认证和授权网站的配置方法。

任务 8.1　Web 服务器简介

8.1.1　Web 服务概述

Web 服务是网络中应用最广泛的服务之一，在网络日益普及的今天，政府、企事业单位都有自己的网站，这些网站要发布到网络上供大家浏览，那么就需要将网站放到 Web 服务器上，再进行相应的配置。

1. Web 服务器

Web 服务器也称为 WWW（World Wide Web）服务器，主要功能是提供网上信息的浏览服务。WWW 是 Internet 的多媒体信息查询工具，是 Internet 上发展最快和目前用得最广泛的服务。正是因为有了 WWW 工具，近年来 Internet 才会迅速发展，且用户数量飞速增长。

2. WWW

WWW 是环球信息网的缩写，也可以简称为 Web，中文名字为"万维网"。它是一个资源空间。这些资源通过超文本传输协议传送给使用者，而后者通过单击链接来获得资源。

3. Http

Http 协议(Hypertext Transfer Protocol,超文本传输协议)是因特网上目前应用最为广泛的一种网络传输协议,所有的 WWW 文件都必须遵守这个协议。Http 是基于 TCP/IP 通信协议来传递数据(HTML 文件,图片文件,查询结果等)。

Http 协议默认使用 80 号端口,但是也可以配置某个 Web 服务器使用另外一个端口(如:8080)。这样,就能在一台服务器运行多个 Web 服务器,每个服务器监听不同的端口。

4. Web"客户端/服务器"方式

如图 8-1 所示为 Web 客户端与服务器之间的通信,其流程如下:

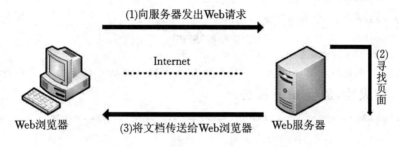

图 8-1　Web 客户端与服务器之间的通信

(1) Web 浏览器使用 Http 命令向服务器发出 Web 请求。

(2) Web 服务器接收到 Web 页面请求后,就发送一个应答,在客户端和服务器之间建立连接,并查找客户端所需的文档。

(3) 若 Web 服务器查找到请求的文档,就会将请求的文档传送给 Web 浏览器。若该文档不存在,Web 服务器会发送一个相应的错误提示文档给客户端。Web 浏览器在接收到文档后,就将它解释并显示在屏幕上。客户端浏览完成后,就断开与 Web 浏览器的连接。

8.1.2　Apache 服务器简介

Apache Http Server(简称 Apache)是 Apache 软件基金会维护开发的一个开放源码的网页服务器,可以在大多数计算机操作系统中运行,由于其多平台和安全性被广泛使用,是最流行的 Web 服务器端软件之一。它快速、可靠并且可通过简单的 API 扩展将 Perl/Python 等解释器编译到服务器中。

Apache 支持许多特性,这些特性大部分都是通过编译的模块实现。这些特性从服务器端的编程语言支持到身份认证方案等。

一些通用的语言接口支持 Perl,Python,Tcl 和 PHP。流行的认证模块包括 mod_access,mod_auth 和 mod_digest,还有 SSL 和 TLS 支持(mod_ssl),代理服务器(proxy)模块,很有用的 URL 重写(由 mod_rewrite 实现),定制日志文件(mod_log_config),以及过滤支持(mod_include 和 mod_ext_filter)。Apache 日志可以通过网页浏览器使用免费的脚本 AWStats 或 Visitors 来进行分析。

任务 8.2 部署 Apache 服务器

8.2.1 安装 Apache 服务器

1. 查询系统是否已经安装 Apache 软件包

```
[root@localhost ~]# rpm -qa|grep httpd
```

2. 安装 Apache 软件包

```
[root@localhost ~]# yum -y install httpd
[root@localhost ~]# rpm -qa|grep httpd
httpd-tools-2.4.6-67.el7.x86_64
httpd-2.4.6-67.el7.x86_64
```

3. 配置防火墙，放行 http 服务

```
[root@localhost ~]# firewall-cmd --permanent --add-service=http
success
[root@localhost ~]# firewall-cmd --reload
success
[root@localhost ~]# firewall-cmd --list-all
public (active)
  target: default
  icmp-block-inversion: no
  interfaces: ens33
  sources:
  services: ssh dhcpv6-client dns http
……
```

4. 启动、测试 httpd 服务

```
[root@localhost ~]# systemctl start httpd
[root@localhost ~]# firefox http://127.0.0.1
```

这时，浏览器会显示 httpd 服务默认页面，如图 8-2 所示。

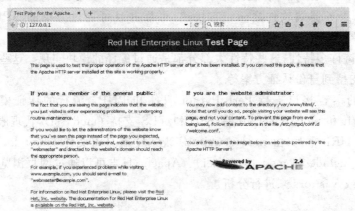

图 8-2 httpd 服务默认页面

8.2.2 认识 Apache 服务器的配置文件

httpd 服务程序的主要配置文件及其存放位置如表 8-1 所示。

表 8-1　httpd 服务程序的配置文件及其存放位置

配置文件名称	存放位置
服务目录	/etc/httpd
主配置文件	/etc/httpd/conf/httpd.conf
网站数据目录	/var/www/html
访问日志	/var/log/httpd/access_log
错误日志	/var/log/httpd/error_log

在 Apache 服务器的主配置文件 httpd.conf 中,有三种类型的信息:注释行信息、全局配置、区域配置。httpd.conf 文件常见的参数及其功能如表 8-2 所示。

表 8-2　httpd.conf 文件常见的参数及其功能

参　数	功　能
ServerRoot	服务目录
ServerAdmin	管理员邮箱
User	运行服务的用户
Group	运行服务的用户组
ServerName	网站服务器的域名
DocumentRoot	网站文档根目录
Directory	网站数据目录的权限
Listen	监听的 IP 地址与端口号
DirectoryIndex	默认的索引页面(主页文件的名称)
ErrorLog	错误日志文件
LogLevel	日志级别
CustomLog	访问日志文件
TimeOut	网页超时时间(默认为 300 s)

网站文档根目录默认的路径是/var/www/html,网站默认的主页文件的名称是 index.html。此时,可以直接修改 index.html 文件的内容,然后再次查看系统的默认页面,如图 8-3 所示,已发生改变。

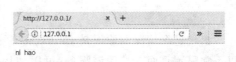

图 8-3　默认页面已发生改变

```
[root@localhost~]# echo "ni hao">/var/www/html/index.html
[root@localhost~]# firefox http://127.0.0.1
```

接下来,将网站文档根目录默认的路径修改为/web,并创建首页文件,同时将主配置文件进行修改(DocumentRoot、Directory 两个参数),重启 httpd 服务后,查看结果,浏览器显示的是 httpd 服务默认页面,如图 8-4 所示。

```
[root@localhost~]# mkdir /web
[root@localhost~]# echo "Welcome">/web/index.html
[root@localhost~]# vim /etc/httpd/conf/httpd.conf
……
DocumentRoot "/web"
#
# Relax access to content within /var/www.
#
<Directory "/web">
    AllowOverride None
    # Allow open access:
    Require all granted
</Directory>
……
[root@localhost~]# systemctl restart httpd
[root@localhost~]# firefox http://127.0.0.1
```

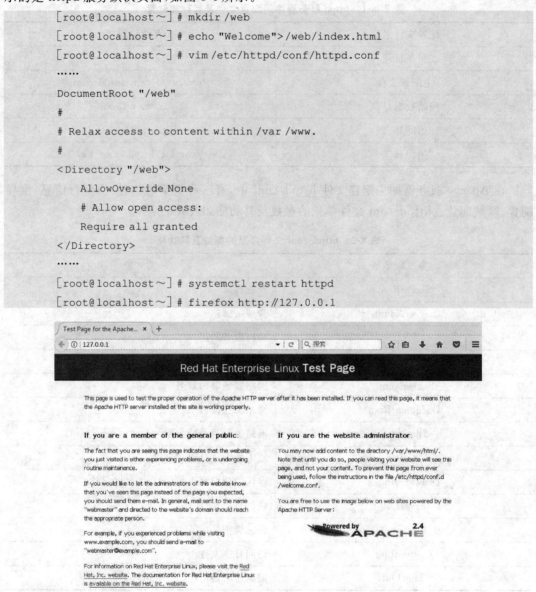

图 8-4 httpd 服务默认页面

出现这个情况的原因是 Linux 系统的 SELinux 设置有问题。只需要禁用该服务,就可以正常显示页面。禁用该服务,可以使用命令 setenforce[0|1]进行临时性设置,也可以修改/etc/sysconfig/selinux 文件,修改 SELinux 相应的值进行永久性设置。

临时性设置:

```
[root@localhost~]# getenforce      //查看当前SELinux服务的运行模式
Enforcing
[root@localhost~]# setenforce 0    //禁用SELinux服务,"1"为启用,"0"为禁用
[root@localhost~]# getenforce
Permissive
```

永久性设置：

```
[root@localhost~]# vim /etc/sysconfig/selinux
# This file controls the state ofSELinux on the system.
# SELINUX=can take one of these three values:
#     enforcing-SELinux security policy is enforced.
#     permissive-SELinux prints warnings instead of enforcing.
#     disabled-NoSELinux policy is loaded.
SELINUX=disable
# SELINUXTYPE=can take one of three two values:
#     targeted-Targeted processes are protected,
#     minimum-Modification of targeted policy.Only selected processes are protected.
#     mls-Multi Level Security protection.
SELINUXTYPE=targeted
```

enforcing：应用 SELinux 所设定的 Policy，所有违反 Policy 的规则(Rules)都会被 SELinux 拒绝；

permissive：：应用 SELinux 所设定的 Policy，但是对于违反规则的操作只会予以记录而并不会拒绝操作；

disabled：完全禁用 SELinux。

禁用 SELinux 服务后，再次查看结果，浏览器已显示正常的页面，如图 8-5 所示。

图 8-5　页面内容已按照设置显示

8.2.3　个人用户主页功能

现在很多网站都支持用户拥有自己的主页空间，用户可以非常容易地管理自己的主页空间。通过 httpd 服务程序提供的个人用户主页功能，可以很容易地满足这个需求。

默认情况下，httpd 服务程序没有启用个人用户主页功能，需要修改配置文件（路径为/etc/httpd/conf.d/userdir.conf)的第 17 行，在 UserDir disabled 参数前面添加♯号；第 24 行，取消 UserDir public_html 参数前面的♯号。

```
[root@localhost~]# vim /etc/httpd/conf.d/userdir.conf
……
17      # UserDir disabled       //让httpd服务程序开启个人用户主页功能
……
24      UserDir public_html      //表示网站数据在用户家目录中的保存目录名称,即public
                                  _html目录
……
```

接下来,创建用户,并在该用户家目录中创建个人主页空间目录 public_html,同时创建默认网页。

```
[root@localhost ~]# useradd jack
[root@localhost ~]# passwd jack
[root@localhost ~]# mkdir /home/jack/public_html
[root@localhost ~]# chmod -Rf 755 /home/jack/
[root@localhost ~]# cd /home/jack/public_html/
0[root@localhost public_html]# echo "Welcome,this is jack's web."> index.html
[root@localhost public_html]# systemctl restart httpd
[root@localhost public_html]# setenforce 0
```

使用格式"http://网址(IP 地址)/~用户名"查看结果,如图 8-6 所示。

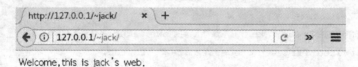

图 8-6 个人用户主页访问效果

8.2.4 相关常规设置

1. 根目录设置

在配置文件中,ServerRoot 参数原来设置 Apache 的配置文件、错误文件、日志文件的存放目录,同时该目录也是整个目录树的根节点,如果下面的字段设置中出现相对路径,就是相对这个路径而言。默认情况下,根目录的路径为/etc/httpd,可以根据实际需求进行修改。

```
ServerRoot "/etc/httpd"
```

2. 客户端连接数限制

客户端连接数限制是指在某一个时间段内,Web 服务器允许多少客户端同时访问。允许同时访问的最大数值就是客户端连接数限制。可以通过在主配置文件 httpd.conf 中添加 prefork 模块进行设置。

```
<IfModule prefork.c>
StartServers 8            //启动 httpd 时,唤醒几个 PID 来处理服务
MinSpareServers 5         //最小的预备使用的 PID 数量
MaxSpareServers 20        //最大的预备使用的 PID 数量
ServerLimit 256           //服务器的限制
MaxClients 256            //最多可以允许多少个客户端同时访问 Web 服务器
MaxRequestsPerChild 4000  //每个程序能够提供的最大传输次数
</IfModule>
```

需要注意的是:MaxClients 的数量不是越大越好,因为它会消耗系统的内存资源,如果数量过大,反而会降低系统的运行速度。

3. 网页编码设置

目前的因特网传输数据的编码以使用万国码(UTF-8)为主,如果出现服务器端的网页编码和客户端的网页编码不一致,就会导致客户端显示的网页内容为乱码。虽然可以调整客户端浏览器的编码来解决这个问题,但是,一般是通过修改服务器端的网页编码来彻底解决这个问题。

```
[root@localhost~]# vim /etc/httpd/conf/httpd.conf
……
AddDefaultCharset UTF-8   //汉字编码一般是 GB2312,可以将 UTF-8 修改为 GB2312
……
```

4. 主机名称设置

ServerName 参数用于定义服务器的名称和端口号。如果没有注册域名,那么可以使用 IP 地址。

```
[root@localhost~]# vim /etc/httpd/conf/httpd.conf
……
ServerName www.jack.com:80
……
```

5. 管理员邮箱设置

当客户端访问服务器发生错误时,服务器通常会将含有错误提示信息的网页反馈给客户端,并提供管理员邮箱地址,以便改正错误。可以通过修改 ServerAdmin 参数进行设置。

```
[root@localhost~]# vim /etc/httpd/conf/httpd.conf
……
ServerAdmin jack@ kk.com
……
```

6. 网站首页和目录相关权限设置

网站首页目录默认的路径是/var/www/html,可以通过修改 DocumentRoot 参数进行设置,网站目录可以使用<Directory></Directory>容器进行设置。在前面的内容中,已经介绍过修改网站首页默认的路径以及目录的修改设置。常见的目录权限及其功能如表 8-3 所示。

表 8-3 Apache 常见目录权限及其功能

权限	功能
Options	设置特定目录中的服务器特性
AllowOverride	设置如何使用访问控制文件.htaccess
Order	设置 Apache 默认的访问权限以及 Allow 和 Deny 语句的处理顺序
Allow	设置允许访问 Apache 服务器的主机
Deny	设置拒绝访问 Apache 服务器的主机

注意:Allow 与 Deny 参数之间的处理顺序是非常重要的,主要有两种处理顺序:

(1) Deny,Allow:Deny 优先处理,但没有写入规则的则默认为 Allow。常用于拒绝所有,开放特定的条件。

(2) Allow,Deny：Allow 优先处理，但没有写入规则的则默认为 Deny。常用于开放所有，拒绝特定的条件。

如果 Allow、Deny 的规则当中有重复的，则以默认情况（Order 的规范）为主。

例如：① 允许所有客户端访问。

```
Order allow,deny
Allow from all
```

② 拒绝 192.168.1.100 和来自.kk.com 域的客户端访问，其他客户端均可以正常访问。

```
Order deny,allow
Deny from 192.168.1.100
Deny from.kk.com
```

③ 仅允许 192.168.1.0/24 网段的客户端访问，但 192.168.0.200 不能访问。

```
Order allow,deny
Allow from 192.168.0.0/24
Deny from 192.168.0.200
```

④ 只允许 192.168.1.0/24 网段的客户端访问。

```
Order deny,allow
Deny from all
Allow from 192.168.1.0/24
```

Options 主要参数及功能如表 8-4 所示。

表 8-4 Option 主要参数及功能

参数	功能
Indexes	允许目录浏览，如果在此目录下找不到首页文件，就显示整个目录下的文件清单
FollowSymLinks	可以在该目录中使用符号链接去访问其他目录
ExecCGI	让此目录具有执行 CGI 程序的权限
Multiviews	允许内容协商的多重视图
All	支持除 Multiviews 以外的所有参数
Includes	允许服务器端使用 SSI(Server—Side Include)技术
IncludesNoExec	允许服务器端使用 SSI(Server—Side Include)技术，并禁止执行 CGI 程序

AllowOverride 主要参数及功能如表 8-5 所示。

表 8-5 AllowOverride 主要参数及功能

参数	功能
ALL	全部权限均可覆盖
AuthConfig	仅有网页认证可覆盖
Indexes	仅允许 Indexes 方面可覆盖
Limits	允许用户使用 Allow、Deny 与 Order 管理可浏览的权限
None	不可覆盖，让.htaccess 文件失效

8.2.5 虚拟主机

虚拟主机是在一台 Web 服务器上,可以为多个独立的 IP 地址、域名或端口号提供不同的 Web 站点,实现多个 Web 站点的访问。

1. 基于 IP 地址的虚拟主机

实现步骤如下:

(1) 为网卡添加多个 IP 地址。

```
[root@localhost ~]# vim /etc/sysconfig/network-scripts/ifcfg-ens33
……
IPADDR1=192.168.1.1
NETMASK=255.255.255.0
IPADDR2=192.168.1.10
NETMASK=255.255.255.0
……
```

(2) 分别创建/www/web1 和/www/web2 两个目录和默认网页。

```
[root@localhost ~]# mkdir -p /www/web1 /www/web2
[root@localhost ~]# echo "wo shi web1" > /www/web1/index.html
[root@localhost ~]# echo "wo shi web2" > /www/web2/index.html
```

(3) 修改主配置文件 httpd.conf,在文件的最后添加以下内容。

```
[root@localhost ~]# vim /etc/httpd/conf/httpd.conf
<VirtualHost 192.168.1.1>
DocumentRoot /www/web1
ServerName www1.kcorp.com
<Directory "/www/web1">
    AllowOverride None
    Require all granted
</Directory>
</VirtualHost>
<VirtualHost 192.168.1.10>
DocumentRoot /www/web2
ServerName www2.kcorp.com
<Directory "/www/web2">
    AllowOverride None
    Require all granted
</Directory>
</VirtualHost>
```

(4) 重启 httpd 服务,记得禁用 SELinux 服务或者配置防火墙放行 http 服务。分别通过两个 IP 地址查看网页,结果如图 8-7、图 8-8 所示。

```
[root@localhost ~]# systemctl restart httpd
```

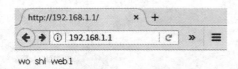
wo shi web1

图 8-7　访问 192.168.1.1 的主页

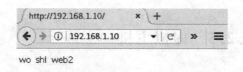
wo shi web2

图 8-8　访问 192.168.1.10 的主页

2. 基于域名的虚拟主机

（1）配置域名 www1.kcorp.com、www2.kcorp.com 分别与 IP 地址 192.168.1.1、192.168.1.2 相互绑定。

```
[root@localhost~]# nslookup www1.kcorp.com
Server: 192.168.1.1
Address: 192.168.1.1# 53
Name: www1.kcorp.com
Address: 192.168.1.1
[root@localhost~]# nslookup www2.kcorp.com
Server: 192.168.1.1
Address: 192.168.1.1# 53
Name: www2.kcorp.com
Address: 192.168.1.10
```

（2）重启 httpd 服务，分别通过两个域名查看网页，结果如图 8-9、图 8-10 所示。

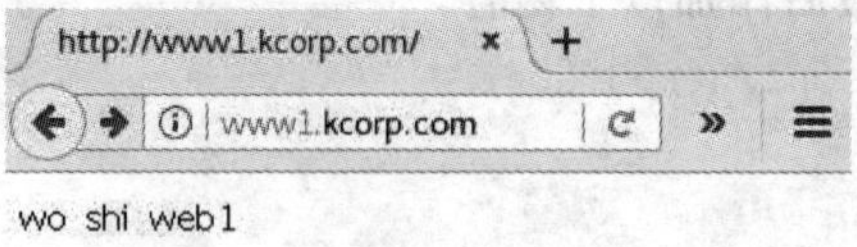

图 8-9　访问 www1.kcorp.com 的主页 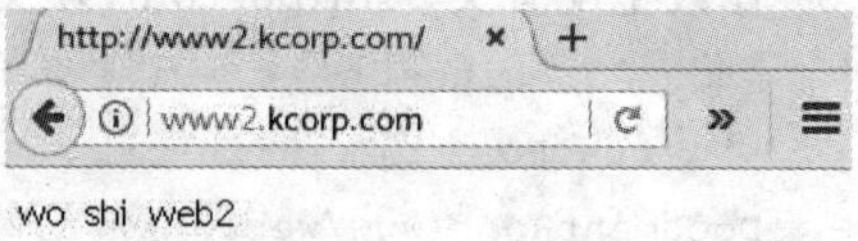　　　图 8-10　访问 www2.kcorp.com 的主页

3. 基于端口号的虚拟主机

（1）基于端口号的虚拟主机服务器只需要一个 IP 地址，各虚拟主机之间通过不同的端口号进行区分。这里使用 IP 地址 192.168.1.1，然后修改主配置文件 httpd.conf，在文件中添加端口号 8080、9090 分别对应两个网页，并监听这两个端口。

```
[root@localhost~]# vim /etc/httpd/conf/httpd.conf
Listen 8080
Listen 9090
<VirtualHost 192.168.1.1:8080>
DocumentRoot/www/web1
ServerName www1.kcorp.com
<Directory "/www/web1">
    AllowOverride None
    Require all granted
</Directory>
</VirtualHost>
```

```
<VirtualHost 192.168.1.1:9090>
DocumentRoot /www/web2
ServerName www2.kcorp.com
<Directory "/www/web2">
    AllowOverride None
    Require all granted
</Directory>
</VirtualHost>
```

(2)重启 httpd 服务,分别通过两个端口号查看网页,结果如图 8-11、图 8-12 所示。

图 8-11　访问 www1.kcorp.com:8080 的主页　　　图 8-12　访问 www1.kcorp.com:9090 的主页

4. 虚拟目录的设置

设置虚拟目录/kk,对应真实的目录/test/web,在主配置文件 httpd.conf 中添加 Alias/kk "/test/web"实现虚拟目录与真实目录的对应。

```
[root@localhost ~]# mkdir -p /test/web
[root@localhost ~]# echo "wo shi web" >/test/web/index.html
[root@localhost ~]# vim /etc/httpd/conf/httpd.conf
<VirtualHost 192.168.1.1>
DocumentRoot /test/web
ServerName www1.kcorp.com
Alias /kk "/test/web"
<Directory "/test/web">
    Options IndexesMultiViews FollowSymLinks
    AllowOverride None
    Order allow,deny
    Allow from all
    Require all granted
</Directory>
</VirtualHost>
```

重启 httpd 服务后,在浏览器中通过 http://192.168.1.1/kk,就可以浏览虚拟目录中的内容,如图 8-13 所示。

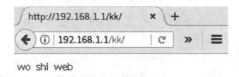

图 8-13　访问虚拟目录

8.2.6 HTTPS 安全认证访问网站

实现步骤如下：
(1) 安装 SSL 服务，并创建证书存放的目录(/etc/httpd/.ssl)。

```
[root@localhost~]# yum -y install mod_ssl
[root@localhost~]# mkdir /etc/httpd/.ssl
```

(2) 生成网站私钥文件。

```
[root@localhost~]# cd /etc/httpd/.ssl
[root@localhost.ssl]# openssl genrsa -out server.key 1024
Generating RSA private key,1024 bit long modulus
..........+ + + + +
..................+ + + + +
e is 65537 (0x10001)
```

openssl 是一个安全套接字层密码库。genrsa 命令用于生成 RSA 私钥，其参数如下：

① -numbits：指定要生成的私钥的长度，默认为 1024。该项必须为命令行的最后一项参数。

② -out：将生成的私钥保存至 filename 文件。

(3) 建立网站证书。

```
[root@localhost.ssl]# openssl req -new -x509 -key server.key -out server.crt
You are about to be asked to enter information that will be incorporated
into your certificate request.
What you are about to enter is what is called a Distinguished Name or a DN.
There are quite a few fields but you can leave some blank
For some fields there will be a default value,
If you enter '.',the field will be left blank.
-----
Country Name (2 letter code)[XX]:cd
State or Province Name (full name)[]:js
Locality Name (eg,city)[Default City]:yf
Organization Name (eg,company)[Default Company Ltd]:yf
Organizational Unit Name (eg,section)[]:sc
Common Name (eg,your name or your server's hostname)[]:
Email Address[]:
```

req 命令用于生成证书请求文件，查看验证证书请求文件，以及生成自签名证书，其参数如下：

① -new：生成证书请求文件。

② -x509：生成自签名证书。

③ -key：证书私钥文件的来源，仅与生成证书请求选项 -new 配合使用。

④ -out：指定生成的证书请求或者自签名证书名称，或者公钥文件名称。

(4) 修改主配置文件 httpd.conf。

```
[root@localhost ~]# vim /etc/httpd/conf/httpd.conf
<VirtualHost 192.168.1.1:443>
DocumentRoot/test/web
ServerName www1.kcorp.com
SSLEngine on
SSLCertificateFile/etc/httpd/.ssl/server.crt
SSLCertificateKeyFile/etc/httpd/.ssl/server.key
<Directory "/test/web">
    AllowOverride None
    Require all granted
</Directory>
</VirtualHost>
[root@localhost.ssl]# systemctl restart httpd
```

重启 httpd 服务后,在浏览器中通过 https://192.168.1.1 进行访问,需要加载网站的安全证书,如图 8-14、图 8-15 所示。

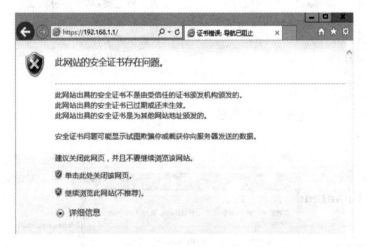

图 8-14　单击"继续浏览此网站(不推荐)"

图 8-15　成功访问

8.2.7　用户认证和授权

实现步骤如下:
(1) 创建登录网站的用户和密码。

```
[root@localhost ~]# htpasswd -c  /etc/pwd jc
New password:
Re-type new password:
Adding password for userjc
```

htpasswd 命令是用来创建 .htaccess 文件身份认证使用的密码,其参数如下:
① -c:新创建一个密码文件。
② -b:用批处理方式创建用户。
③ -D:删除一个用户。
④ -m:采用 MD5 编码加密。
⑤ -p:采用明文格式的密码。
(2) 修改主配置文件 httpd.conf。

```
<VirtualHost 192.168.1.1>
DocumentRoot /test/web
ServerName www1.kcorp.com
<Directory "/test/web">
    AllowOverride AuthConfig
    Orderdeny,allow
    Allow from all
    AuthType Basic                  //认证的类型:基本身份验证
    AuthName "Please Login:"        //设置认证提示的信息和内容
    AuthUserFile /etc/pwd           //设置认证文件的位置
    Require  valid-user             //设置允许访问的用户
</Directory>
</VirtualHost>
```

重启 httpd 服务后,在浏览器中通过 http://192.168.1.1 进行访问,需要进行用户认证,如图 8-16、图 8-17 所示。

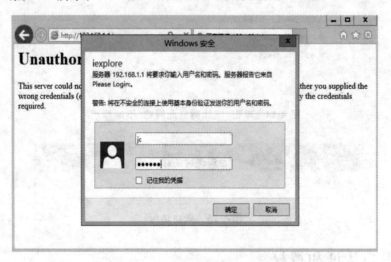

图 8-16　输入用户名和密码

图 8-17　成功访问

自我测试

一、填空题

1. Web 服务器主要功能是提供_____服务。
2. WWW 是_____的缩写。
3. HTTP 协议默认使用_____号端口。
4. Apache HTTP Server 是最流行的_____服务器端软件之一。

二、简答题

1. 简述 HTTP 协议的功能。
2. 简述 Web"客户端/服务器"方式。

三、实训题

1. 在 Apache 服务器上创建 Web 站点/var/www/html/myweb，在 myweb 站点下创建网站首页 myweb.html，网页内容为 Hello。配置后，能在浏览器上输入 http://localhost 查看网页内容。

2. 在 192.168.88.120 地址上配置 web1 站点，域名为 www.web1.com，默认站点根目录为/var/www/html/web1，使用 8080 端口；在 192.168.88.220 地址上配置 web2 站点，域名为 www.web2.com，默认站点根目录为/var/www/html/web2，使用 9090 端口。配置完成后，使用浏览器对两个站点进行测试。

项目 9 Vsftpd 服务器

项目综述

随着公司业务发展,很多文件需要在网络上进行传输,为了方便文件的上传和下载,公司网络管理员将搭建 FTP 服务器来实现这些功能。

项目目标

- 了解 FTP 服务相关知识及工作原理;
- 掌握 vsftpd 服务器的配置方法;
- 掌握基于虚拟用户的 FTP 服务器的配置方法;
- 掌握基于 FTPS 服务器的配置方法。

◀ 任务 9.1 FTP 服务器简介 ▶

9.1.1 FTP 服务器基本概念

1. FTP 协议

FTP 协议(文件传输协议)用于 Internet 上的控制文件的双向传输。同时,它也是一个应用程序。基于不同的操作系统有不同的 FTP 应用程序,而所有这些应用程序都遵守同一种协议传输文件。

2. FTP 工作过程

FTP 采用"客户端/服务器"方式,用户端要在自己的本地计算机上安装 FTP 客户程序。FTP 的工作过程如图 9-1 所示。

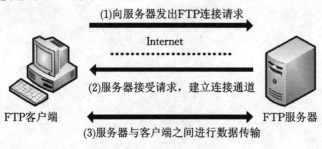

图 9-1 FTP 工作过程

FTP 采用三次握手的可信传输,其过程如下:

(1) 客户端主动向服务器发送 FTP 连接请求

为连接到服务器,客户端将随机使用一个 1024 以上的端口主动连接到 FTP 服务器的 21 端口,在连接数据包中会包含握手信号(SYN)标志。

(2) 服务器接收到信号后发出响应

当服务器接收到客户端的请求后,向客户端发出响应,同时服务器建立等待连接的资源,将带有 SYN 与确认的封包发给客户端。

(3) 客户端回应确认封包

客户端接收到服务器的封包后,会再次发送一个确认封包给服务器,这时,双方就正式建立了传输通道。

以上创建的通道只能用于执行 FTP 命令,如果要进行数据的传送,就需要再建立一条数据传输的通道。

(1) 客户端发送数据传输请求的命令给服务器

客户端使用另外一个高于 1024 的端口进行准备连接,并主动使用指令信道(21 端口)发送一个命令给服务器,表示已经准备好一个数据传输的端口,准备进行传输。

(2) 服务器以 ftp-data 端口主动连接到客户端

服务器收到命令后,会以 ftp-data 端口(一般为 20 端口)通知客户端已经与那个高于 1024 的端口进行连接。

(3) 客户端响应服务器,完成三次握手,建立数据传输的通道。

FTP 主要用到的端口有以下两个:

① 命令通道的 FTP 端口(默认为 21 端口),主要用于命令传输;

② 数据传输的 ftp-data 端口(默认为 20 端口),主要用于数据传输。

FTP 有两种工作模式:主动传输模式(Active FTP,FTP 服务器主动向客户端发起连接请求)和被动传输模式(Passive FTP,FTP 服务器等待客户端发起连接请求,FTP 默认的工作模式)。

FTP 客户程序有字符界面和图形界面两种。字符界面的 FTP 的命令复杂、繁多。图形界面的 FTP 客户程序,操作上要简捷方便得多。

3. FTP 的使用

在 FTP 的使用当中,用户经常遇到两个概念:"下载(Download)"和"上传(Upload)"。"下载"文件就是从远程主机拷贝文件至自己的计算机上;"上传"文件就是将文件从自己的计算机中拷贝至远程主机上。用 Internet 的语言来说,用户可通过客户端程序向(从)远程主机上传(下载)文件。两个权限是读取和写入。

4. FTP 服务器的分类

FTP 服务器有两类:

(1) 匿名 FTP 服务器。它有公共的用户名和密码(就像是没有用户名和密码)。公共用户名为:anonymous;密码为空。匿名 FTP 是这样一种机制,用户可通过它连接到远程主机上,并从其下载文件,而无须成为其注册用户。

(2) 普通 FTP 服务器。它需要用户名和密码。使用普通 FTP 时必须首先登录,在远程主机上获得相应的权限以后,方可上传或下载文件。

9.1.2 FTP 服务的用户分类

使用 FTP 的用户需要经过验证后才能登录,FTP 服务的用户分为三类:

(1) 本地用户(Real 用户)

本地用户也就是实体用户,是 Linux 系统中的用户,即系统本机的用户。本地用户可以通过输入自己的账号和密码进行授权登录。当授权访问的本地用户登录系统后,其登录目录为用户自己的家目录,本地用户可以进行上传和下载。通常不建议使用本地用户通过 FTP 方式远程访问系统。

(2) 虚拟用户(Guest 用户)

只能采用 FTP 方式使用系统的用户,不能直接使用 shell 登录系统,即虚拟用户。访问服务器时需要验证,其登录目录是 vsftpd 为其指定的目录,通常情况下,该用户可以进行上传和下载,大多数 FTP 用户都属于这类用户。

(3) 匿名用户(Anonymous 用户)

如果用户在远程 FTP 服务器上没有账号,则称此用户为匿名用户。当匿名用户登录系统后,其登录目录为匿名 FTP 服务器的根目录(默认为/var/ftp)。一般情况下,应尽可能多的对该用户进行限制,仅允许该用户进行下载。

◀ 任务 9.2 部署 Vsftpd 服务器 ▶

9.2.1 安装 Vsftpd 服务器

1. 查询系统是否已经安装 vsftpd 软件包

```
[root@localhost ~]# rpm -qa|grep vsftpd
```

2. 安装 vsftpd 软件包

```
[root@localhost ~]# yum -y install vsftpd
[root@localhost ~]# rpm -qa|grep vsftpd
vsftpd-3.0.2-22.el7.x86_64
```

3. 配置防火墙,放行 ftp 服务

```
[root@localhost ~]# firewall-cmd --permanent --add-service=ftp
success
[root@localhost ~]# firewall-cmd --reload
success
[root@localhost ~]# firewall-cmd --list-all
public (active)
  target: default
```

```
        icmp-block-inversion: no
        interfaces: ens33
        sources:
        services: ssh dhcpv6-client dns http ftp
    ……
```

也可以使用以下命令,将 SELinux 中的防火墙设置为 on,并设置 ftp 服务完全访问为允许,使得所有通过验证的用户都可以进行上传、下载。

```
[root@localhost~]# setsebool -P ftpd_full_access=on
[root@localhost~]# getsebool -a |grep ftp
ftpd_anon_write --> off
ftpd_connect_all_unreserved --> off
ftpd_connect_db --> off
ftpd_full_access --> on
ftpd_use_cifs --> off
ftpd_use_fusefs --> off
ftpd_use_nfs --> off
ftpd_use_passive_mode --> off
```

4. 启动、测试 httpd 服务

```
[root@localhost~]# systemctl start vsftpd
[root@localhost~]# firefox ftp://127.0.0.1
```

这时,因为是使用匿名用户登录 vsftpd 服务器,浏览器会显示服务器默认的根目录/var/ftp/pub,如图 9-2 所示。

图 9-2 服务器默认的根目录

9.2.2 认识 Vsftpd 服务器的配置文件

vsftpd 服务程序的主要配置文件及其功能,如表 9-1 所示。

表 9-1 vsftpd 服务程序的主要配置文件及其功能

配置文件名称	存放位置	功能
主配置文件	/etc/vsftpd/vsftpd.conf	vsftpd 的主配置文件
PAM 配置文件	/etc/pam.d/vsftpd	用于用户认证的 PAM 配置文件
用户文件	/etc/vsftpd/ftpusers	指定不能登录 vsftpd 用户的文件
用户文件	/etc/vsftpd/user_list	限制用户登录 vsftpd 的配置文件
匿名用户主目录	/var/ftp	默认情况下,除了 root 用户具有读写权限,其他的目录都只有只读权限
匿名用户下载目录	/var/ftp/pub	

vsftpd 服务程序主配置文件 vsftpd.conf 常见的参数及其功能如表 9-2 所示。

表 9-2 vsftpd 服务程序主配置文件常见参数及其功能

参数	功能
background=<YES/NO>	是否运行在后台监听模式,默认值为 YES
listen=<YES/NO>	是否以独立运行的方式监听服务
listen_port=21	设置 FTP 服务的监听端口
listen_address=<IP 地址>	设置需要监听的 IP 地址
download_enable=<YES/NO>	是否允许下载文件
write_enable=<YES/NO>	是否开放本地用户的写权限
userlist_enable=<YES/NO> userlist_deny=<YES/NO>	设置用户列表为"允许"还是"禁止"操作。userlist_enable 决定/etc/vsftpd/userlist 文件是否启用,YES 为启用。userlist_deny 决定/etc/vsftpd/userlist 文件中的用户是否允许访问 FTP 服务器,若为 YES,则该文件中的用户不能访问服务器;若为 NO,则该文件中的用户可以访问服务器
pasv_enbale=<YES/NO>	是否使用被动模式的数据连接
pasv_min_port=<n> pasv_max_port=<m>	设置被动模式数据连接的端口范围
ftpd_banner=<message>	设置用户连接服务器后的显示信息
banner_file=<filename>	设置用户连接服务器后的显示信息,信息存放在指定的 filename 文件中
connect_timeout=<n>	如果客户尝试连接 vsftpd 服务器超过 n 秒,则强制断开,默认为 60
accept_timeout=<n>	当使用者以被动模式进行数据传输时,服务器发出 passive port 指令后,等待客户超过 n 秒就强制断开,默认为 60
data_connection_timeout=<n>	设置空闲的数据连接在 n 秒后中断,默认为 120
idle_session_timeout=<n>	设置空闲的用户会话在 n 秒后中断,默认为 600
max_clients=<n>	最大客户连接数,0 为不限制

续表 9-2

参数	功能
max_per_ip=<n>	同一 IP 地址的最大连接数,0 为不限制
local_enable=<YES/NO>	设置是否允许本地用户访问
guest_enbale=<YES/NO>	当设置为 YES 时,所有非匿名用户都视为虚拟用户
wirte_enable=<YES/NO>	是否让本地用户具有写权限
local_umask=<nnn>	设置本地用户上传文件的 umask 掩码值,默认值为 022(新建的目录权限是 755,代表文档的所有者具有读写执行权限,组内用户具有读和执行权限,其他用户具有读和执行权限)
local_max_rate=<n>	设置本地用户的最大传输速率,单位为 bytes/sec,0 为不限制
local_root=<filename>	设置本地用户的 FTP 根目录
chroot_local_user=<YES/NO>	设置为 YES 时,所有的本地用户将执行 chroot
chroot_list_enable=<YES/NO> chroot_list_file=<filename>	当 chroot_local_user 设置为 NO 且 chroot_list_enable 设置为 YES 时,只有 filename 文件中指定的用户才能执行 chroot,chroot_list_file 默认值为/etc/vsftpd/chroot_list
anonymous_enable=<YES/NO>	设置是否允许匿名用户访问
anon_max_rate=<n>	设置匿名用户的最大传输速率,单位为 bytes/sec,0 为不限制
anon_upload_enabel=<YES/NO>	设置是否允许匿名用户上传文件
anon_umask=<nnn>	设置匿名用户上传文件的 umask 掩码值,默认值为 022
anon_root=<filename>	设置匿名用户的 FTP 根目录
anon_upload_enable=<YES/NO>	设置是否允许匿名用户上传文件
anon_mkdir_write_enable=<YES/NO>	设置是否允许匿名用户创建目录
anon_other_write_enable=<YES/NO>	设置是否开放匿名用户的其他写入权限(包括重命名、删除等操作权限)
tcp_wrappers=<YES/NO>	设置服务器是否支持 tcp_wrappers
pam_service_name=vsftpd	设置 PAM 模块的名称,该文件存放在/etc/pam.d/vsftpd
xferlog_enable=<YES/NO>	设置是否启用 FTP 的上传和下载日志
xferlog_file=/var/log/xferlog	设置日志记录文件存放的位置以及文件的名称
xferlog_std_format=<YES/NO>	当设置为 YES 时,日志文件将以标准 xferlog 的格式书写

9.2.3 配置匿名 FTP 服务器

在公司搭建一台 FTP 服务器,允许匿名用户上传、下载文件,匿名用户的 FTP 根目录为/var/ftp。实现步骤如下:

(1) 创建测试文件,设置文件权限,将文件所有者设为 ftp,或者将 pub 目录赋予其他用户写的权限。

```
[root@localhost~]# touch /var/ftp/pub/ceshi.tar
[root@localhost~]# ll -ld /var/ftp/pub
drwxr-xr-x. 3 root root 34 6月  27 10:16 /var/ftp/pub
[root@localhost~]# chown ftp /var/ftp/pub
[root@localhost~]# ll -ld /var/ftp/pub
drwxr-xr-x. 4 ftp root 58 6月  27 10:31 /var/ftp/pub
[root@localhost~]# chmod o+w /var/ftp/pub
[root@localhost~]# ll -ld /var/ftp/pub
drwxr-xrwx. 3 ftp root 34 6月  27 10:16 /var/ftp/pub
```

（2）编辑主配置文件 vsftpd.conf，修改以下内容。

```
[root@localhost~]# vim /etc/vsftpd/vsftpd.conf
……
anonymous_enable=YES
anon_root=/var/ftp
anon_upload_enable=YES
anon_mkdir_write_enable=YES
anon_other_write_enable=YES
……
```

（3）重启 vsftpd 服务，记得禁用 SELinux 服务或者配置防火墙放行 ftp 服务。

```
[root@localhost~]# systemctl restart vsftpd
[root@localhost~]# firewall-cmd --permanent --add-service=ftp
[root@localhost~]# firewall-cmd --reload
```

（4）在 Windows 10 客户端的"我的电脑"地址栏中输入 ftp://192.168.1.1，打开 pub 目录，将文件 ceshi.tar 下载到系统桌面，将系统桌面的 windows ceshi.txt 文件上传到服务器，如图 9-3、图 9-4 所示。

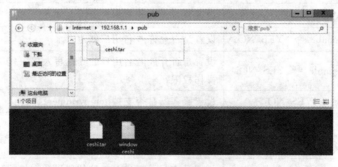

图 9-3　客户端访问服务器

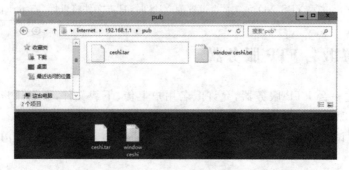

图 9-4　将文件 ceshi.tar 下载到系统桌面

也可以在命令行下,使用 ftp 命令连接服务器(在输入密码处直接按回车键)。通过 cd 命令切换目录、dir 命令显示列出的文件及目录、mkdir 命令创建目录、rmdir 删除目录,如图 9-5 所示。

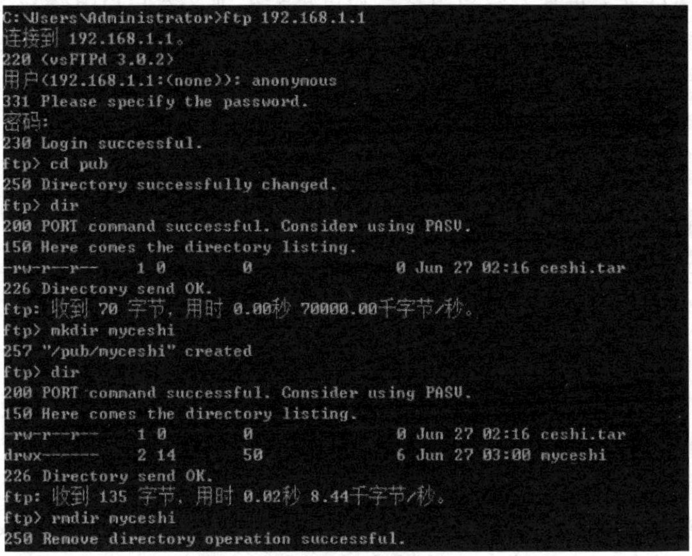

图 9-5　使用 ftp 命令连接服务器

9.2.4　配置普通 FTP 服务器

为公司用户创建 FTP 服务器登录账户 ftp1、ftp2,允许这两个账户登录 FTP 服务器,但不能登录本地系统,这两个账户的根目录限制在/myftp/ftp 目录下。实现步骤如下:

(1) 创建 FTP 服务器账户 ftp1、ftp2,并禁止这两个用户登录本地系统,同时创建用户 user1。

```
[root@localhost~]# useradd -s /sbin/nologin ftp1
[root@localhost~]# useradd -s /sbin/nologin ftp2
[root@localhost~]# useradd -s /sbin/nologin user1
[root@localhost~]# passwd ftp1
[root@localhost~]# passwd ftp2
[root@localhost~]# passwd user1
```

(2) 编辑主配置文件 vsftpd.conf,修改以下内容。

```
[root@localhost~]# vim /etc/vsftpd/vsftpd.conf
……
anonymous_enable=NO
local_enable=YES
write_enable=YES
local_root=/myftp/ftp
chroot_local_user=NO
chroot_list_enable=YES
chroot_list_file=/etc/vsftpd/chroot_list
pam_service_name=vsftpd
allow_writeable_chroot=YES
```

注意，只要启用了 chroot，参数 allow_writeable_chroot＝YES 必须添加，否则客户端在登录时，会出现错误信息"500 OOPS：vsftpd：refusing to run with writable root inside chroot（ ）"。这是因为从 2.3.5 之后，vsftpd 增强了安全检查，如果用户被限定在其主目录下，则该用户的主目录不再具有写权限了。如果检查发现还有写权限，就会报该错误。要修复这个错误，可以用命令去除用户主目录的写权限"chmod a-w/myftp/ftp"，也可以在 vsftpd 的配置文件中添加一行配置：allow_writeable_chroot＝YES。

（3）创建/etc/vsftpd/chroot_list 文件，添加账户 ftp1、ftp2。

```
[root@localhost~]# vim /etc/vsftpd/chroot_list
ftp1
ftp2
```

（4）创建 FTP 根目录/myftp/ftp，设置文件权限。

```
[root@localhost~]# touch /myftp/ftp/ceshi.tar
[root@localhost~]# ls-dl/myftp/ftp
drwxr-xr-x. 2 root root 23 6月  27 11:44 /myftp/ftp
[root@localhost~]# chmod-R o+w /myftp/ftp
[root@localhost~]# ls-dl /myftp/ftp/
drwxr-xrwx. 2 root root 23 6月  27 11:44 /myftp/ftp
```

注意：需要重启 vsftpd 服务，记得禁用 SELinux 服务或者配置防火墙放行 ftp 服务。

（5）在 Windows 10 客户端上进行测试。

① 使用账户 ftp1、ftp2 登录服务器后，这两个账户不能切换目录，但能建立文件夹和上传、下载文件。如图 9-6、图 9-7 所示。

图 9-6　账户 ftp1 登录服务器

```
ftp> put 1.txt
200 PORT command successful. Consider using PASV.
150 Ok to send data.
226 Transfer complete.
ftp> ls
200 PORT command successful. Consider using PASV.
150 Here comes the directory listing.
1.txt
ceshi.tar
myceshi
226 Directory send OK.
ftp: 收到 30 字节，用时 0.00秒 30000.00千字节/秒。
ftp> cd /var
550 Failed to change directory.
```

图 9-7　不能切换目录

② 使用账户 user1 登录服务器后，该账户能切换目录，建立文件夹，上传、下载文件。如图 9-8、图 9-9 所示。

```
C:\Users\Administrator>ftp 192.168.1.1
连接到 192.168.1.1。
220 (vsFTPd 3.0.2)
用户(192.168.1.1:(none)): user1
331 Please specify the password.
密码：
230 Login successful.
ftp> pwd
257 "/myftp/ftp"
ftp> mkdir userr1file
257 "/myftp/ftp/userr1file" created
ftp> ls
200 PORT command successful. Consider using PASV.
150 Here comes the directory listing.
1.txt
ceshi.tar
myceshi
userr1file
226 Directory send OK.
ftp: 收到 42 字节，用时 0.00秒 42000.00千字节/秒。
ftp> get ceshi.tar c:\user1\cs.tar
200 PORT command successful. Consider using PASV.
150 Opening BINARY mode data connection for ceshi.tar (0 bytes).
226 Transfer complete.
ftp> lcd c:\user1
目前的本地目录 C:\user1。
ftp> !dir
 驱动器 C 中的卷没有标签。
 卷的序列号是 CE3D-CB3B

 C:\user1 的目录

2020/06/27  13:19    <DIR>          .
2020/06/27  13:19    <DIR>          ..
2020/06/27  13:02                 0 2.txt
2020/06/27  13:19                 0 cs.tar
               2 个文件              0 字节
               2 个目录 20,571,824,128 可用字节
```

图 9-8　账户 user1 登录服务器

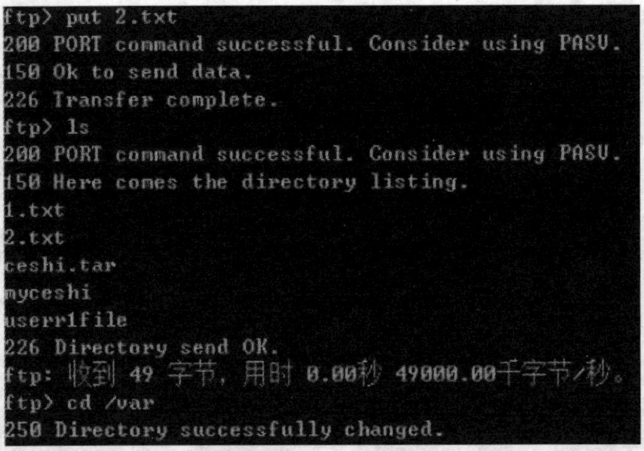

```
ftp> put 2.txt
200 PORT command successful. Consider using PASV.
150 Ok to send data.
226 Transfer complete.
ftp> ls
200 PORT command successful. Consider using PASV.
150 Here comes the directory listing.
1.txt
2.txt
ceshi.tar
myceshi
userr1file
226 Directory send OK.
ftp: 收到 49 字节,用时 0.00秒 49000.00千字节/秒。
ftp> cd /var
250 Directory successfully changed.
```

图 9-9 可以切换目录

9.2.5 配置使用虚拟用户的 FTP 服务器

在实际使用 FTP 服务器时,一般都会关闭匿名访问服务,而使用普通用户(实体用户)访问。这时,FTP 用户其实是在使用服务器中真实的用户账户,会对服务器产生严重的安全隐患,甚至可能导致黑客攻击服务器。为此,可以使用虚拟用户模式(将虚拟的账户映射为服务器的实体账户,客户端使用虚拟账户登录 FTP 服务器),提高系统的安全性。

为公司用户创建虚拟账户 ftp3、ftp4,用于登录 FTP 服务器。访问的主目录为/vmyftp/vftp,用户可以查看文件,但是不具有修改、删除等权限。

实现步骤如下:

(1) 创建用户文件,其中奇数行为账户名,偶数行为密码。

```
[root@localhost~]# mkdir /vftp
[root@localhost~]# vim /vftp/vuser.txt
ftp3
123456
ftp4
123456
```

(2) 生成数据库文件,修改数据库文件权限,为了保证系统安全,将明文的用户信息文件删除。需要使用 db_load 命令生成 db 数据库文件,以提供给系统使用虚拟用户信息。

```
[root@localhost~]# db_load -T -t hash -f /vftp/vuser.txt /vftp/vuser.db
[root@localhost~]# ls /vftp/
vuser.db  vuser.txt
[root@localhost~]# chmod 600 /vftp/vuser.db
[root@localhost~]# rm -f /vftp/vuser.txt
```

db_load 命令的作用是将用户信息文件转换为数据库并使用 hash 加密,其参数如下:

① -T:允许应用程序将文本文件转译载入数据库。

② -t:用来指定转译载入的数据库类型,参数 hash 表示使用 hash 码加密。

③ -f:指定包含用户名和密码的文本文件,文件的内容是:奇数行是用户名、偶数行是该用户对应的密码。

(3) 建立用于支持虚拟用户的 PAM 文件。

PAM 的全称为可插拔认证模块(Pluggable Authentication Modules),它是一种认证机制,使用它,不需要重新安装应用程序。通过修改指定的配置文件,就能够让系统管理员根据需求灵活地调整服务程序的不同认证方式。PAM 模块配置文件位于/etc/pam.d 目录下。

```
[root@localhost~]# vim /etc/pam.d/vsftpd
auth      required pam_userdb.so   db=/vftp/vuser
account   required pam_userdb.so   db=/vftp/vuser
```

注意,这两行必须位于该文件的前面两行,可以将文件原来的内容全部删除或进行注释。另外,不需要写数据库文件的后缀。

(4) 创建虚拟用户对应的系统用户。

```
[root@localhost~]# mkdir -p /vmyftp/vftp
[root@localhost~]# touch /vmyftp/vftp/1.txt
[root@localhost~]# useradd -d /vmyftp/vftp vuser
[root@localhost~]# chown vuser.vuser /vmyftp/vftp
[root@localhost~]# chmod 555 /vmyftp/vftp
[root@localhost~]# ls -ld /vmyftp/vftp
dr-xr-xr-x. 2 vuser vuser 6 6月  27 17:06 /vmyftp/vftp
```

(5) 编辑主配置文件 vsftpd.conf,修改以下内容。

```
[root@localhost~]# vim /etc/vsftpd/vsftpd.conf
……
anonymous_enable=NO                  //禁止匿名用户访问
local_enable=YES                     //允许本地用户访问
guest_enable=YES                     //开启虚拟用户访问
guest_username=vuser                 //设置虚拟用户对应的系统用户为 vuser
allow_writeable_chroot=YES
pam_service_name=vsftpd              //配置 vsftpd 使用的 PAM 模块为 vsftpd
……
```

注意:需要重启 vsftpd 服务,记得禁用 SELinux 服务或者配置防火墙放行 ftp 服务。

(6) 在 Windows 10 客户端上进行测试。

使用账户 ftp3、ftp4 登录服务器后,这两个账户不能切换目录,可以查看、下载文件,但是不具有修改、删除权限。如图 9-10、图 9-11、图 9-12 所示。

图 9-10 账户 ftp3 登录服务器

```
C:\Users\Administrator>ftp 192.168.1.1
连接到 192.168.1.1。
220 (vsFTPd 3.0.2)
用户(192.168.1.1:(none)): ftp4
331 Please specify the password.
密码:
230 Login successful.
ftp> ls
200 PORT command successful. Consider using PASV.
150 Here comes the directory listing.
1.txt
226 Directory send OK.
ftp: 收到 10 字节，用时 0.00秒 10000.00千字节/秒。
ftp> cd /var
550 Failed to change directory.
ftp> mkdir ceshi
550 Create directory operation failed.
ftp> rm 1.txt
550 Remove directory operation failed.
```

图 9-11　账户 ftp4 登录服务器

```
C:\Users\Administrator>ftp 192.168.1.1
连接到 192.168.1.1。
220 (vsFTPd 3.0.2)
用户(192.168.1.1:(none)): ftp3
331 Please specify the password.
密码:
230 Login successful.
ftp> ls
200 PORT command successful. Consider using PASV.
150 Here comes the directory listing.
1.txt
226 Directory send OK.
ftp: 收到 10 字节，用时 0.00秒 10000.00千字节/秒。
ftp> get 1.txt c:\user1\1.txt
200 PORT command successful. Consider using PASV.
150 Opening BINARY mode data connection for 1.txt (0 bytes).
226 Transfer complete.
ftp> lcd c:\user1
目前的本地目录 C:\user1。
ftp> !dir
 驱动器 C 中的卷没有标签。
 卷的序列号是 CE3D-CB3B

 C:\user1 的目录

2020/06/27  17:33    <DIR>          .
2020/06/27  17:33    <DIR>          ..
2020/06/27  17:33                 0 1.txt
2020/06/27  13:02                 0 2.txt
2020/06/27  13:19                 0 cs.tar
               3 个文件              0 字节
               2 个目录 20,554,461,184 可用字节
ftp> put 2.txt
200 PORT command successful. Consider using PASV.
550 Permission denied.
```

图 9-12　不具有修改、删除权限

9.2.6　配置基于独立配置文件的 FTP 服务器

根据公司的需求，对使用 FTP 服务的用户要进行权限限制，需要创建两个虚拟账户（公共账户 ftppub，私有账户 ftppri），并对两个虚拟账户设置不同的权限以及限制客户端的链接数、下载速度。

实现步骤如下：

(1) 创建用户文件，其中奇数行为账户名，偶数行为密码。

```
[root@localhost~]# mkdir /vftpweb
[root@localhost~]# vim /vftpweb/vuser.txt
ftppub
123456
ftppri
123456
```

（2）生成数据库文件，修改数据库文件权限，为了保证系统安全，将明文的用户信息文件删除。

```
[root@localhost~]# db_load -T -t hash -f /vftpweb/vuser.txt /vftpweb/vuser.db
[root@localhost~]# ls /vftpweb/
vuser.db  vuser.txt
[root@localhost~]# chmod 600 /vftpweb/vuser.db
[root@localhost~]# rm -f /vftpweb/vuser.txt
```

（3）建立用于支持虚拟用户的 PAM 文件。

```
[root@localhost~]# vim /etc/pam.d/vsftpd
auth      required pam_userdb.so   db=/vftpweb/vuser
account   required pam_userdb.so   db=/vftpweb/vuser
```

注意，这两行必须位于该文件的前面两行，可以将文件原来的内容全部删除或进行注释。另外，不需要写数据库文件的后缀。

（4）创建虚拟用户对应的系统用户。

由于用户要进行权限限制，可以将两个账户的目录隔离，从而控制用户的文件访问。公共账户 ftppub 对应系统账户 ftppubuser，访问的主目录为 /myvftp/public，公共账户只允许下载。私有账户 ftppri 对应系统账户 ftppriuser，访问的主目录为 /myvftp/private，私人账户可以上传、下载。

```
[root@localhost~]# mkdir /myvftp
[root@localhost~]# useradd -d /myvftp/public ftppubuser
[root@localhost~]# chown ftppubuser:ftppubuser /myvftp/public/
[root@localhost~]# chmod o=r /myvftp/public/
[root@localhost~]# useradd -d /myvftp/private ftppriuser
[root@localhost~]# chown ftppriuser:ftppriuser /myvftp/private/
[root@localhost~]# chmod o=rw /myvftp/private/
[root@localhost~]# touch /myvftp/public/pub.txt
[root@localhost~]# touch /myvftp/private/pri.txt
```

（5）编辑主配置文件 vsftpd.conf，修改以下内容。

```
[root@localhost~]# vim /etc/vsftpd/vsftpd.conf
……
anonymous_enable=NO            //禁止匿名用户访问
local_enable=YES               //允许本地用户访问
allow_writeable_chroot=YES
max_clients=200                //设置 FTP 服务器的最大客户连接数为 200
max_per_ip=5                   //同一 IP 地址的最大连接数为 5
pam_service_name=vsftpd        //配置 vsftpd 使用的 PAM 模块为 vsftpd
user_config_dir=/ftpconfig
                               //设账户配置目录，在该目录里创建与虚拟账户同名的配置文件
……
```

(6) 创建用户账户配置文件。
① 公共账户 ftppub

```
[root@localhost ~]# mkdir /ftpconfig
[root@localhost ~]# vim /ftpconfig/ftppub
guest_enable=YES                    //开启虚拟用户访问
guest_username=ftppubuser           //设置虚拟用户对应的系统用户为 ftppubuser
anon_world_readable_only=YES        //设置虚拟账号可读,其可以进行下载
anon_max_rate=30000                 //限定虚拟账号最大传输速率为 30 kB/s
```

② 私人账户 ftppri

```
[root@localhost ~]# vim /ftpconfig/ftppri
guest_enable=YES                    //开启虚拟用户访问
guest_username=ftppriuser           //设置虚拟用户对应的系统用户为 ftppriuser
anon_world_readable_only=NO         //设置虚拟用户可以浏览 FTP 目录和下载文件
write_enable=YES                    //该项目与 anon_upload_enable、anon_mkdir_
                                      write_enable 配合使用,才能使它们生效
anon_upload_enable=YES              //设置虚拟用户具有上传权限
anon_mkdir_write_enable=YES         //设置虚拟用户可以创建目录
allow_writeable_chroot=YES
anon_max_rate=50000                 //限定虚拟账号最大传输速率为 50 kB/s
```

注意:在配置文件中,一定要删除不必要的空格。否则,客户端登录时,会出现错误信息"500 OOPS:bad bool value in config file for:anon_world_readable_only",同时,需要重启 vsftpd 服务,记得禁用 SELinux 服务或者配置防火墙放行 ftp 服务。

(7) 在 Windows 10 客户端上进行测试。

① 使用公共账户 ftppub 登录服务器后,可以浏览下载文件,但不能创建文件夹、上传文件。如图 9-13、图 9-14 所示。

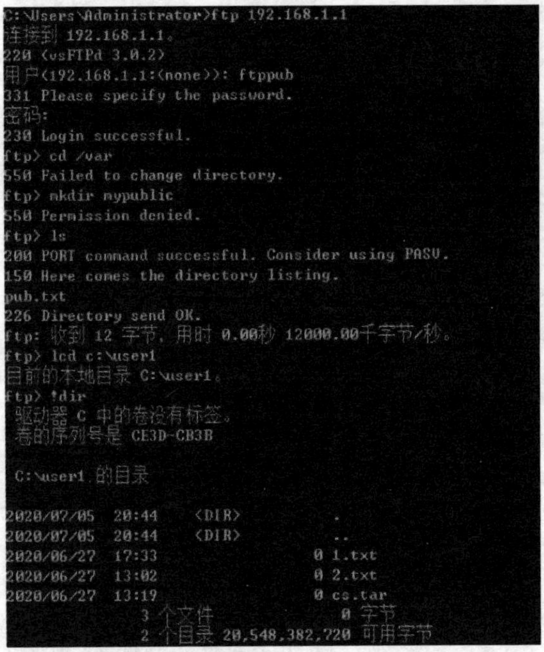

图 9-13 公共账户 ftppub 登录服务器

图 9-14 可以浏览下载文件，但不能创建文件夹、上传文件

② 使用私有账户 ftppri 登录服务器后，可以创建文件夹，上传、下载文件。如图 9-15、图 9-16 所示。

图 9-15 私有账户 ftppri 登录服务器

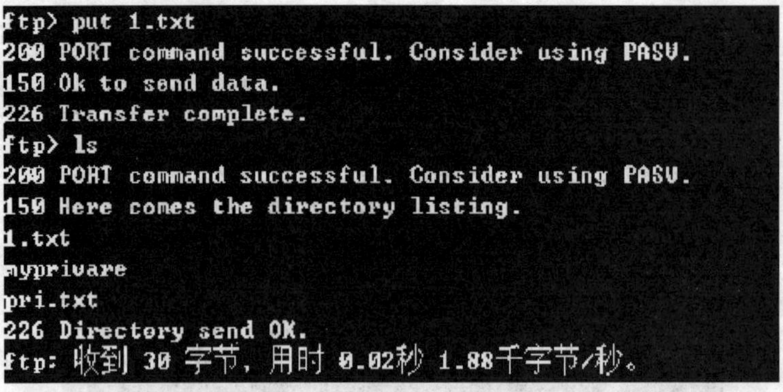

```
ftp> put 1.txt
200 PORT command successful. Consider using PASV.
150 Ok to send data.
226 Transfer complete.
ftp> ls
200 PORT command successful. Consider using PASV.
150 Here comes the directory listing.
1.txt
myprivare
pri.txt
226 Directory send OK.
ftp: 收到 30 字节，用时 0.02秒 1.88千字节/秒。
```

图 9-16　可以创建文件夹，上传、下载文件

9.2.7　配置基于 SSL 的 FTP 服务器

当前公司使用 FTP 服务，属于明文传输，可以使用加密连接（TLS/SSL）的方式，提高数据传输时的安全性。vsftpd 默认支持基于 STARTTLS 的 TLS/SSL。

vsftpd 与 TLS/SSL 常见的参数及功能如表 9-3 所示。

表 9-3　vsftpd 与 TLS/SSL 常见的参数及功能

参数	功能
ssl_enable=＜YES/NO＞	设置是否开启 SSL 连接
validate_cert=＜YES/NO＞	若设置为 YES，要求接收的所有 SSL 客户端证书必须经过验证
rsa_cert_file=＜filename＞	设置 RSA 证书的文件位置
rsa_private_key_file=＜filename＞	设置 RSA 私钥的文件位置
ssl_sslv2=＜YES/NO＞	设置是否支持 SSL V2 版本协议，默认为 NO
ssl_sslv3=＜YES/NO＞	设置是否支持 SSL V3 版本协议，默认为 NO
ssl_tlsv1=＜YES/NO＞	设置是否支持 TLS V1 版本协议，默认为 YES
allow_anon_ssl=＜YES/NO＞	设置是否为匿名用户启用 SSL 连接，默认为 NO
force_anon_logins_ssl=＜YES/NO＞	若设置为 YES，所有匿名用户登录都强制使用 SSL 连接发送用户口令
force_anon_data_ssl=＜YES/NO＞	若设置为 YES，所有匿名用户登录都强制使用 SSL 连接发送和接收数据
force_local_logins_ssl=＜YES/NO＞	若设置为 YES，所有本地用户登录都强制使用 SSL 连接发送用户口令
force_local_data_ssl=＜YES/NO＞	若设置为 YES，所有本地用户登录都强制使用 SSL 连接发送和接收数据
require_cert=＜YES/NO＞	若设置为 YES，要求所有 SSL 客户端连接都提供客户端证书
ca_certs_file=＜filename＞	设置 CA 证书文件位置

实现步骤如下:

(1) 安装 SSL 服务,并创建证书存放的目录(/etc/vsftpd/ssl)。

```
[root@localhost ~]# yum -y install mod_ssl
[root@localhost ~]# mkdir -p /etc/vsftpd/ssl
```

(2) 生成私钥文件。

```
[root@localhost ~]# cd /etc/vsftpd/ssl
[root@localhost ssl]# openssl genrsa -out server.key 1024
Generating RSA private key,1024 bit long modulus
..........+ + + + +
............+ + + + + +
e is 65537 (0x10001)
```

(3) 建立证书。

```
[root@localhost ssl]# openssl req -new -x509 -key server.key -out server.crt
You are about to be asked to enter information that will be incorporated
into your certificate request.
What you are about to enter is what is called a Distinguished Name or a DN.
There are quite a few fields but you can leave some blank
For some fields there will be a default value,
If you enter '.',the field will be left blank.
......
Country Name (2 letter code)[XX]:cn
State or Province Name (full name)[]:jj
Locality Name (eg,city)[Default City]:kc
Organization Name (eg,company)[Default Company Ltd]:yh
Organizational Unit Name (eg,section)[]:sc
Common Name (eg,your name or your server's hostname)[]:
Email Address[]:
```

(4) 编辑主配置文件 vsftpd.conf,添加以下内容。

```
[root@localhost ~]# vim /etc/vsftpd/vsftpd.conf
......
ssl_enable=YES
ssl_sslv3=YES
force_local_logins_ssl=YES
force_local_data_ssl=YES
ssl_ciphers=HIGH        //加密方法,HIGH 代表密钥长度值大于 128 位
rsa_cert_file=/etc/vsftpd/ssl/server.crt
rsa_private_key_file=/etc/vsftpd/ssl/server.key
......
```

(5) 在 Windows 10 客户端上,使用 FileZilla 进行测试,如图 9-17 ~ 图 9-20 所示。

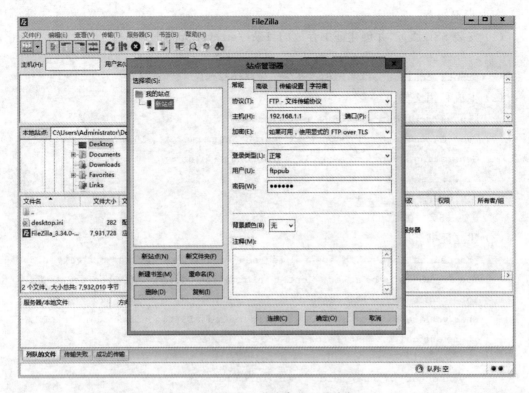

图 9-17　添加公共账户 ftppub 信息

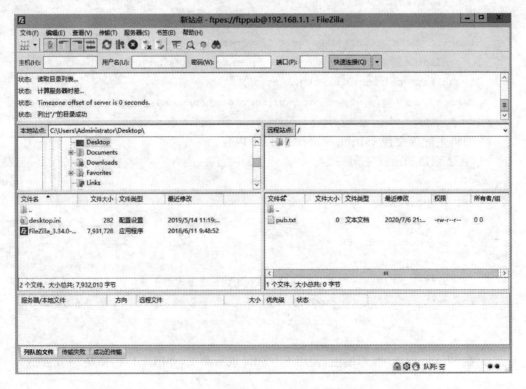

图 9-18　公共账户 ftppub 登录服务器

项目 9　Vsftpd 服务器

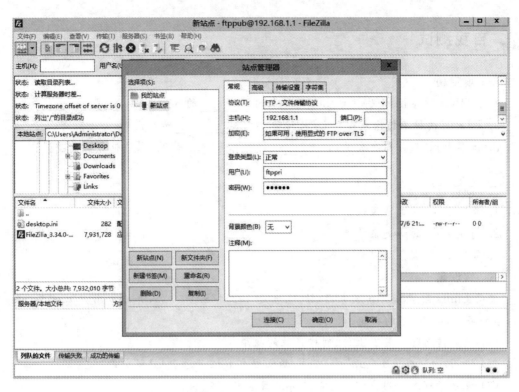

图 9-19　添加私有账户 ftppri 信息

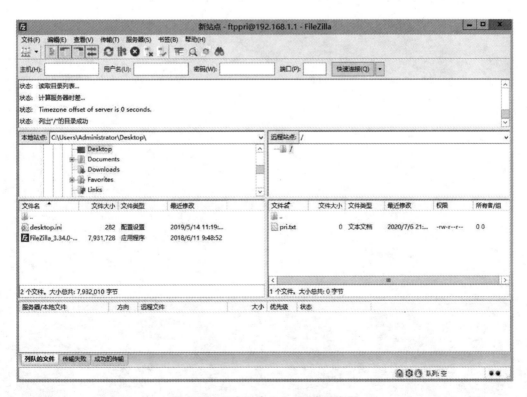

图 9-20　私有账户 ftppri 登录服务器

自我测试

一、填空题

1. FTP（文件传输协议）用于 Internet 上的_____的双向传输。
2. FTP 有两种工作模式：_____传输模式和_____传输模式。
3. 匿名 FTP 服务器有公共的_____和_____，普通 FTP 服务器需要_____和_____。

二、简答题

1. 简述 FTP 的工作原理。
2. 简述 FTP 的两类服务器。
3. 简述 FTP 服务的用户分类。

三、实训题

使用 Linux 配置 Vsftp 服务器，IP 地址是 192.168.160.25，并添加用户 tom 和 lucy。Vsftp 服务器具有以下的功能：

(1) 设置匿名账号具有上传、创建目录的权限。
(2) 使用 /etc/vsftpd/ftpusers 文件，禁止用户 tom 登录 FTP 服务器。
(3) 设置所有本地用户都锁定在 /home 目录。
(4) 配置基于主机的访问控制，拒绝 192.168.160.100 主机访问。

项目 10　邮件服务器

📚 项目综述

随着公司业务发展,需要搭建公司自己的邮件服务器,以满足日常办公的需要。公司网络管理员将搭建邮件服务器来实现这些功能。

📚 项目目标

- 了解电子邮件服务相关知识及工作原理;
- 掌握 Postfix 服务器的配置方法;
- 掌握 Sendmail 服务器的配置方法;
- 掌握 Dovexot 服务器的配置方法;
- 掌握具有认证功能的电子邮件系统的配置方法。

◀ 任务 10.1　电子邮件服务器简介 ▶

电子邮件(Electronic Mail,E-mail)服务是互联网中重要的服务。在当今信息时代,人们在互联网上经常使用该服务,通过电子邮件来提高学习、工作的效率。电子邮件地址的格式是"用户名@接收服务器域名(主机地址)"。第一部分用户名代表用户邮箱账号,对于同一个邮件接收服务器而言,这个账号必须是唯一的;第二部分"@"是分隔符;第三部分是用户邮箱的邮件接收服务器的域名,用来标识其所在的位置。如我们常使用的 QQ 邮箱地址就是"QQ 号@QQ.COM"。

10.1.1　电子邮件系统的组成

Linux 系统中的电子邮件系统包括 MUA(Mail User Agent,邮件用户代理)、MTA(Mail Transfer Agent,邮件传送代理)、MDA(Mail Dilivery Agent,用户投递代理)三个组件。

1. MUA

MUA 是电子邮件系统的客户端程序,它是用户和电子邮件系统的接口,主要负责邮件的发送和接收,以及提供用户浏览、编写邮件。除非用户直接使用类似 Telnet 的软件登录邮件服务器来主动发出邮件,否则就需要通过 MTA 来帮助用户发送邮件到邮件服务器上。目前常见的用户代理软件有基于 Windows 平台的 Outlook、Foxmail,基于 Linux 平台的 Kmail、mail、elm、pine 等。

2. MTA

MTA 是电子邮件系统的服务器端程序,主要负责邮件的接收和转发。MTA 有以下两种功能:

(1) 接收邮件

使用简单邮件传输协议(SMTP),将来自客户端或者其他 MTA 的邮件收下。

(2) 转发邮件

如果该邮件的目的地址并不是本地的用户,那么 MTA 就会将该邮件转发给其他 MTA,这就是邮件转发功能(转发就是要求服务器向其他服务器传递邮件的一种请求)。

目前常见的邮件传送代理软件有基于 Windows 平台的 Exchange,基于 Linux 平台的 Sendmail、Postfix 等。

3. MDA

MTA 把邮件投递到邮件接收者所在的邮件服务器,MDA 负责分析由 MTA 所收到的邮件表头或内容等数据,然后将邮件按照接收者的用户名投递到邮箱中。同时,MDA 还有分析与过滤邮件的功能,如过滤垃圾邮件、自动回复功能。

4. MUA、MTA、MDA 协同工作

总体来说,当使用 MUA 程序写信(例如 elm、pine 或 mail)时,应用程序把信件传给 Sendmail 或 Postfix 这样的 MTA 程序。如果信件是寄给局域网或本地主机的,那么 MTA 程序从地址上就可以确定这个信息。如果信件是发给远程系统的用户,那么 MTA 程序必须能够进行路由,与远程邮件服务器建立连接并发送邮件。MTA 程序还必须能够处理发送邮件时产生的问题,并且向发信人报告出错的信息。例如,当邮件没有填写地址或者收信人不存在时,MTA 程序要向发信人报错。MTA 程序还支持别名机制,使用户能够方便地用不同的名字与其他用户、主机、网络通信。MDA 主要是把 MTA 收到的邮件信息投递到相应的邮箱中。

10.1.2 电子邮件协议

电子邮件服务是基于邮件服务协议来完成电子邮件的传输,常见的邮箱协议有:

① 简单邮件传输协议(Simple Mail Transfer Protocol,SMTP):属于客户端/服务器模型,用来发送和中转发出的电子邮件,默认工作在 TCP 的 25 端口。

② 邮局协议版本 3(Post Office Protocol 3,POP3):属于客户端/服务器模型,用于将电子邮件存储到本地主机,默认工作在 TCP 的 110 端口。POP3 允许从服务器上将邮件存储到本地计算机上,同时删除保存在邮件服务器上的邮件。

③ Internet 消息访问协议版本 4(Internet Message Access Protocol 4,IMAP4):用于在本地主机上访问邮件,默认工作在 TCP 的 143 端口。

需要注意的是，POP3 与 IMAP4 都用于处理电子邮件的接收任务，但是，POP3 是将邮件下载到本地计算机上进行处理，而 IMAP4 是需要持续访问邮件服务器，在服务器上处理。

④ 多用途互联网邮件扩充（Multipurpose Internet Mail Extensions，MIME）：是为了帮助协调和统一为二进制数据而发明的多种编码方案。MIME 使用简单的字符串组成，最初是为了标识邮件 Email 附件的类型，在 html 文件中可以使用 content-type 属性表示，描述了文件类型的互联网标准。MIME 类型能包含视频、图像、文本、音频、应用程序等数据。在没有 MIME 之前，电子邮件系统只能处理纯文本文件。

10.1.3 电子邮件的传输过程

电子邮件与普通邮件（快递）有着类似的地方，发信者注明收件人的姓名、地址，发送服务器把邮件传送到接收方服务器，收件方服务器再把邮件发送到收件人的邮箱。如图 10-1 所示。

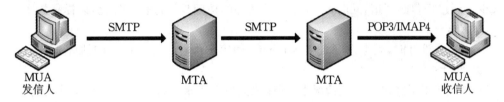

图 10-1　电子邮件发送示意图

一封邮件的传递过程如图 10-2 所示。

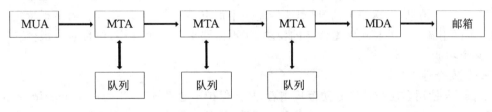

图 10-2　电子邮件传递过程

（1）邮件用户在客户端使用 MUA 撰写邮件，然后将撰写好的邮件提交到本地 MTA 上的缓存区；

（2）MTA 每隔一段时间发送一次缓存区中的邮件队列。MTA 根据邮件的接收者地址，使用 DNS 服务器的 MX 解析邮件地址的域名，从而决定将邮件投递到哪一个目标主机；

（3）目标主机上的 MTA 收到邮件后，根据邮件地址中的用户名判断用户的邮箱，并使用 MDA 将邮件投递到该用户的邮箱；

（4）该邮件的接收者可以使用常用的 MUA 软件登录邮箱查阅、处理新邮件。

任务 10.2 配置电子邮件系统

10.2.1 Postfix 简介

Postfix 是由 Wietse Zweitze Venema 博士所发展的,是 Venema 博士在 1988 年利用他在 IBM 公司的第一个休假年进行的计划,Postfix 是在 IBM 的 GPL 协议之下开发的 MTA (邮件传输代理)软件。在 RHEL7 的 Linux 系统中默认使用发件服务是由 Postfix 服务程序提供,而在之前的 RHEL5、RHEL6 等版本中,使用的是 Sendmail 服务程序。Postfix 相对于 Sendmail 减少了许多没有必要的配置,且稳定性、并发性有了大幅提升。

1. Postfix 设计目标

(1) 高性能

Postfix 比同类的服务器产品速度快 3 倍以上,Postfix 采用了 Web 服务器设计原理,从而减少了创建进程的开销,同时,对其他一些文件的访问技术进行了优化,以提高效率。

(2) 兼容性

Postfix 在设计时,就考虑到了与 Sendmail 兼容的问题,从而使移植更加容易。

(3) 健壮性

Postfix 实现了在过量负载情况下,程序依然可以正常运行。当本地文件系统没有可用空间或者没有可用内存时,Postfix 会自动放弃,而不是尝试再次运行,从而导致系统崩溃。

(4) 灵活性

Postfix 结构上由十多个小的子模块组成,每个子模块完成特定的任务,如通过 SMTP 协议接收一个消息,发送一个消息,本地传递一个消息,重写一个地址等。当出现特定的需求时,可以用新版本的模块来替代老的模块,而不需要更新整个程序,而且它也很容易实现关闭某个功能。

(5) 安全性

Postfix 使用多层防护措施防范攻击者,从而保护本地系统,几乎每一个 Postfix 守护进程都能运行在固定低权限的 chroot 之下,在网络和安全敏感的本地投递程序之间没有直接的路径(一个攻击者必须首先突破若干个其他的程序,才有可能访问本地系统)。Postfix 甚至不绝对信任自己的队列文件或 IPC 消息中的内容以防止被欺骗。Postfix 在输出发送者提供的消息之前会首先过滤消息,而且 Postfix 程序没有 Set-UID。

(6) 开放性

Postfix 遵从 IBM 的开放源代码版权许可证,用户可以自由地分发该软件,进行二次开发。其唯一的限制就是必须将对 Postfix 做的修改返回给 IBM 公司。

2. Postfix 的特点

(1) 配置容易

配置一个基本的 SMTP 服务器,只需要修改配置文件的几个参数。

(2) 支持虚拟域

在大多数通用情况下,增加对一个虚拟域的支持仅仅需要改变一个 Postfix 查找信息表。其他的邮件服务器则通常需要多个级别的别名或重定向才能实现。

（3）UCE（Unsolicited Commercial Email）控制

Postfix 能限制哪个主机允许通过自身转发邮件，并且支持限定什么邮件允许接收。Postfix 实现通常的控制功能：黑名单列表、RBL 查找、HELO/发送者 DNS 核实，实现了邮件头和邮件内容的过滤。Postfix 利用正则表达式模式映射表，其提供了高效的过滤电子邮件功能，默认使用 POSIX 格式的正则表达式。

（4）表查询

Postfix 没有实现地址重写语言，而是使用了一种扩展的表查询来实现地址重写功能。表可以是本地的纯文件、dbm 或 db 等格式的文件，或者 LDAP、NIS、SQL 关系数据库等。

3. Postfix 服务程序的配置文件

Postfix 服务程序的主要配置文件及功能如表 10-1 所示。

表 10-1　Postfix 服务程序的主要配置文件及功能

配置文件名称	存放位置	功能
配置文件	/etc/Postfix/main.cf	主配置文件
	/etc/Postfix/master.cf	主控守护进程配置文件
	/etc/pam.d/smtp.Postfix	Postfix 的 PAM 配置文件
守护进程	/usr/libexec/Postfix/master	Postfix 常驻内存的主控守护进程
	/usr/libexec/Postfix/*	被 Postfix 主守护进程控制的其他组件进程
systemd 的服务配置单元	/usr/lib/systemd/system/Postfix.service	Postfix 服务单元配置文件
Postfix 的管理工具	/usr/sbin/Postfix	Postfix 的控制程序
	/usr/sbin/postconf	显示、编辑 /etc/Postfix/master.cf 的配置工具
	/usr/sbin/postalias	建立、修改、查询别名
	/usr/sbin/postmap	建立、修改、查询映射表
	/usr/sbin/postqueue	邮件队列管理工具
	/usr/sbin/postcat	打印队列文件的内容
	/usr/sbin/postsuper	系统管理员的邮件队列管理工具
	/usr/sbin/postlog	向邮件日志直接写入信息的工具
与 Sendmail 兼容的工具	/usr/sbin/Sendmail	与 Sendmail 兼容的邮件发送替代工具
	/usr/bin/newaliases	与 Sendmail 兼容的别名数据库生成替代工具
	/usr/bin/mailq	与 Sendmail 兼容的邮件队列查询替代工具
邮件队列目录	/var/spool/Postfix/	Postfix 的邮件队列目录
用户邮箱目录	/var/spool/mail/	用户邮箱目录（mdir 格式）
文档	/usr/share/doc/Postfix-2.10.1/	Postfix 的文档目录

4. 常用的 Postfix 配置工具

可以使用 postconf 工具对 Postfix 的配置文件进行显示、更新。如表 10-2 所示。

表 10-2　postconf 命令常见参数及其功能

命令	功能	命令	功能
postconf -df	显示所有参数的默认值	postconf -dfx	显示所有参数的默认值，并替换变量的值
postconf -df<参数>	显示指定参数的默认值	postconf -e<参数=值>	编辑 main.cf 中的指定参数
postconf -#<参数>	将指定的参数设置为 Postfix 编译时使用的默认值	postconf -m	显示支持的映射表类型
postconf -t	显示内置的投递状态通知消息	postconf -b	显示内置的投递状态通知消息，并替换模板中变量的值
postconf -nf	显示 main.cf 中明确指定的非默认值	postconf -nfx	显示 main.cf 中明确指定的非默认值，并替换变量的值
postconf -f<参数>	显示指定参数的当前值	postconf -Mf	显示 main.cf 中所有服务的设置
postconf -Mf<参数>	显示 main.cf 中指定类型的服务设置	postconf -l	显示支持的邮箱锁机制类型
postconf -a	显示支持的 SASL 服务插件类型	postconf -A	显示支持的 SAS 客户插件类型

10.2.2　配置常规的 Postfix 服务器

1. 查询系统是否已经安装 Postfix 软件包

```
[root@localhost ~]# rpm -qa|grep Postfix
```

2. 安装 Postfix 软件包

```
[root@localhost ~]# yum -y install Postfix
[root@localhost ~]# rpm -qa |grep Postfix
Postfix-2.10.1-6.el7.x86_64
```

3. 配置防火墙，放行 smtp 服务，并启动 Postfix 服务

```
[root@localhost ~]# firewall-cmd --permanent --add-service=smtp
success
[root@localhost ~]# firewall-cmd --reload
success
[root@localhost ~]# firewall-cmd --list-all
public (active)
```

```
        target: default
        icmp-block-inversion: no
        interfaces: ens33
        sources:
        services: ssh dhcpv6-client dns http ftp smtp
    ……
    [root@localhost~]# systemctl start Postfix
```

4. Postfix 服务程序主配置文件的重要参数

虽然 Postfix 服务程序主配置文件(/etc/Postfix/main.cf)有 679 行左右,但是在配置过程中,仅需要修改几个参数,表 10-3 列出了其主要的配置参数。

表 10-3 Postfix 服务程序主配置文件主要参数及其功能

参数	功能
myhostname	邮局系统的主机名(FQDN 名)
mydomain	邮局系统的域名
myorigin	从本机发出邮件的域名名称
inet_interfaces	监听的网卡接口
inet_protocols	设置监听的 IP 协议类型,可以是 IPv4 或者是 IPv6。all 代表两种协议
mydestination	可接收邮件的主机名或域名,只有当发出来的邮件的收件人地址与该参数值匹配时,Postfix 才会将该邮件接收下来
mynetworks	设置可转发哪些主机的邮件
relay_domains	设置可转发哪些网域的邮件
home_mailbox	指定邮箱相对用户根目录的路径,以及采用的信箱格式(以斜线结尾为 Maildir 格式,反之为 mbox 格式)

在 Postfix 服务程序主配置文件中,需要修改 5 个地方的参数。

(1)第 76 行,修改 myhostname 为邮局系统的主机名,这里我们使用项目 7 所配置的主机名(mail.kcorp.com),如图 10-3、图 10-4 所示。

```
$TTL 1D
@       IN SOA      @ root.kcorp.com. (
                                0       ; serial
                                1D      ; refresh
                                1H      ; retry
                                1W      ; expire
                                3H )    ; minimum
@       IN  NS      dns.kcorp.com.
@       IN  MX  10  mail.kcorp.com.
dns     IN  A       192.168.1.1
www     IN  A       192.168.1.1
ftp     IN  A       192.168.1.1
mail    IN  A       192.168.1.1
```

图 10-3 正向解析文件

```
$TTL 1D
@       IN SOA          @ root.kcorp.com. (
                        0       ; serial
                        1D      ; refresh
                        1H      ; retry
                        1W      ; expire
                        3H )    ; minimum
@       IN      NS      dns.kcorp.com.
@       IN      MX      10      mail.kcorp.com.
1       IN      PTR     dns.kcorp.com.
1       IN      PTR     www.kcorp.com.
1       IN      PTR     ftp.kcorp.com.
1       IN      PTR     mail.kcorp.com.
```

图 10-4 反向解析文件

```
myhostname=mail.kcorp.com
```

（2）第 83 行，修改 mydomain 为邮局系统的域名，这里我们使用项目 7 配置的域名（kcorp.com）。

```
mydomain=kcorp.com
```

（3）第 99 行，修改 myorigin 的值为直接调用 mydomai 的值。

```
myorigin=$ mydomain
```

（4）第 116 行，修改 inet_interfaces 的值为 all。

```
inet_interfaces=all
```

（5）第 164 行，修改 mydestination 的值，直接调用前面定义的 myhostname、mydomain 为变量。

```
mydestination=$ myhostname,$ mydomain
```

5．配置别名和群发

可以通过配置 aliases 文件来设定邮件的别名，在这个文件中，左边是别名，右边是实际存在的用户账号。如为账号 jack 设置别名为 jerry；设置组寄信功能，让 stugroup 包含账号 stu1、stu2、stu3。

```
[root@localhost~]# vim /etc/aliases
# Basic system aliases -- these MUST be present.
mailer-daemon:  postmaster
postmaster:     root
# General redirections for pseudo accounts.
bin: root
daemon:root
adm:root
lp:             root
jack:           jerry
stugroup:       stu1,stu2,stu3
……
```

设置完成后，需要使用 newaliases 命令生成 aliases.db 数据库。

```
[root@localhost~]# newaliases
[root@localhost~]# ls -al /etc/
-rw-r--r--.   1 root root    12288 8月   4 11:10 aliases.db
```

6. 邮件过滤

可以使用 access 文件来进行邮件的过滤，其用法为：

规则的范围或规则(IP/部分 IP/主机名等)　　　　处理方式

如允许 192.168.1.0/24 网段、kcorp.com 发送邮件，拒绝 192.168.50.0/24、kc.kcorp.com 发送邮件。

```
[root@localhost ~]# vim /etc/Postfix/access
192.168.1        OK
.kcorp.com       OK
192.168.50       REJECT
kc.kcorp.com REJECT
[root@localhost ~]# vim /etc/Postfix/main.cf
……
smtpd_client_restrictions=check_client_access hash:/etc/Postfix/access
                              //需添加本行信息，access 规则才生效
……
```

使用 postmap 生成新的 access.db 数据库。

```
[root@localhost ~]# postmap hash:/etc/Postfix/access
[root@localhost ~]# ls -l /etc/Postfix/access*
-rw-r--r--. 1 root root 20957 8月   4 11:29 /etc/Postfix/access
-rw-r--r--. 1 root root 12288 8月   4 11:47 /etc/Postfix/access.db
```

配置完成后，不需要重启 Postfix 服务，配置文件已经生效。

7. 限制邮箱容量

（1）设置用户邮箱的大小限制

如限制用户的邮箱最大为 4 M。

```
[root@localhost ~]# vim /etc/Postfix/main.cf
……
message_size_limit=4000000
……
```

（2）使用磁盘配额限制用户邮箱的大小

① 将 1/var 挂载到/dev/sdb1 上，如图 10-5 所示。

```
[root@localhost ~]# df -hT
文件系统              类型       容量   已用   可用  已用% 挂载点
/dev/mapper/rhel-root xfs        17G   3.3G   14G   20%  /
devtmpfs              devtmpfs  897M     0  897M    0%  /dev
tmpfs                 tmpfs     912M     0  912M    0%  /dev/shm
tmpfs                 tmpfs     912M  9.0M  903M    1%  /run
tmpfs                 tmpfs     912M     0  912M    0%  /sys/fs/cgroup
/dev/sda1             xfs      1014M  179M  836M   18%  /boot
tmpfs                 tmpfs     183M   20K  183M    1%  /run/user/0
/dev/sr0              iso9660   3.8G  3.8G     0  100%  /run/media/root/RHEL-7.4 Server.x86_64
/dev/sdb1             ext4      9.8G   37M  9.2G    1%  /var
```

图 10-5　查看系统挂载信息

② 修改 fstab 文件，实现开机自动挂载。

```
[root@localhost ~]# sudo blkid
/dev/sda1: UUID="93abc0c5-f930-430c-ab0a-e2db81e2b643" TYPE="xfs"
/dev/sda2: UUID="MClI8p-w32z-KOGo-Wu8I-hyCR-Cckb-3hzpr8" TYPE="LVM2_member"
/dev/sdb1: UUID="4bc69e28-e86d-42a6-b6e4-8f9e0c54f0f7" TYPE="ext4"
/dev/mapper/rhel-root: UUID="4296eb59-b0e6-4e95-af35-c7d798818ac7" TYPE="xfs"
/dev/mapper/rhel-swap: UUID="a5b5bf7b-f096-4d7c-b133-28dc99d4040e" TYPE="swap"
[root@localhost ~]# vim /etc/fstab
# /etc/fstab
# Created by anaconda on Tue Mar 3 00:32:39 2020
# Accessible filesystems,by reference,are maintained under '/dev/disk'
# See man pages fstab(5),findfs(8),mount(8) and/or blkid(8) for more info
#
/dev/mapper/rhel-root      /                   xfs     defaults 0 0
UUID=93abc0c5-f930-430c-ab0a-e2db81e2b643/boot xfs     defaults       0 0
/dev/mapper/rhel-swap      swap                swap    defaults       0 0
UUID=4bc69e28-e86d-42a6-b6e4-8f9e0c54f0f7/var ext4 defaults,usrquota,grpquota
                 0 0
```

③ 设置磁盘配额

```
[root@localhost ~]# quotacheck -cmvug /var
[root@localhost ~]# ls -al /var/
……
-rw-------. 1 root root  7168 8月   4 21:32 aquota.group
-rw-------. 1 root root  7168 8月   4 21:32 aquota.user
……
[root@localhost ~]# quotaon -auvg
/dev/sdb1 [/var]: group quotas turned on
/dev/sdb1 [/var]: user quotas turned on
[root@localhost ~]# quotaon -p /var
group quota on /var (/dev/sdb1) is on
user quota on /var (/dev/sdb1) is on
[root@localhost ~]# edquota -u jack    //设置用户 jack 的磁盘配额硬限制为 100M
Disk quotas for user jack (uid 1000):
  Filesystem    blocks       soft       hard     inodes     soft      hard
  /dev/sdb1      0            0        100000      0         0         0
```

10.2.3 配置 Dovecot 服务程序

Dovecot 是一个邮件访问代理(Mail Access Agent,MAA),用于将用户连接到系统邮件库,为 MUA 提供用户认证,为 MUA 使用 POP 或 IMAP 协议,从用户邮箱读取邮件做准备。Linux 系统要使用 POP3/POP3S、IMAP/ IMAPS 协议接收邮件,就需要安装配置 Dovecot 服务程序。

1. Dovecot 的特点

（1）采用模块化设计；
（2）完全兼容 UW IMAP 和 Courier IMAP；
（3）包含内置的 LDA 和 LMTP 服务，并提供可选的 Sieve 过滤支持；
（4）支持标准的 mbox、Maildir 及其开发的高性能 dbox 邮箱格式；
（5）支持多账户存储方式，如 PAM、口令文件、SQL 等；
（6）支持 SASL 和 TLS。

2. Dovecot 服务程序的配置文件

Dovecot 服务程序的主要配置文件及其功能如表 10-4 所示。

表 10-4　Dovecot 服务程序的主要配置文件及其功能

配置文件名称	存放位置	功能
配置文件	/etc/dovecot/dovecot.conf	主配置文件
	/etc/dovecot/conf.d/*-*.conf	不同进程组件的配置文件
	/etc/dovecot/conf.d/auth-*.conf.ext	不同验证模块的配置文件
	/etc/pam.d/dovecot	Dovecot 的 PAM 配置文件
守护进程	/usr/sbin/dovecot	主守护进程
	/usr/libexec/dovecot/{pop3,imap}-login	登录进程
	/usr/libexec/dovecot/auth	验证进程
	/usr/libexec/dovecot/{pop3,imap}	登录后，处理邮件的进程
systemd 的服务配置单元	/usr/lib/systemd/system/dovecot.service	Dovecot 服务单元配置文件
管理工具	/usr/bin/doveadm	Dovecot 的控制程序
	/usr/bin/doveconf	显示 Dovecot 的配置选项
	/usr/bin/dsync	Dovecot 的邮箱同步工具
文档	/usr/share/doc/dovecot-2.2.10/	Dovecot 的文档目录

3. Dovecot 常用的配置工具

可以使用 doveconf 工具对 Dovecot 的配置文件进行显示、更新，如表 10-5 所示。

表 10-5　doveconf 命令常见参数及其功能

命令	功能	命令	功能
doveconf -d	显示所有参数的默认值	doveconf -d<参数>	显示指定参数的默认值
doveconf -a	显示所有参数的当前值	doveconf <参数>	显示指定参数的当前值
doveconf -n	显示所有修改了默认值的参数	doveconf -N	显示所有修改了默认值的参数以及明确设置时默认值的参数

4. 查询系统是否已经安装 Dovecot 软件包

```
[root@localhost ~]# rpm -qa |grep dovecot
```

5. 安装 Dovecot 软件包

```
[root@localhost ~]# yum -y install dovecot
[root@localhost ~]# rpm -qa |grep dovecot
dovecot-2.2.10-8.el7.x86_64
```

6. 配置防火墙

放行 POP3 服务对应的 TCP 端口 110，IMAP 服务对应的 TCP 端口 143，并启动 Dovecot 服务。

```
[root@localhost ~]# firewall-cmd --permanent --add-port=110/tcp
[root@localhost ~]# firewall-cmd --permanent --add-port=143/tcp
[root@localhost ~]# firewall-cmd --reload
success
[root@localhost ~]# systemctl start dovecot
```

7. 使用 netstat 命令检测是否开启了 POP3、IMAP 服务对应的端口

```
[root@localhost ~]# netstat -an|grep :110
tcp        0      0 0.0.0.0:110             0.0.0.0:*               LISTEN
tcp6       0      0 :::110                  :::*                    LISTEN
[root@localhost ~]# netstat -an|grep :143
tcp        0      0 0.0.0.0:143             0.0.0.0:*               LISTEN
tcp6       0      0 :::143                  :::*                    LISTEN
```

8. 配置 Dovecot 服务程序

（1）第 24 行，修改 protocols 的值为 imap、pop3、lmtp。

```
[root@localhost ~]# vim /etc/dovecot/dovecot.conf
protocols = imap pop3 lmtp
```

（2）第 48 行，修改允许登录的网络地址为任意地址。

```
login_trusted_networks=0.0.0.0/0
```

（3）因 Dovecot 服务程序为了保证电子邮件系统的安全，默认强制用户使用加密的方式登录，而当前并没有将系统进行加密，所以，需要添加下列参数来允许用户进行明文的登录。

```
disable_plaintext_auth=no
```

9. 指定邮件格式与存储路径

在 Dovecot 服务程序中，配置邮件格式和存储路径的配置文件是 10-mail.conf（文件路径是 /etc/dovecot/conf.d/10-mail.conf）。仅需要修改该文件的第 24 行，将注释符号（#）取消即可。

```
[root@localhost ~]# vim /etc/dovecot/conf.d/10-mail.conf
……
mail_location=mbox:~/mail:INBOX=/var/mail/%u
……
```

10. 创建用户及其邮件目录

建立用户 mail1、mail2，并指定用户保存邮件的目录，重启 dovecot 服务。

　　[root@localhost~]# useradd mail1
　　[root@localhost~]# echo "12345678"|passwd--stdin mail1
　　更改用户 mail1 的密码。
　　passwd:所有的身份验证令牌已经成功更新。
　　[root@localhost~]# useradd mail2
　　[root@localhost~]# echo "12345678"|passwd--stdin mail2
　　更改用户 mail2 的密码。
　　passwd:所有的身份验证令牌已经成功更新。
　　[root@localhost~]# mkdir-p /home/mail1/mail/.imap/INBOX
　　[root@localhost~]# mkdir-p /home/mail2/mail/.imap/INBOX
　　[root@localhost~]# systemctl restart dovecot

10.2.4　电子邮件系统的测试

1. 在 Windows 2012 系统中安装 office2007，利用 Outlook 软件进行邮件的测试。
实现步骤如下：

（1）配置 Windows 2012 系统的 TCP/IP 相关参数，并测试是否能正常解析邮件域名。如图 10-6、图 10-7 所示。

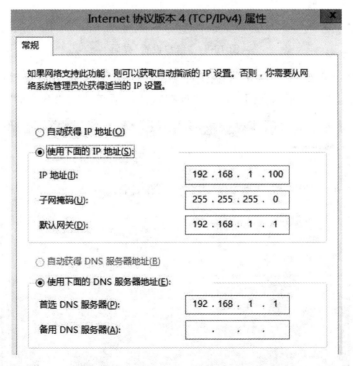

图 10-6　"TCP/IPv4"属性界面

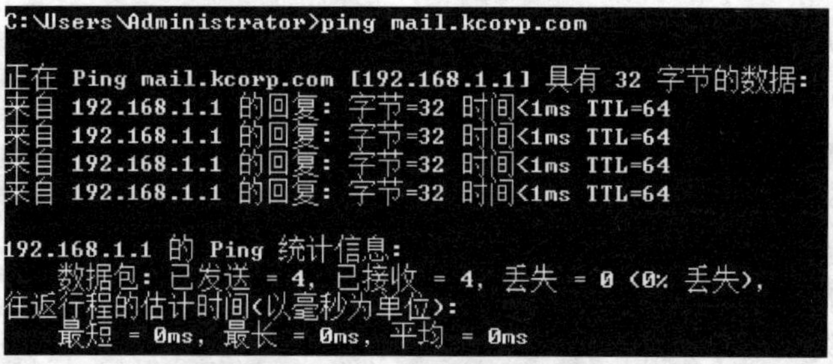

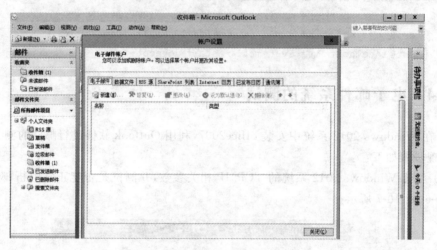

图 10-7　测试解析邮件域名

(2) 启动 Outlook 软件,依次单击"工具"→"账户设置",在图 10-8 中单击"新建"。

图 10-8　"账户设置"界面

(3) 在图 10-9 中单击"下一步",在图 10-10 中,勾选"手动配置服务器设置或其他服务器类型"选项,单击"下一步",在图 10-11 中保持默认值不变,单击"下一步"。

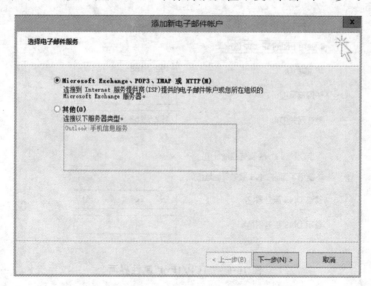

图 10-9　单击"下一步"

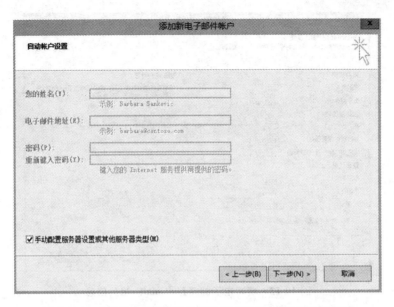

图 10-10　勾选"手动配置服务器设置或其他服务器类型"选项

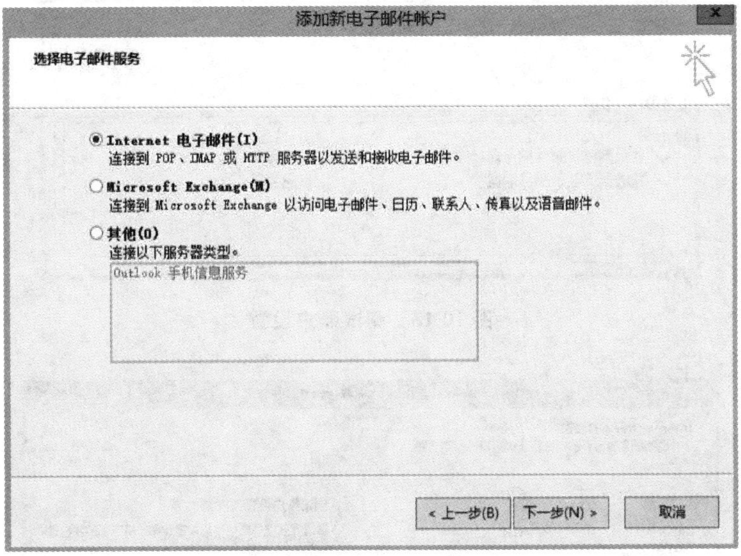

图 10-11　保持默认值不变

（4）在图 10-12 中填写电子邮箱账户 mail1 相关信息。填写完成后,单击"测试账户设置",进行电子邮件服务登录验证,如图 10-13 所示。

需要注意的是,"接收邮件服务器"、"发送邮件服务器"除填写服务器的 IP 地址外,也可以添加 DNS 主机记录,使用主机地址进行填写"pop3.kcorp.com、smtp.kcorp.com"。

（5）重复邮件 mail1 的设置步骤,添加电子邮件账户 mail2,并进行电子邮件服务登录验证,如图 10-14、图 10-15 所示。

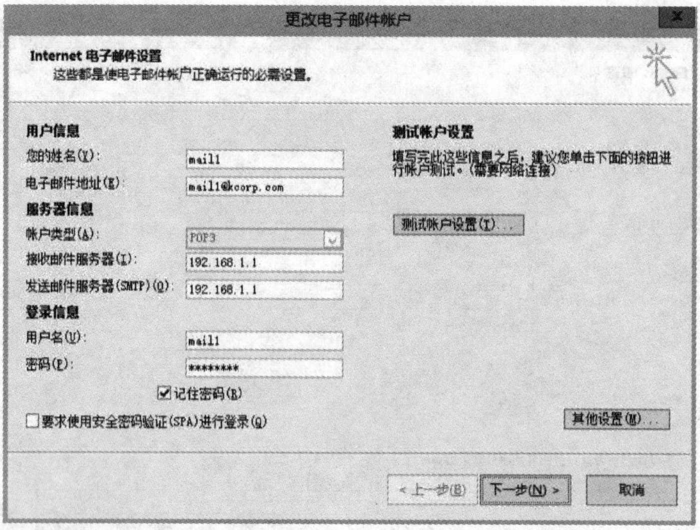

图 10-12　填写电子邮箱账户 mail1 相关信息

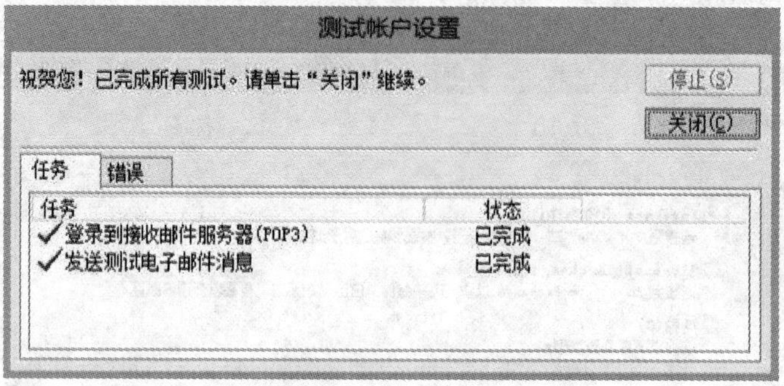

图 10-13　测试账户设置

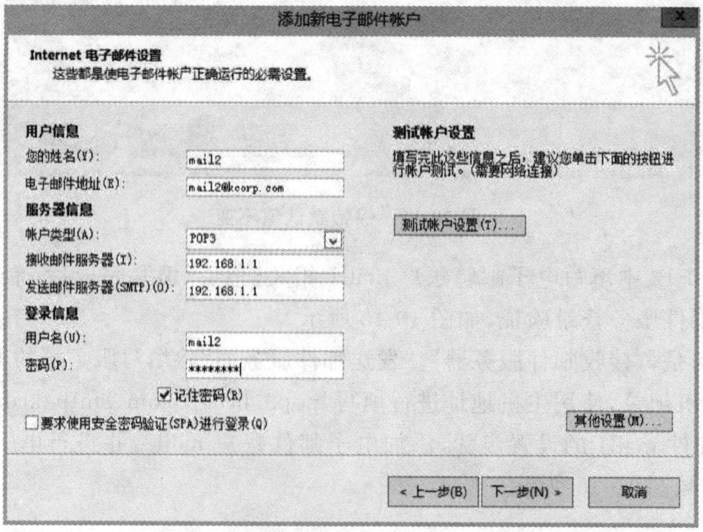

图 10-14　填写电子邮箱账户 mail2 相关信息

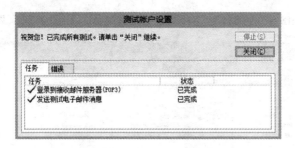

图 10-15　测试账户设置

（6）完成电子邮件账户 mail、mail2 的设置后，这时使用的账户为 mail1，可以选择账户 mail2，设置为默认账户，单击"设为默认值"，如图 10-16 所示。此操作主要是进行账户直接的切换，从而验证邮件的收发。

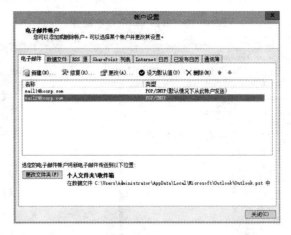

图 10-16　设置默认电子邮件账户

（7）测试邮件的发送与接收

① 账户 mail1 发送邮件给账户 mail2，如图 10-17 所示。

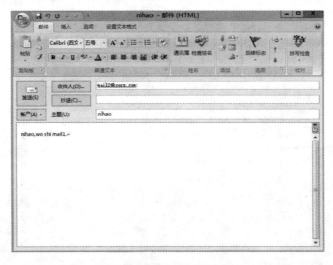

图 10-17　账户 mail1 发送邮件给账户 mail2

② 换到账户 mail2，接收邮件，并查看接收到的邮件。如图 10-18、图 10-19 所示。

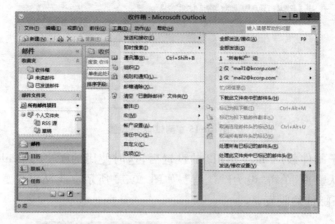

图 10-18　账户 mail2 接收邮件

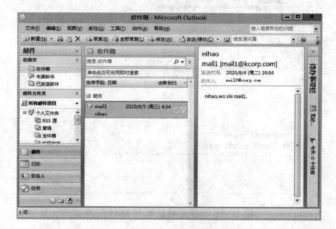

图 10-19　查看接收到的邮件

③ 账户 mail2 回信给 mail1，切换到 mail1，查看邮件内容，如图 10-20、图 10-21 所示。

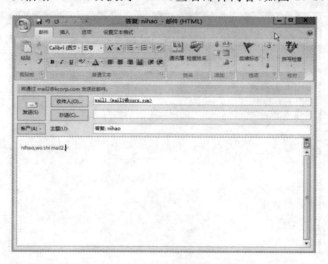

图 10-20　账户 mail2 回信给 mail1

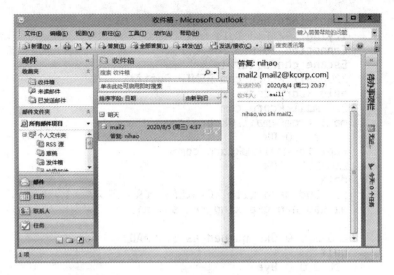

图 10-21　mail1 查看邮件内容

2. 在 RHEL7 中安装 Telnet 服务,进行邮件的测试。

实现步骤如下:

(1) 安装 Telnet 服务。

```
[root@localhost~]# yum-y install telnet
[root@localhost~]# yum-y install telnet-server
```

(2) 账户 mail1 发送邮件给账户 mail2,如图 10-22 所示。

```
[root@localhost~]# telnet 192.168.1.1 25        //连接服务器
Trying 192.168.1.1…
Connected to 192.168.1.1.
Escape character is '^]'.
220 mail.kcorp.com ESMTP Postfix
helo kcorp.com                                  //告诉服务器你是谁
250 mail.kcorp.com
mail from:<mail1@ kcorp.com>                    //谁发送邮件
250 2.1.0 Ok
rcpt to:<mail2@ kcorp.com>                      //谁接收邮件
250 2.1.5 Ok
data
354 End data with <CR> <LF> .<CR> <LF>
ni hao,hen gao xing ren shi ni.                 //邮件内容
.                                               //.代表邮件内容结束
250 2.0.0 Ok: queued as 5C42AAD23
quit                                            //退出
221 2.0.0 Bye
Connection closed by foreign host.
```

(3) 账户 mail2 查看账户 mail1 发来的邮件,如图 10-23 所示。

```
[root@localhost ~]# telnet 192.168.1.1 25
Trying 192.168.1.1...
Connected to 192.168.1.1.
Escape character is '^]'.
220 mail.kcorp.com ESMTP Postfix
helo kcorp.com
250 mail.kcorp.com
mail from: <mail1@kcorp.com>
250 2.1.0 Ok
rcpt to: <mail2@kcorp.com>
250 2.1.5 Ok
data
354 End data with <CR><LF>.<CR><LF>
ni hao,hen gao xing ren shi ni.
.
250 2.0.0 Ok: queued as 5C42AAD23
quit
221 2.0.0 Bye
Connection closed by foreign host.
```

图 10-22 账户 mail1 发送邮件给账户 mail2

```
[root@localhost ~]# telnet 192.168.1.1 110        //连接服务器
Trying 192.168.1.1...
Connected to 192.168.1.1.
Escape character is '^]'.
+ OK[XCLIENT]Dovecot ready.
user mail2                                         //邮件账户名
+ OK
pass 12345678                                      //邮件账户密码
+ OK Logged in.
list                                               //查看该用户的邮件列表,有 1 封邮件
+ OK 1 messages:
1 427
.
retr 1                                             //查看编号为 1 的邮件信息
+ OK 427 octets
Return-Path: <mail1@kcorp.com>
X-Original-To: mail2@kcorp.com
Delivered-To: mail2@kcorp.com
Received: from kcorp.com (slave.kcorp.com[192.168.1.1])
by mail.kcorp.com (Postfix) with SMTP id 5C42AAD23
for <mail2@kcorp.com> ;Wed,5 Aug 2020 04:50:23 + 0800 (CST)
Message-Id: <20200804205031.5C42AAD23@mail.kcorp.com>
Date: Wed,5 Aug 2020 04:50:23 + 0800 (CST)
From: mail1@kcorp.com
ni hao,hen gao xing ren shi ni.
```

```
[root@localhost ~]# telnet 192.168.1.1 110
Trying 192.168.1.1...
Connected to 192.168.1.1.
Escape character is '^]'.
+OK [XCLIENT] Dovecot ready.
user mail2
+OK
pass 12345678
+OK Logged in.
list
+OK 1 messages:
1 427
.
retr 1
+OK 427 octets
Return-Path: <mail1@kcorp.com>
X-Original-To: mail2@kcorp.com
Delivered-To: mail2@kcorp.com
Received: from kcorp.com (slave.kcorp.com [192.168.1.1])
        by mail.kcorp.com (Postfix) with SMTP id 5C42AAD23
        for <mail2@kcorp.com>; Wed,  5 Aug 2020 04:50:23 +0800 (CST)
Message-Id: <20200804205031.5C42AAD23@mail.kcorp.com>
Date: Wed,  5 Aug 2020 04:50:23 +0800 (CST)
From: mail1@kcorp.com

ni hao,hen gao xing ren shi ni.
.
```

图 10-23　账户 mail2 查看账户 mail1 发来的邮件

10.2.5　配置 Sendmail 服务程序

要使用 Sendmail 服务程序，需要关闭 Postfix 服务，并安装 Sendmail 服务。
实现步骤如下：

(1) 安装 Sendmail 服务，关闭 Postfix 服务。

```
[root@localhost ~]# yum -y install Sendmail
[root@localhost ~]# yum -y install Sendmail-cf
[root@localhost ~]# systemctl stop Postfix
```

(2) 配置 Sendmail 服务主配置文件，修改第 118 行的内容。

```
[root@localhost ~]# vim /etc/mail/Sendmail.mc
……
DAEMON_OPTIONS(`Port=smtp,Addr=0.0.0.0,Name=MTA')dnl
……
```

(3) 安装 m4 软件包，并使用 m4 命令生成 Sendmail.cf 文件。

```
[root@localhost ~]# yum -y install m4
[root@localhost mail]# m4 Sendmail.mc> Sendmail.cf
```

(4) 修改 local-host-names 文件，添加域名到该文件。

```
[root@localhost ~]# vim /etc/mail/local-host-names
……
kcorp.com
```

(5) 编辑 access 文件，开启转发权限，加入允许通过本机转发邮件的域名信息，并启用 makemap hash 命令生成数据库文件 access.db，然后启动 Sendmail 服务。

```
[root@localhost ~]# vim /etc/mail/access
Connect:localhost.localdomain RELAY
Connect:localhost    RELAY
Connect:127.0.0.1                        RELAY
192.168.1.0                              RELAY
kcorp.com                                RELAY
[root@localhost ~]# cd /etc/mail/
[root@localhost mail]# makemap hash access.db <  access
[root@localhost mail]# systemctl start Sendmail
```

（6）使用 Outlook 软件进行测试。账户 mail1 写信给账户 mail2，mai2 能正常接收、查阅邮件。如图 10-24、图 10-25 所示。

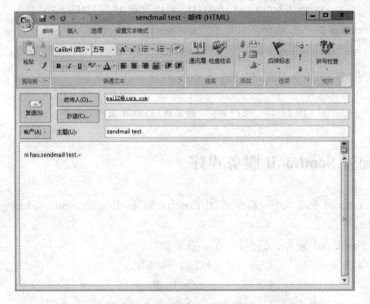

图 10-24　账户 mail1 写信给账户 mail2

图 10-25　账户 mail2 查看账户 mail1 发来的邮件

任务 10.3 部署认证功能的电子邮件系统

10.3.1 使用 Cyrus-SASL 开启 SMTP 的 SASL 认证

为了避免邮件服务器成为垃圾邮件、各类广告的中转站、集散地,可以对转发邮件的客户端进行认证。可以开启 SMTP 的 SASL 认证,使通过身份验证的用户转发邮件。SMTP 认证的机制是通过 Cyrus-SASL 软件包来实现的。

SASL(Simple Authentication and Security Layer)的机制是对协议执行验证。如果有某种服务(SMTP 或 IMAP)使用了 SASL,使用这种协议的应用程序之间将会共享代码。

实现步骤如下:

(1) 安装 cyrus-sasl 软件包。

```
[root@localhost ~] # yum -y install cyrus-sasl
```

(2) 配置、测试 saslauthd 服务。

```
[root@localhost ~] # saslauthd -v        //查看 saslauthd 支持的密码管理机制
saslauthd 2.1.26
authentication mechanisms: getpwent kerberos5 pam rimap shadow ldap httpform
//机制类型
[root@localhost ~] # vim /etc/sysconfig/saslauthd
……
MECH=shadow    //将默认的密码管理机制 pam 修改为 shadow,本地用户认证
……
[root@localhost ~] # setsebool -P allow_saslauthd_read_shadow on
//让 SELinux 允许 saslauthd 程序读取 /etc/shadow 文件
[root@localhost ~] # systemctl start saslauthd      //启动 saslauthd 服务
[root@localhost ~] # testsaslauthd -u mail1 -p '12345678'
//测试 saslauthd 认证功能
0: OK "Success."
```

(3) 配置 smtp.conf 文件,开启 SMTP 的 SASL 认证。

```
[root@localhost ~] # vim /etc/sasl2/smtpd.conf
pwcheck_method: saslauthd
mech_list: plain login        //支持的机制
log_level:3        //登录文件信息等级,设置为 3
mech_list: /run/saslauthd/mux    //设置 smtp 查找 cyrus-sasl 的路径
```

(4) 配置 main.cf 文件,开启 Postfix 支持 SMTP 认证。

```
[root@localhost ~] # vim /etc/Postfix/main.cf
smtpd_sasl_auth_enable=yes    //启用 SASL 认证
smtpd_sasl_security_options=noanonymous    //禁止使用匿名方式登录
broken_sasl_auth_clients=yes    //兼容早期的非标准 SMTP 认证协议
```

```
        smtpd_recipient_restrictions=permit_sasl_authenticated,reject_unauth_
destination  //认证网络允许，没有认证的拒绝
    [root@localhost~]# systemctl restart Postfix
```

（5）测试。

① 使用 printf 命令计算用户名和密码对应编码。

```
    [root@localhost~]# printf "mail1" | openssl base64
bWFpbDE=
    [root@localhost~]# printf "12345678" | openssl base64
MTIzNDU2Nzg=
```

② mail1 使用 telnet 发送邮件。

```
    [root@localhost~]# printf "mail1" | openssl base64
bWFpbDE=           //mail1 的 Base64 编码
    [root@localhost~]# printf "12345678" | openssl base64
MTIzNDU2Nzg=       //密码 12345678 的 Base64 编码
    [root@localhost~]# telnet 192.168.1.1 25
Trying 192.168.1.1…
Connected to 192.168.1.1.
Escape character is '^]'.
220 mail.kcorp.com ESMTP Postfix
ehlo kcorp.com
250-mail.kcorp.com
250-PIPELINING
250-SIZE 10240000
250-VRFY
250-ETRN
250-AUTH PLAIN LOGIN
250-AUTH=PLAIN LOGIN
250-ENHANCEDSTATUSCODES
250-8BITMIME
250 DSN
auth login        //使用 SMTP 认证登录
334 VXNlcm5hbWU6
bWFpbDE=          //输入 mail1 的 Base64 编码
334 UGFzc3dvcmQ6
MTIzNDU2Nzg=      //输入密码 12345678 的 Base64 编码
235 2.7.0 Authentication successful
mail from:<mail1@ kcorp.com>
250 2.1.0 Ok
rcpt to:<mail2@ kcorp.com>
250 2.1.5 Ok
data
354 End data with <CR><LF>.<CR><LF>
```

```
ni hao,SASL.
.
250 2.0.0 Ok: queued asB5ECB226B94
quit
221 2.0.0 Bye
Connection closed by foreign host.
```

③ mail2 使用 Outlook 接收邮件,需要勾选"我的发送服务器(SMTP)要求验证"选项。如图 10-26、图 10-27 所示。

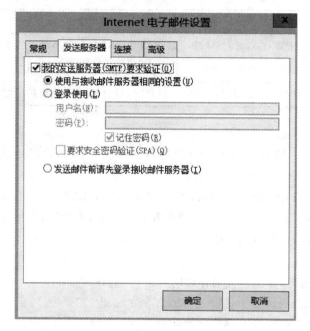

图 10-26　勾选"我的发送服务器(SMTP)要求验证"选项

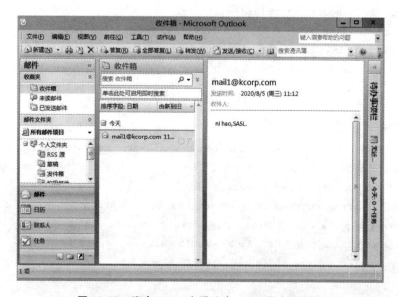

图 10-27　账户 mail2 查看账户 mail1 发来的邮件

10.3.2 基于 TLS/SSL 的邮件服务

Postfix、Dovecot 可以使用 OpenSSL 提供的库，实现基于 TLS/SSL 的连接。通过这样的连接，可以对通信数据进行加密，以及基于用户 TLS 证书的认证。表 10-6 列出了邮件协议与 TLS/SSL 直接的支持关系。

表 10-6　邮件协议与 TLS/SSL 的支持关系

协议	端口	功能	协议	端口	功能
SMTPS	465/TCP	SMTP over TLS	SMTP	25/TCP	通过 SMTP 协议扩展指令 STARTTLS 实现
POP3S	995/TCP	POP3 over TLS	POP3	110/TCP	通过 POP3 协议扩展指令 STLS 实现
IMAPS	993/TCP	IMAP4 over TLS	IMAP4	143/TCP	通过 IMAP4 协议扩展指令 STARTTLS 实现

说明：

① SMTP/POP3/IMAP4 over TLS 使用与 SMTP/POP3/IMAP4 独立的端口作为连接。客户端连接 465/995/993 端口直接进行加密传输。

② 通过 STARTTLS 将纯文本协议 SMTP/POP3/IMAP4 连接升级到 TLS/SSL 加密连接。客户端使用 25/110/143 端口，直到服务器发送 STARTTLS 指令后，如果客户端支持 STARTTLS，在经过协商后，才能开始进行加密传输。

实现步骤如下：

（1）安装 SSL 服务，并创建证书存放的目录（/etc/mail/.ssl）。

```
[root@localhost~]# yum-y install mod_ssl
[root@localhost~]# mkdir-p cd/etc/mail/.ssl
```

（2）生成邮件私钥文件。

```
[root@localhost~]# cd /etc/mail/.ssl
[root@localhost.ssl]# openssl genrsa-out server.key 1025
Generating RSA private key,1025 bit long modulus
...................+ + + + + +
.............+ + + + + +
e is 65537 (0x10001)
```

（3）建立邮件证书。

```
[root@localhost.ssl]# openssl req-new-x509-key server.key  -out server.crt
You are about to be asked to enter information that will be incorporated
into your certificate request.
What you are about to enter is what is called a Distinguished Name or a DN.
There are quite a few fields but you can leave some blank
For some fields there will be a default value,
If you enter '.',the field will be left blank.
……
Country Name (2 letter code)[XX]:NC
```

```
State or Province Name (full name)[]:GH
Locality Name (eg,city)[Default City]:GK
Organization Name (eg,company)[Default Company Ltd]:SC
Organizational Unit Name (eg,section)[]:
Common Name (eg,your name or your server's hostname)[]:
Email Address[]:
```

（4）配置基于 TLS 的 Postfix，添加以下内容到 main.cf 文件中。（将之前的 SASL 认证配置注释或删除）

```
[root@localhost.ssl]# vim /etc/Postfix/main.cf
……
smtpd_tls_cert_file=/etc/mail/.ssl/server.crt    //指定服务器证书文件的位置
smtpd_tls_key_file=/etc/mail/.ssl/server.key     //指定服务器私钥文件的位置
smtpd_tls_session_cache_database=btree:/etc/Postfix/smtpd_scache
                                                 //指定 TLS 会话缓存数据库的位置
smtpd_tls_protocols=!SSLv2,!SSLv3                //指定服务器使用的 SSL/TLS 协议的版本
……
```

（5）修改 master.cf 文件，将第 26、27、28 的注释号（#）取消。

```
[root@localhost.ssl]# vim /etc/Postfix/master.cf
……
smtps     inet  n    -    n    -    -    smtpd
 -o syslog_name=Postfix/smtps
 -o smtpd_tls_wrappermode=yes
……
[root@localhost.ssl]# systemctl restart Postfix
```

（6）使用 telnet 查看 Postfix 是否已支持 TLS。

```
[root@localhost.ssl]# telnet 192.168.1.1 25
Trying 192.168.1.1…
Connected to 192.168.1.1.
Escape character is '^]'.
220 mail.kcorp.com ESMTP Postfix
ehlo kcorp.com
250-mail.kcorp.com
250-PIPELINING
250-SIZE 10240000
250-VRFY
250-ETRN
250-STARTTLS
250-ENHANCEDSTATUSCODES
250-8BITMIME
250 DSN
starttls
220 2.0.0 Ready to start TLS           //已开启 TLS
```

(7) 配置 Dovecot 服务程序的 10-ssl.conf 文件,让其支持 TLS。

```
[root@localhost.ssl]# vim /etc/dovecot/conf.d/10-ssl.conf
ssl=yes    //启用 SSL/TLS
ssl_cert=</etc/mail/.ssl/server.crt       //指定服务器证书文件的位置
ssl_key=</etc/mail/.ssl/server.key        //指定服务器私钥文件的位置
ssl_protocols=! SSLv2 ! SSLv3             //指定服务器使用的 SSL/TLS 协议的版本
[root@localhost.ssl]# systemctl restart dovecot
```

(8) 使用 netstat 命令查看 Dovecot 服务是否已支持 SSL/TLS,如图 10-28 所示。

```
[root@localhost .ssl]# netstat -ltp|grep dovecot
tcp    0    0 0.0.0.0:pop3      0.0.0.0:*         LISTEN    10744/dovecot
tcp    0    0 0.0.0.0:imap      0.0.0.0:*         LISTEN    10744/dovecot
tcp    0    0 0.0.0.0:imaps     0.0.0.0:*         LISTEN    10744/dovecot
tcp    0    0 0.0.0.0:pop3s     0.0.0.0:*         LISTEN    10744/dovecot
tcp6   0    0 [::]:pop3         [::]:*            LISTEN    10744/dovecot
tcp6   0    0 [::]:imap         [::]:*            LISTEN    10744/dovecot
tcp6   0    0 [::]:imaps        [::]:*            LISTEN    10744/dovecot
tcp6   0    0 [::]:pop3s        [::]:*            LISTEN    10744/dovecot
```

图 10-28　查看 Dovecot 服务是否已支持 SSL/TLS

(9) 使用 Outlook 软件进行邮件的测试,这时,因为要使用 SSL/TLS 进行邮件的发送与接收,需要在"Internet 电子邮件设置"→"高级"里,设置协议以及协议所使用的端口号(账户 mail1、mail2 均需要设置。同时,之前对于 SASL 认证的勾选"我的发送服务器(SMTP)要求验证"选项,要取消,不勾选),如图 10-29、图 10-30 所示。

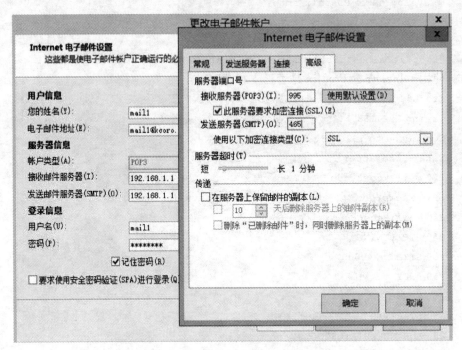

图 10-29　账户 mail1 设置接收、发送服务器的端口号

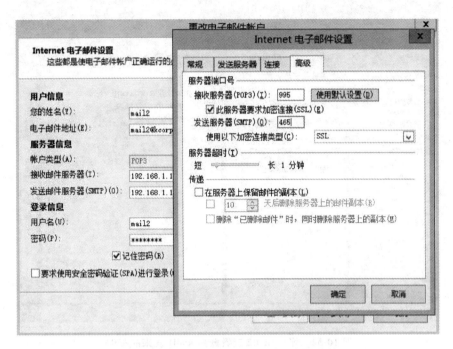

图 10-30　账户 mail2 设置接收、发送服务器的端口号

现在账户 mail1、mail2 就可以正常进行邮件的发送与接收,账户 mail1 写信给账户 mail2,mai2 能正常接收、查阅邮件。如图 10-31、图 10-32 所示。

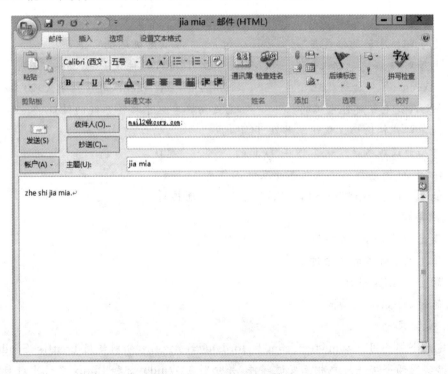

图 10-31　账户 mail1 写信给账户 mail2

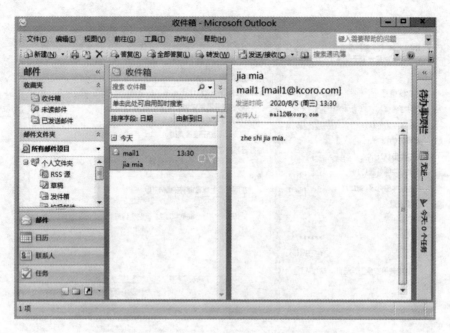

图 10-32 账户 mail2 查看账户 mail1 发来的邮件

自我测试

一、填空题

1. 电子邮件地址的格式是"_____@_____（主机地址）"。
2. 电子邮件系统包括_____（Mail User Agent，邮件用户代理）、_____（Mail Transfer Agent，邮件传送代理）、_____（Mail Dilivery Agent，用户投递代理）三个组件。
3. MUA 是电子邮件系统的_____，它是用户和电子邮件系统的接口。
4. 简单邮件传输协议默认工作在 TCP 的_____端口。
5. 邮局协议版本 3 默认工作在 TCP 的_____端口。
6. Internet 消息访问协议版本 4 默认工作在 TCP 的_____端口。
7. Dovecot 是一个邮件访问代理，用于将用户连接到_____。

二、简答题

1. 简述 MTA 的两种功能。
2. 简述电子邮件的传输过程。
3. 简述 Postfix 设计目标。
4. 简述 Dovecot 的特点。

三、实训题

创建两个邮件用户 tom@est.com 和 Jordan@test.com，分别使用 Postfix、Sendmail 两个服务器实现两个邮件用户相互发送邮件，并分别在 Windows 和 Linux 系统中进行测试。

项目 11　代理服务器

项目综述

随着公司业务发展,为了确保公司的办公电脑在访问因特网上的资源的时候,过滤、屏蔽掉一些不安全的信息,网络管理员将在公司内部搭建代理服务器,来实现这些功能。

项目目标

- 了解代理服务器相关知识及工作原理;
- 掌握正向代理服务器的配置方法;
- 掌握反向代理服务器的配置方法。

任务 11.1　代理服务器概述

11.1.1　代理服务器基本概念

代理服务器是指可以代理局域网内的计算机向 Internet 获取网页或者其他数据的一种服务器。通过使用代理服务器,可以节省网络带宽,加快内部网络访问因特网 WWW 的速度,同时通过在代理服务器上设置相关访问限制,可以过滤或屏蔽掉一些不安全的信息,提高网络访问的安全性。目前,人们在使用因特网时,经常会用到一些网络加速器,如腾讯网游加速器、网易 UU 加速器等,其实,都是使用代理服务器,来提高网络访问速度。如图 11-1 所示,客户端通过代理服务器去访问 Internet 上的服务器,获取需要的信息。

一般而言,代理服务器会部署在整个局域网的单点对外防火墙上,而局域网内部的计算机都通过代理服务器来访问 Internet 上的服务器。在实际的应用中,我们可以去设置浏览器的相关选项,以使用代理服务器。如 Windows 系统的 Internet Explorer 浏览器,通过"Internet 选项"→"连接"→"局域网(LAN)设置",输入代理服务器的地址及端口号,如图 11-2 所示。

当使用了代理服务器后,客户端就会使用代理服务器的 IP 地址去访问服务器,在服务器端,也会认为是与代理服务器的 IP 的主机进行通信,从而隐藏了客户端真实的 IP 地址。所以当外部计算机入侵客户端时,会误认为代理服务器就是客户端,导致攻击对象错误,因此,代理服务器还具有防火墙的作用。

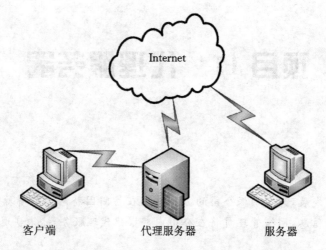

图 11-1 客户端使用代理服务器访问 Internet 上的服务器

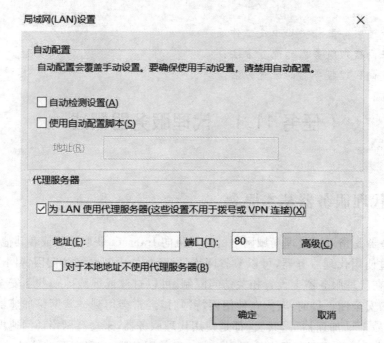

图 11-2 设置代理服务器的地址及端口号

11.1.2 代理服务器的工作原理

如图 11-3 所示,说明了代理服务器的工作原理。
(1) 客户端 A 通过代理服务器向服务器 A 提出访问请求;
(2) 代理服务器查询自己的缓存中是否有客户端 A 需要访问的数据;
(3) 如果代理服务器没有客户端 A 需要的数据,那么代理服务器就向服务器 A 提出访问请求;

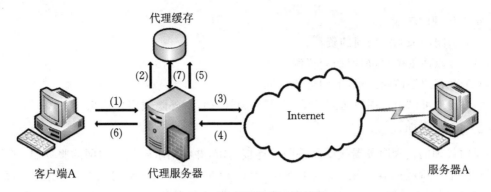

图 11-3　代理服务器工作原理

（4）服务器 A 响应代理服务器的访问请求；

（5）代理服务器获取服务器 A 的数据后，将数据信息保存到自己的缓存中，以便以后再次访问时使用；

（6）代理服务器将数据发送给客户端 A；

（7）当客户端 A 再次访问服务器 A 的数据时，代理服务器直接查找缓存，将缓存中已经存在的符合客户端 A 需要的数据取出，发给客户端 A。不再向 Internet 上的服务器 A 请求数据，这样就极大地提高了客户端访问数据的速度。

11.1.3　代理服务器的分类

在代理服务器为用户提供代理服务时，可以分为正向代理和反向代理两种模式。

1. 正向代理

正向代理分为标准正向代理、透明代理两种模式。标准正向代理，用户需在客户端的浏览器等软件中手动设置代理服务器的地址和端口，由代理服务器代替客户端去请求数据，否则默认不使用代理服务。透明代理，用户不需要在客户端上指定代理服务器的地址和端口，而是直接通过默认路由、防火墙策略等，将需要访问的服务器重定向给代理服务器处理。正向代理即客户端代理，服务端不知道实际发起请求的客户端。

正向代理示意图如图 11-4 所示。

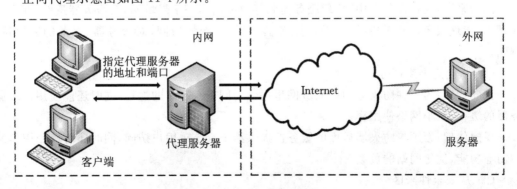

图 11-4　正向代理示意图

正向代理的功能为：
（1）访问原来无法访问的资源；
（2）可以使用缓存，加速访问资源；
（3）对客户端访问授权，上网进行认证；
（4）代理可以记录用户访问记录（上网行为管理），对外隐藏用户信息。

2. 反向代理

反向代理的代理服务器接受因特网的连接请求，并将请求转发给内网的服务器，获得数据后返回给因特网的客户端，类似于使用 iptables 的 DNAT 策略的发布服务器。也就是说，反向代理不需要用户进行任何设置，直接访问服务器真实 IP 地址或者域名，但是服务器内部会自动根据访问内容进行跳转及返回，用户不知道它最终访问的是哪些机器。反向代理即服务端代理，客户端不知道实际提供服务的服务端。

反向代理示意图如图 11-5 所示。

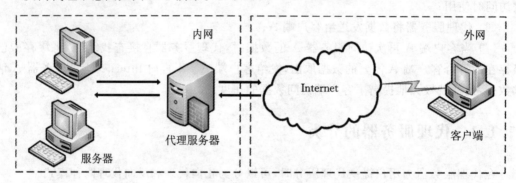

图 11-5　反向代理示意图

反向代理的功能为：
（1）保证内网的安全，阻止 Web 攻击，大型网站通常将反向代理服务器作为公网访问地址，Web 服务器在内网；
（2）负载均衡，通过反向代理服务器来优化网站的负载。

3. 正向代理与反向代理的区别

总的来说，在正向代理中，代理服务器和客户端同属一个局域网，对服务器透明；在反向代理中，代理服务器和客户端（之前的服务器）同属一个局域网，对服务器（之前的客户端）透明。

（1）从用途上来区分

正向代理：为了在防火墙内的局域网里提供访问 Internet 的途径，同时还能使用代理服务器的缓存，减小网络使用率。

反向代理：为了将防火墙后面的服务器提供给 Internet 用户访问，同时还可以使用负载均衡等功能，优化网站的负载。

（2）从安全性来区分

正向代理：允许客户端通过它访问任意网站并且隐蔽客户端自身，因此必须采取安全措施来确保，仅为经过授权的客户端提供服务。

反向代理:对外是透明的,访问者并不知道自己访问的是代理服务器,对访问者而言,认为访问的就是真实的服务器。

◀ 任务 11.2 配置代理服务器 ▶

11.2.1 安装代理服务器

在 Linux 系统中,使用 Squid 服务程序来实现代理功能。对于 Web 用户而言,Squid 是一个高性能的代理缓存服务软件,能够代替用户向网站服务器请求数据,并自动处理所下载的数据。当一个用户想要访问网站服务器上的页面、图片等数据时,可以向 Squid 发出一个申请,要 Squid 代替其进行下载,然后 Squid 连接所申请网站服务器并请求相关数据,并将这些数据传送给用户,同时将这些数据存储在自己的服务器上。当其他的用户再次申请同样的页面、图片等数据时,Squid 就把存储在本地服务器的数据传送给用户,从而节省了用户访问数据的时间,并且减小了网站服务器的负载压力。

(1)查询系统是否已经安装 Squid 软件包。

[root@localhost~]# rpm-qa|grepsquid

(2)安装 Squid 软件包。

[root@localhost~]# yum-y install squid
[root@localhost~]# rpm-qa|grep squid
squid-3.5.20-10.el7.x86_64
squid-migration-script-3.5.20-10.el7.x86_64

(3)配置防火墙,放行 Squid 服务。

[root@localhost~]# firewall-cmd --permanent --add-service=squid
[root@localhost~]# firewall-cmd --reload
[root@localhost~]# firewall-cmd --list-all

默认情况下,Squid 服务使用的是 TCP 协议的 3128 端口。

[root@localhost~]# semanage port -l | grep squid_port_t
squid_port_t   tcp   3128,3401,4827
squid_port_t   udp   3401,48274,启动 squid 服务

(4)启动 Squdi 服务。

[root@localhost~]# systemctl start squid

11.2.2 认识 Squid 服务程序的配置文件

1. Squid 服务程序的配置文件

Squid 服务程序的主要配置文件及其功能如表 11-1 所示。

表 11-1　Squid 服务程序的主要配置文件及其功能

配置文件名称	存放位置	功能
配置文件	/etc/squid/squid.conf	Squid 的主配置文件
	/etc/squid/mime.conf	Squid 支持的 Internet 的文件格式
Squid 的管理工具	/usr/sbin/squid	Squid 的控制程序
	/var/spool/squid	Squid 默认的缓存存储位置
	/user/lib64/squid	Squiid 额外的控制模块

2. Squid 服务程序主配置文件的重要参数

Squid 服务程序主配置文件的重要参数及其功能如表 11-2 所示。

表 11-2　Squid 服务程序主配置文件的重要参数及其功能

参数	功能
http_port 3128	默认监听客户端要求的端口号
cache_mem 8M	设置内存缓冲区的大小
cache_dir ufs/var/spool/squid 100 16 256	设置硬盘缓冲区的大小,100 代表磁盘使用量仅用掉该文件系统的 100 MB,16 代表第一层次目录共有 16 个,256 代表每层次目录内部再分为 256 个次目录
cache_effective_user squid	设置启动 squid PID 的拥有者
cache_effective_group squid	设置启动 squid PID 的组
dns_nameservers<IP 地址>	默认使用服务器的 DNS 地址
cache_log/var/log/squid/cache.log	缓存日志文件的存放位置
cache_access_log/var/log/squid/access.log	访问日志文件的存放位置
visible_hostname <名称>	设置 squid 服务器的名称

如图 11-6 所示,就显示"cache_dir ufs/var/spool/squid 100 16 256"这段参数的意思,第一层次目录为 16 个,第二层次"00"的目录为 256 个。

```
[root@localhost ~]# ls /var/spool/squid/
00  01  02  03  04  05  06  07  08  09  0A  0B  0C  0D  0E  0F  swap.state
[root@localhost ~]# ls /var/spool/squid/00
00  0D  1A  27  34  41  4E  5B  68  75  82  8F  9C  A9  B6  C3  D0  DD  EA  F7
01  0E  1B  28  35  42  4F  5C  69  76  83  90  9D  AA  B7  C4  D1  DE  EB  F8
02  0F  1C  29  36  43  50  5D  6A  77  84  91  9E  AB  B8  C5  D2  DF  EC  F9
03  10  1D  2A  37  44  51  5E  6B  78  85  92  9F  AC  B9  C6  D3  E0  ED  FA
04  11  1E  2B  38  45  52  5F  6C  79  86  93  A0  AD  BA  C7  D4  E1  EE  FB
05  12  1F  2C  39  46  53  60  6D  7A  87  94  A1  AE  BB  C8  D5  E2  EF  FC
06  13  20  2D  3A  47  54  61  6E  7B  88  95  A2  AF  BC  C9  D6  E3  F0  FD
07  14  21  2E  3B  48  55  62  6F  7C  89  96  A3  B0  BD  CA  D7  E4  F1  FE
08  15  22  2F  3C  49  56  63  70  7D  8A  97  A4  B1  BE  CB  D8  E5  F2  FF
09  16  23  30  3D  4A  57  64  71  7E  8B  98  A5  B2  BF  CC  D9  E6  F3
0A  17  24  31  3E  4B  58  65  72  7F  8C  99  A6  B3  C0  CD  DA  E7  F4
0B  18  25  32  3F  4C  59  66  73  80  8D  9A  A7  B4  C1  CE  DB  E8  F5
0C  19  26  33  40  4D  5A  67  74  81  8E  9B  A8  B5  C2  CF  DC  E9  F6
```

图 11-6　硬盘缓冲区的大小

3. 访问控制列表

在 Squid 服务程序中，使用访问控制列表（ACL）来控制哪一台客户端可以访问 Web 服务器，并指定访问的方式。Squid 服务器通过检查 ACL 中配置的策略（规则）来决定是否允许某客户端访问，可以基于源地址、目的地址、域名、时间和日期等，来使用 ACL 定义策略（规则）。

ACL 的基本语法为：

acl　＜自定义的 acl 名称＞　＜acl 控制类型＞　＜控制的内容＞

在定义 acl 名称的时候，要尽量使用意义明确的名称。设置好 ACL 之后，还需要使用 http_access 命令去设置拒绝或允许某个 ACL 的访问请求，其格式为：

http_access　〔allow/deny〕　ACL 名称

需要注意的是，配置 Squid 的 ACL 的匹配顺序与网络设备（如三层交换机、路由器、防火墙）的 ACL 的匹配顺序规则都是一样的，都是由上到下进行匹配。一旦匹配成功后，就立即执行相应的操作并结束匹配过程。所以，为了避免 ACL 将所有的流量全部拒绝或者全部允许，往往会在 ACL 最下面写 deny all 或者 allow all，以避免安全隐患。

需要注意的是：

① 没有设置任何规则时，Squid 服务将拒绝客户端的请求；

② 有规则但找不到相匹配的项时，Squid 将采用与最后一条规则相反的权限。

如图 11-7 所示，路由器使用 ACL 对数据进行过滤，从而提高网络环境的安全性。

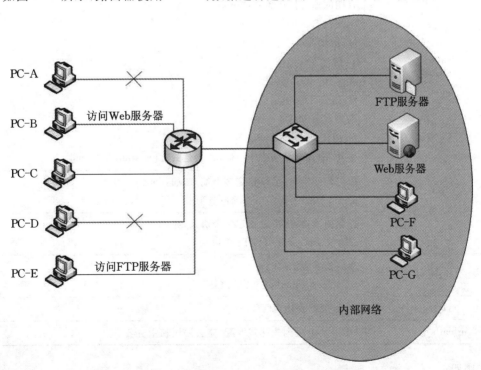

图 11-7　使用 ACL 对数据进行过滤

通过在路由器上使用 ACL，限制了 PC-A、PC-D 访问内部网络，PC-B、PC-E 能够访问内部网络中具体的服务器，PC-C 能完全访问内部网络，不受任何限制。ACL 规则匹配顺序是从上至下的，ACL 会将流量与列表中的每条语句逐一进行比较，直至找到匹配项或比较完所有语句为止，然后执行操作，跳出匹配过程。

常见的 ACL 列表类型及其功能如表 11-3 所示。

表 11-3　常见的 ACL 列表类型及其功能

类型	功能
src ip-address/netmask	客户端源 IP 地址和子网掩码
src addr1 -addr2/netmask	客户端源 IP 地址范围
dst ip-address/netmask	客户端目的 IP 地址和子网掩码
myipip-address/netmask	本地套接字 IP 地址
srcdomain domain	客户端所属的域
dstdomain domain	Internet 服务器所属的域
srcdom_regex expression	对来源 URL 进行正则表达式匹配
dstdom_regex expression	对目的 URL 进行正则表达式匹配
time	指定时间，语法如下： acl aclname time[day -abbrevs][h1:m1 -h2:m2][hh:mm -hh:mm] 日期的缩写指代关系如下： S：指代 Sunday M：指代 Monday T：指代 Tuesday W：指代 Wednesday H：指代 Thursday F：指代 Friday A：指代 Saturday 另外，h1:m1 必须小于 h2:m2，表达式为[hh:mm-hh:mm]
port	指定访问端口，可以指定多个端口，如：port 80 70 21 port 0-1024…指定一个端口范围
proto	指定使用协议，可以指定多个协议，如： acl aclname proto HTTP FTP…
method	指定请求方法，如： acl aclname method GET POST
url_regex	URL 规则表达式匹配
urlpath_regex	URL-path 规则表达式匹配，略去协议和主机名

例如：

（1）拒绝所有客户端的请求。

```
acl deny-all src 0.0.0.0/0.0.0.0
http_access deny deny-all
```

(2) 禁止 192.168.100.0/24 网段的客户端请求。

  acl100 src 192.168.100.0/255.255.255.0

  http_access deny100

(3) 仅允许 192.168.100.20 的客户端请求,其他客户端不允许访问。

  acl20 src 192.168.100.20

  http_accessallow 20

  http_access deny all

(4) 禁止所有客户端访问域名为 www.abc.com 的网站。

  acl abc dstdomain www.abc.com

  http_access deny abc

(5) 禁止 192.168.100.0/24 网络的客户端在周一至周五的 9 点至 18 点进行访问。

  acl 100 src 192.168.100.0/255.255.255.0

  acl bdtime time MTWHF 9:00-18:00

  http_access deny 100 bdtime

(6) 屏蔽 www.abc.com 网站。

  acl abc dstdomain-i www.abc.com

  http_access deny abc

"-i"表示忽略大小写字母,默认情况下 Squid 区分大小写字母。

(7) 禁止所有客户端访问网址中包含 boy 关键词的网站。

  aclboy url_regex-i boy

  http_access deny boy

(8) 禁止所有客户端访问 www.bc.com 网站。

  aclbd_url url_regex http://www.bc.com

  http_access deny bd_url

(9) 禁止所有客户端访问 22、23、25 端口。

  acl bd_port port 22 23 25

  http_access deny bd_port

如果写成 http_access deny ! bd_port 则表示拒绝所有非 bd_port 列表中的端口,"!"符号称为"非",表示取反的意思。

(10) 禁止客户端下载文件后缀为.rar、.avi、.mp3、.exe 类型的文件。

  acl badfile urlpath_regex-i \.rar$ \.avi$ \.mp3$ \.exe$

  http_access deny badfile

11.2.3 配置标准正向代理服务器

如图 11-8 所示,客户端(RHEL1,IP 地址为 192.168.1.1/24,网络连接模式为 VMnet1[仅主机模式]),代理服务器(RHEL2,IP1 地址为 192.168.1.2/24,IP2 地址为 200.200.200.2/24,网络连接模式分别为 VMnet1、VMnet8[NAT 模式]),服务器(RHEL3,IP 地址为 200.200.200.1/24,网络连接模式为 VMnet8)。通过在代理服务器上,配置标准正向代理,实现客户端访问服务器的 Web 服务。

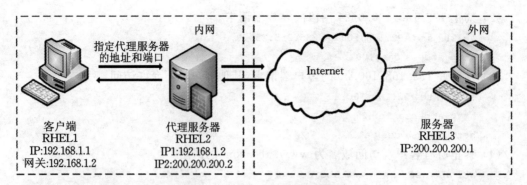

图 11-8 标准正向代理应用环境

实现步骤为：

(1) 配置三台计算机的 IP 地址。

客户端(RHEL1)：

```
[root@RHEL1~]# vim /etc/sysconfig/network-scripts/ifcfg-ens33
……
IPADDR=192.168.1.1
NETMASK=255.255.255.0
GATEWAY=192.168.1.2
……
```

代理服务器(RHEL2)：

```
[root@RHEL2~]# vim /etc/sysconfig/network-scripts/ifcfg-ens33
……
IPADDR=192.168.1.2
NETMASK=255.255.255.0
……
[root@RHEL2~]# vim /etc/sysconfig/network-scripts/ifcfg-ens33
……
IPADDR=200.200.200.2
NETMASK=255.255.255.0
……
```

服务器(RHEL3)：

```
[root@RHEL3~]# vim /etc/sysconfig/network-scripts/ifcfg-ens33
……
IPADDR=200.200.200.1
NETMASK=255.255.255.0
……
```

(2) 配置代理服务器(RHEL2)。

```
[root@RHEL2~]# vim /etc/squid/squid.conf
……
http_port 3128
cache_dir ufs/var/spool/squid 100 16 256
```

```
visible_hostname RHEL2
……
[root@RHEL2~]# systemctl stop firewalld
[root@RHEL2~]# systemctl restart squid
```

(3) 配置服务器(RHEL3)的 Web 服务,如图 11-9 所示。

```
[root@RHEL3~]# yum -y install httpd
[root@RHEL3~]# systemctl start httpd
[root@RHEL3~]# firefox http://200.200.200.1
```

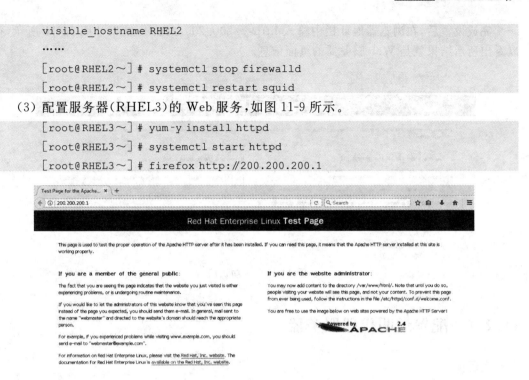

图 11-9　httpd 服务默认页面

(4) 在客户端(RHEL1)上使用 Firefox 浏览器进行测试。在该浏览器上依次操作:按 ALT 键,在菜单中单击"编辑"→"首选项"→"高级"→"网络"→"设置",打开"连接设置"界面,并填写代理服务器相关数据信息,如图 11-10 所示。

图 11-10　手动配置代理服务器信息

完成设置后,在浏览器地址栏中输入 http://200.200.200.1,结果如图 11-11 所示,可以看出所示结果就是 Web 服务器的页面信息。

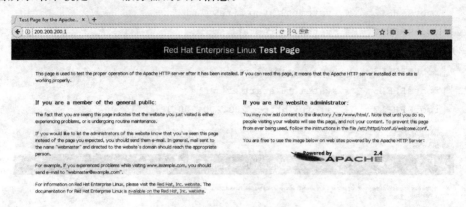

图 11-11 成功访问

11.2.4 配置透明代理服务器

实现步骤:

(1) 配置代理服务器(RHEL2),仅需在前面配置标准正向代理服务器的基础上修改一行参数。

```
[root@RHEL2~]# vim /etc/squid/squid.conf
……
http_port 3128 transparent
```

(2)配置代理服务器(RHEL2)的 iptables 规则,使用 SNAT 数据转发功能,将客户端主机源网络地址为 192.168.1.0/24、TCP 端口为 80 的请求,转发到代理服务器的 3128 端口。

```
[root@RHEL2~]# systemctl stop firewalld
[root@RHEL2~]# systemctl start iptables
[root@RHEL2~]# iptables -F
[root@RHEL2~]# iptables -t nat -I PREROUTING -s 192.168.1.0/24 -p tcp --dport 80 -j REDIRECT --to-ports 3128
[root@RHEL2~]# service iptables save
iptables: Saving firewall rules to /etc/sysconfig/iptables:[确定]
```

(3) 在客户端(RHEL1)上使用 Firefox 浏览器进行测试。删除图 11-10 所设置的代理服务器相关参数(也就是说,在浏览器上不进行任何设置),设置为"不使用代理"。之后,在浏览器地址栏中输入 http://200.200.200.1,结果如图 11-12 所示。

(4) 查看服务器(RHEL3)的 Web 服务访问日志。

```
[root@RHEL3~]# vim /var/log/httpd/access_log
200.200.200.2 - - [09/Aug/2020:18:45:14 +0800] "GET / HTTP/1.1" 403 3985 "-" "Mozilla/5.0 (X11; Linux x86_64; rv:52.0) Gecko/20100101 Firefox/52.0"
200.200.200.2 - - [09/Aug/2020:18:45:14 +0800] "GET /icons/apache_pb2.gif HTTP/1.1" 304 - "http://200.200.200.1/" "Mozilla/5.0 (X11; Linux x86_64; rv:52.0) Gecko/20100101 Firefox/52.0"
```

项目 11 代理服务器

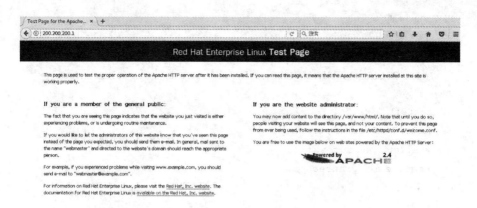

图 11-12 成功访问

11.2.5 配置反向代理服务器

反向代理是实现图 11-8 中的外网服务器(RHEL3)去访问内网客户端(RHEL1)的 Web 服务。

实现步骤为：

(1) 配置代理服务器(RHEL2)。

```
[root@RHEL2~]# vim /etc/squid/squid.conf
……
acl localnet src 200.200.200.0/24
http_access allow localnet
http_access deny all
http_port 200.200.200.2:80 vhost  //指定监听端口，accel 加速模式，vhost 启用反向代理
cache_peer 192.168.1.1 parent 80 0 no-query originserver weight=5 max_conn=30
……
[root@RHEL2~]# systemctl start firewalld
[root@RHEL2~]# firewall-cmd --permanent --add-service=squid
success
[root@RHEL2~]# firewall-cmd --permanent --add-port=80/tcp
success
[root@RHEL2~]# firewall-cmd --reload
success
```

no-query 表示不发送 icp 信息包到此机器；originserver 设置网站源服务器；weight 设置权重；max_conn 设置网站的最大连接数。

(2) 配置客户端(RHEL1)的 Web 服务，如图 11-13 所示。

```
[root@RHEL1~]# yum -y install httpd
[root@RHEL1~]# mkdir /web
[root@RHEL1~]# echo "Welcome"> /web/index.html
[root@RHEL1~]# vim /etc/httpd/conf/httpd.conf
```

```
……
DocumentRoot "/web"
#
# Relax access to content within /var/www.
#
<Directory "/web">
    AllowOverride None
    # Allow open access:
    Require all granted
</Directory>
……
[root@RHEL1~]~]# systemctl restart httpd
[root@RHEL1~]systemctl stop firewalld
[root@RHEL1~]setenforce 0
[root@RHEL1~]firefox http://192.168.1.1
```

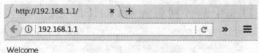

图 11-13 httpd 服务正常

(3) 在服务器(RHEL3)上使用 Firefox 浏览器进行测试。删除图 11-10 所设置的代理服务器相关参数(也就是说,在浏览器上不进行任何设置),设置为"不使用代理"。之后,在浏览器地址栏中输入 http://200.200.200.1,结果如图 11-14 所示。

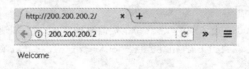

图 11-14 成功访问

自我测试

一、填空题

1. 代理服务器是可以代理局域网内的计算机向 Internet 获取_____的一种服务。
2. 代理服务器为用户提供代理服务时,可以分为_____和_____两种模式。
3. 正向代理分为_____、_____两种模式。
4. 在 Linux 系统中,使用_____服务程序来实现代理功能。
5. 在 Squid 服务程序中,使用_____来控制哪一台客户端可以访问 Web 服务器,并指定访问的方式。

二、简答题

1. 简述代理服务器的工作原理。

2. 简述正向代理的功能。
3. 简述反向代理的功能。

三、实训题

如图 11-15 所示,公司使用 squid 作为代理服务器(内网 IP 地址为 192.168.1.1/24),公司内部 IP 地址段为 192.168.1.0/24,现准备使用 9090 作为代理端口,实现以下功能:

(1) 客户端在设置代理服务器地址和端口后,能够访问互联网的 Web 服务器(201.200.160.1)。

(2) 客户端在不需要设置代理服务器地址和端口的情况下,能够访问互联网的 Web 服务器(201.200.160.1)。

(3) 配置反向代理服务器。

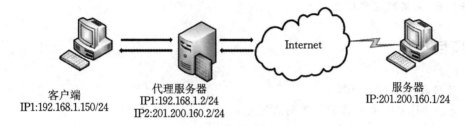

图 11-15 代理服务网络拓扑结构图

项目 12　数据库服务器

项目综述

随着公司业务发展,公司准备使用数据库来存储数据,网络管理员将在公司内部搭建数据库服务器,来实现这些功能。

项目目标

- 了解数据库相关知识;
- 掌握 MySQL 数据库服务器的配置方法;
- 掌握 MariaDB 数据库服务器的配置方法。

◀ 任务 12.1　MySQL 数据库服务器 ▶

12.1.1　MySQL 数据库简介

在 Linux 系统中,可以安装 Oracle、MySQL、PostSQL、DB2、MariaDB 等关系型数据库管理系统(Relational Database Management System,RDBMS)。MySQL 数据库在其中,是较具有竞争力的数据库,它是一个高性能、多线程、多用户,同时建立在客户端/服务器结构上的关系型数据库管理系统。

关系型数据库管理系统将数据保存在不同的表中,而不是将所有数据放在一个大仓库中,这样就增加了对数据库的访问速度,并提高了灵活性。MySQL 所使用的 SQL 语言是用于访问库的最常用的标准化语言。MySQL 分为社区版和商业版,具有体积小、速度快、拥有成本低、开放源代码等特点,一般中小型网站都选择 MySQL 作为网站的数据库。

12.1.2　安装 MySQL 数据库

1. 下载 MySQL 软件包

由于在 RHEL7 系统中,默认没有 MySQL 服务,所以需要到 MySQL 官方网站上(https://dev.mysql.com/downloads/repo/yum/)下载 MySQL 的 YUM 仓库文件,如图 12-1 所示。也可以直接使用 wget 命令下载 MySQL 的 YUM 仓库文件,如图 12-2 所示。

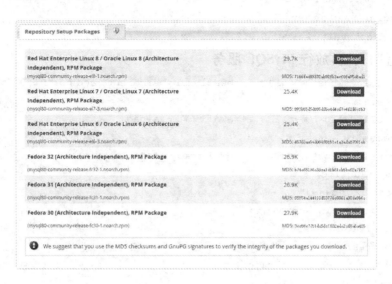

图 12-1　下载 MySQL 的 YUM 仓库文件

```
[root@localhost mysql]# mkdir /mysql
[root@localhost mysql]# cd /mysql
```

wget 命令为：wget https://repo.mysql.com//mysql80-community-release-el7-3.noarch.rpm

图 12-2　使用 wget 命令下载 MySQL 的 YUM 仓库文件

2. 安装 MySQL 软件包

在安装 MySQL 软件包前，先卸载 MariaDB 软件包，避免安装时发生错误。

```
[root@localhost ~]# rpm -qa|grep MariaDB
MariaDB-libs-5.5.56-2.el7.x86_64
MariaDB-5.5.56-2.el7.x86_64
[root@localhost ~]# yum -y remove MariaDB*
```

卸载完成后，就可以安装 MySQL 依赖包及服务。

```
[root@localhost mysql]# yum -y install mysql80-community-release-el7-3.noarch.rpm
[root@localhost mysql]# yum -y install mysql-community-server
[root@localhost mysql]# rpm -qa|grep mysql
mysql-community-client-8.0.21-1.el7.x86_64
mysql-community-common-8.0.21-1.el7.x86_64
```

```
    mysql-community-libs-8.0.21-1.el7.x86_64
    mysql80-community-release-el7-3.noarch
    mysql-connector-java-5.1.25-3.el7.noarch
    mysql-community-server-8.0.21-1.el7.x86_64
```

3. 配置防火墙，放行 MySQL 服务

```
[root@localhost mysql]# firewall-cmd --permanent --add-port=3306/tcp
success
[root@localhost mysql]# firewall-cmd --reload
success
[root@localhost mysql]# firewall-cmd --list-all
public (active)
  target: default
  icmp-block-inversion: no
  interfaces: ens33
  sources:
  services:ssh dhcpv6-client squid
  ports: 80/tcp 3306/tcp
……
```

4. 启动 MySQL 服务

```
[root@localhost mysql]# systemctl start mysqld
```

5. 登录 MySQL 数据库

安装好 MySQL 数据库后，默认情况下，root 用户是没有登录密码的，可以在启动 MySQL 服务后，在 MySQL 命令行模式下修改 root 密码。在安装 MySQL 服务时，系统默认创建了数据库超级管理员"root@localhost"，并设置了该用户的默认密码，可以在/vat/log/mysqld.log 文件中查看，并使用该密码登录 MySQL 数据库，如图 12-3 所示，超级管理员的默认密码为"ebqYw.2％*3#b"。

```
[root@localhost ~]# cat /var/log/mysqld.log| grep password
2020-08-10T04:53:15.632271Z 6 [Note] [MY-010454] [Server] A temporary password is generated for root@localhost: ebqYw.2%*3#b
[root@localhost ~]# mysql -u root -p
Enter password:
Welcome to the MySQL monitor.  Commands end with ; or \g.
Your MySQL connection id is 15
Server version: 8.0.21

Copyright (c) 2000, 2020, Oracle and/or its affiliates. All rights reserved.

Oracle is a registered trademark of Oracle Corporation and/or its
affiliates. Other names may be trademarks of their respective
owners.

Type 'help;' or '\h' for help. Type '\c' to clear the current input statement.

mysql>
```

图 12-3 登录 MySQL 数据库

接下来，就可以修改 root 用户的密码，命令为：ALTER USER 'root'@'localhost' IDENTIFIED BY 'MySQL123!' PASSWORD EXPIRE NEVER。修改后，下次就可以使用修改后的密码登录 MySQL 数据库。

需要注意的是，MySQL 数据库的默认密码策略要求密码必须包含数字、大小写字母和特殊符号。另外，常用 MySQL 命令都以";"结束，只有少量特殊命令不能加";"结束，如备份数据库命令。

```
mysql> ALTER USER 'root'@'localhost' IDENTIFIED BY 'MySQL123!' PASSWORD
EXPIRE NEVER;
    Query OK,0 rows affected (0.01 sec)
```

6. 数据库相关知识

(1) 数据库文件

MySQL 数据库文件有".FRM"".MYD"和".MYI"三类,其中".FRM"是描述表结构的文件,".MYD"是表的数据文件,".MYI"是表数据文件中的索引文件。

(2) 数据类型

数据类型规定了某个字段所允许输入的数据的值的类型,MySQL 常用的数据类型如表 12-1 所示。

表 12-1 MySQL 常用的数据类型

数据类型	系统数据类型
整数型	TINYINT,SMALLINT,MEDIUMINT,INT,BIGINT
精确数值型	DECIMAL(M,D),NUMERIC(M,D)
浮点型	FLOAT,REAL,DOUBLE
位型	BIT
字符型	CHAR,VARCHAR,BLOB,TEXT,ENUM,SET
文本型	TINYTEXT,TEXT,MEDIUMTEXT,LONGTEXT
BLOB 类型	TINYBLOB,BLOB,MEDIUMBLOB,LONGBLOB
日期时间型	DATETIME,DATE,TIMESTAMP,TIME,YEAR
Unicode 字符型	NCHAR,NVARCHAR
二进制型	BINARY,VARBINARY

(3) 表的类型

MySQL 支持多个存储引擎作为对不同表的类型的处理器,存储引擎也称为表的类型。表的类型指明了表中数据的存储格式,MyISAM 为 MySQL 默认的存储引擎。表 12-2 列出了常用的表的存储引擎。另外,可以使用命令 show engines 查看 MySQL 服务实例支持的存储引擎。

表 12-2 常用的表的存储引擎

表的存储引擎	说明
MyISAM	二进制存储引擎
BDB	带页面锁定的事务安全表
CSV	数据以逗号分隔的表
ARCHIVE	档案存储引擎
EXAMPLE	示例引擎

（4）约束

约束是 MySQL 提供的自动保证数据完整性的一种机制，是数据库服务器强制用户必须遵从的业务逻辑。它通过限制字段中的数据、记录数据和表之间的数据来保证数据的完整性。表 12-3 列出了常用的约束类型。

表 12-3 常用的约束类型

类型	说明
primary key	指定主键
foreign key	指定外键
index	指定索引
unique	指定唯一索引
check	限制输入到一个或多个属性值的范围

（5）常用的 SQL 操作语句

结构化查询语言（Structured Query Language, SQL）是一种专门用来与数据库通信的语言，它可以帮助用户操作关系型数据库。常见的 SQL 操作语句及其功能如表 12-4 所示。

表 12-4 常见的 SQL 操作语句及其功能

语句	功能
create database［数据库名］	创建数据库
show databases［数据库名］	查看数据库
alter database［数据库名］［选项］	修改数据库
drop database［数据库名］	删除数据库
use［数据库名］	使用指定的数据库
show tables	查看表
create table［表名］	创建表
alter table［表名］［add/change/modify/drop］［字段名］/［数据类型］	修改表的结构 add：添加字段 change：修改字段 modify：修改字段；类型 drop：删除字段
alter table［原表名］rename to［新表名］	修改表名
create table［原表名］like［新表名］	复制表
drop table［表名］	删除表
insert into［表名］［字段 1,字段 2,…］value［字段 1 的值,字段 2 的值,…］	插入记录
update［表名］set［字段 1＝值,字段 2＝值,…］where［匹配条件］	更新记录
select［＊/字段名］from［表名］where［匹配条件］	查询记录，"＊"代表表中所有的字段
delete from［表名］where［匹配条件］	删除记录

12.1.3 数据库的创建与使用

员工信息表(staff)和工资表(salary)的结构如表12-5、表12-6所示。

表 12-5 员工信息表

字段名	数据类型	是否为空	长度	备注	说明
number	char	否	9	主键	员工编号
name	varchar	是	10		员工姓名
sex	char	是	4		员工性别
department	varchar	是	15		工作部门

表 12-6 工资表

字段名	数据类型	是否为空	长度	备注	说明
number	char	否	9	主键	员工编号
name	varchar	是	10		员工姓名
salary	decimal	是	4	小数位1位	员工工资

1. 查看已有的数据库

MySQL 数据库主要分为系统数据库、示例数据库和用户数据库。可以使用命令 show databases 命令查看系统默认的数据库，如图 12-4 所示。

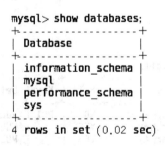

图12-4 查看系统默认的数据库

(1) information_schema 数据库类似"数据字典"，主要提供访问数据库元数据的方式。元数据是关于数据的数据，如数据库名、数据表名、列的数据类型以及访问权限等。

(2) mysql 是 MySQL 的核心数据库，它记录了用户及其访问权限等 MySQL 所需的控制和管理信息，一旦该数据库被损坏，MySQL 将无法正常运行。

(3) performance_schema 数据库主要用于收集数据库服务器的性能参数。

(4) sys 数据库包含了一系列的存储过程、自定义函数以及视图，可以帮助用户快速地了解系统的元数据信息。该数据库还结合了 information_schema、performance_schema 数据库的相关数据，以便让用户快速地检索元数据。

2. 创建数据库

```
mysql> create database company;
Query OK,1 row affected (0.05 sec)
```

3. 创建数据表

```
mysql> use company;
Database changed
mysql> create table staff(number char(9) primary key,name varchar(10),sex char(4),department varchar(15));
Query OK,0 rows affected (0.06 sec)
mysql> create table salary(number char(9) primary key,name varchar(10),salary decimal(4,1));
Query OK,0 rows affected (0.05 sec)
```

4. 显示所有表

显示数据库中的所有表,如图 12-5 所示。

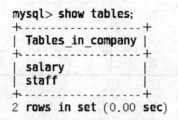

图 12-5 显示数据库中的所有表

5. 查看表结构

查看表结构如图 12-6 所示。

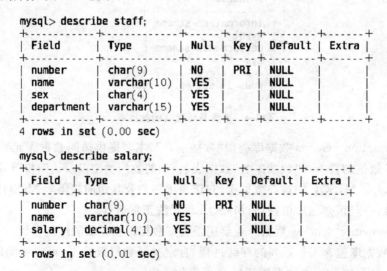

图 12-6 查看表结构

6. 修改表结构

(1) 在员工信息表中增加一个字段 age(年龄),数据类型为 int(2),如图 12-7 所示。

```
mysql> alter table staff add age int(2);
Query OK, 0 rows affected, 1 warning (0.07 sec)
Records: 0  Duplicates: 0  Warnings: 1

mysql> describe staff;
+------------+-------------+------+-----+---------+-------+
| Field      | Type        | Null | Key | Default | Extra |
+------------+-------------+------+-----+---------+-------+
| number     | char(9)     | NO   | PRI | NULL    |       |
| name       | varchar(10) | YES  |     | NULL    |       |
| sex        | char(4)     | YES  |     | NULL    |       |
| department | varchar(15) | YES  |     | NULL    |       |
| age        | int         | YES  |     | NULL    |       |
+------------+-------------+------+-----+---------+-------+
5 rows in set (0.00 sec)
```

图 12-7 增加字段

(2) 将工资表中的字段 salary 的名称改为 wages,字段类型改为 decimal(18,2),如图 12-8 所示。

```
mysql> alter table salary change salary wages decimal(18,2);
Query OK, 0 rows affected (0.06 sec)
Records: 0  Duplicates: 0  Warnings: 0

mysql> describe salary;
+--------+---------------+------+-----+---------+-------+
| Field  | Type          | Null | Key | Default | Extra |
+--------+---------------+------+-----+---------+-------+
| number | char(9)       | NO   | PRI | NULL    |       |
| name   | varchar(10)   | YES  |     | NULL    |       |
| wages  | decimal(18,2) | YES  |     | NULL    |       |
+--------+---------------+------+-----+---------+-------+
3 rows in set (0.00 sec)
```

图 12-8 修改字段名称

(3) 将员工信息表中的字段 age 的类型改为 char(4),如图 12-9 所示。

```
mysql> alter table staff modify age char(4);
Query OK, 0 rows affected (0.03 sec)
Records: 0  Duplicates: 0  Warnings: 0

mysql> describe staff;
+------------+-------------+------+-----+---------+-------+
| Field      | Type        | Null | Key | Default | Extra |
+------------+-------------+------+-----+---------+-------+
| number     | char(9)     | NO   | PRI | NULL    |       |
| name       | varchar(10) | YES  |     | NULL    |       |
| sex        | char(4)     | YES  |     | NULL    |       |
| department | varchar(15) | YES  |     | NULL    |       |
| age        | char(4)     | YES  |     | NULL    |       |
+------------+-------------+------+-----+---------+-------+
5 rows in set (0.01 sec)
```

图 12-9 修改字段类型

(4) 删除员工信息表中的字段 age,如图 12-10 所示。

```
mysql> alter table staff drop age;
Query OK, 0 rows affected (0.05 sec)
Records: 0  Duplicates: 0  Warnings: 0

mysql> describe staff;
+------------+-------------+------+-----+---------+-------+
| Field      | Type        | Null | Key | Default | Extra |
+------------+-------------+------+-----+---------+-------+
| number     | char(9)     | NO   | PRI | NULL    |       |
| name       | varchar(10) | YES  |     | NULL    |       |
| sex        | char(4)     | YES  |     | NULL    |       |
| department | varchar(15) | YES  |     | NULL    |       |
+------------+-------------+------+-----+---------+-------+
4 rows in set (0.00 sec)
```

图 12-10 删除字段

(5) 将员工信息表的名称修改为 staffs,修改成功后,再将该表名称修改为 staff,如图 12-11 所示。

```
mysql> alter table staff rename to staffs;
Query OK, 0 rows affected (0.00 sec)

mysql> describe staffs;
+------------+-------------+------+-----+---------+-------+
| Field      | Type        | Null | Key | Default | Extra |
+------------+-------------+------+-----+---------+-------+
| number     | char(9)     | NO   | PRI | NULL    |       |
| name       | varchar(10) | YES  |     | NULL    |       |
| sex        | char(4)     | YES  |     | NULL    |       |
| department | varchar(15) | YES  |     | NULL    |       |
+------------+-------------+------+-----+---------+-------+
4 rows in set (0.00 sec)

mysql> alter table staffs rename to staff;
Query OK, 0 rows affected (0.02 sec)

mysql> describe staff;
+------------+-------------+------+-----+---------+-------+
| Field      | Type        | Null | Key | Default | Extra |
+------------+-------------+------+-----+---------+-------+
| number     | char(9)     | NO   | PRI | NULL    |       |
| name       | varchar(10) | YES  |     | NULL    |       |
| sex        | char(4)     | YES  |     | NULL    |       |
| department | varchar(15) | YES  |     | NULL    |       |
+------------+-------------+------+-----+---------+-------+
4 rows in set (0.00 sec)
```

图 12-11 修改表的名称

7. 复制、删除表

将 staff 表进行复制,并将表名命名为 sta,完成后,删除表 sta,如图 12-12 所示。

8. 向表中插入记录

向员工信息表、工资表中插入记录,如图 12-13 所示。

```
mysql> create table sta like staff;
Query OK, 0 rows affected (0.02 sec)

mysql> show tables;
+-------------------+
| Tables_in_company |
+-------------------+
| salary            |
| sta               |
| staff             |
+-------------------+
3 rows in set (0.01 sec)

mysql> drop table sta;
Query OK, 0 rows affected (0.01 sec)

mysql> show tables;
+-------------------+
| Tables_in_company |
+-------------------+
| salary            |
| staff             |
+-------------------+
2 rows in set (0.00 sec)
```

图 12-12　复制、删除表

```
mysql> insert into staff values('1001','jack','male','jinglishi');
Query OK, 1 row affected (0.02 sec)

mysql> insert into salary values('1001','jack','3500');
Query OK, 1 row affected (0.00 sec)

mysql> select * from staff;
+--------+------+------+------------+
| number | name | sex  | department |
+--------+------+------+------------+
| 1001   | jack | male | jinglishi  |
+--------+------+------+------------+
1 row in set (0.00 sec)

mysql> select * from salary;
+--------+------+---------+
| number | name | wages   |
+--------+------+---------+
| 1001   | jack | 3500.00 |
+--------+------+---------+
1 row in set (0.00 sec)
```

图 12-13　向表中插入记录

9. 更新表中数据

更新工资表中 jack 的工资为 4500，如图 12-14 所示。

```
mysql> update salary set wages='4500' where name='jack';
Query OK, 1 row affected (0.00 sec)
Rows matched: 1  Changed: 1  Warnings: 0

mysql> select * from salary;
+--------+------+---------+
| number | name | wages   |
+--------+------+---------+
| 1001   | jack | 4500.00 |
+--------+------+---------+
1 row in set (0.00 sec)
```

图 12-14 更新表中数据

10. 删除表中数据

删除工资表中的信息,如图 12-15 所示。

```
mysql> delete from salary;
Query OK, 1 row affected (0.00 sec)

mysql> select * from salary;
Empty set (0.00 sec)
```

图 12-15 删除表中数据

可以在删除数据信息时进行条件匹配,再删除表中数据信息,如图 12-16 所示。

```
mysql> insert into salary values('1002','lucy','2500');
Query OK, 1 row affected (0.00 sec)

mysql> insert into salary values('1003','jerry','5500');
Query OK, 1 row affected (0.03 sec)

mysql> select * from salary;
+--------+-------+---------+
| number | name  | wages   |
+--------+-------+---------+
| 1002   | lucy  | 2500.00 |
| 1003   | jerry | 5500.00 |
+--------+-------+---------+
2 rows in set (0.00 sec)

mysql> delete from salary where number='1003';
Query OK, 1 row affected (0.00 sec)

mysql> select * from salary;
+--------+------+---------+
| number | name | wages   |
+--------+------+---------+
| 1002   | lucy | 2500.00 |
+--------+------+---------+
1 row in set (0.00 sec)
```

图 12-16 依据条件删除表中数据

12.1.4 备份与恢复数据库

1. 备份数据库

mysqldump 命令用于备份数据库,其用法为:

mysqldump -u 用户名 -p 用户密码 数据库名＞备份文件名.sql

如图 12-17 所示,将数据库 company 备份到/mysql 下,并命名为 company.sql。

```
[root@localhost ~]# mysqldump -u root -p company>/mysql/company.sql;
Enter password:
[root@localhost ~]# ls /mysql
company.sql   mysql80-community-release-el7-3.noarch.rpm
```

图 12-17 备份数据库

2. 恢复数据库

恢复数据库同样可以使用 mysqldump 命令,其用法为:

mysqldump -u 用户名 -p 用户密码 数据库名＜备份文件名.sql,如图 12-18 所示(仅显示部分结果)。

```
[root@localhost ~]# mysqldump -u root -p company</mysql/company.sql;
Enter password:
-- MySQL dump 10.13  Distrib 8.0.21, for Linux (x86_64)
--
-- Host: localhost    Database: company
-- ------------------------------------------------------
-- Server version       8.0.21

/*!40101 SET @OLD_CHARACTER_SET_CLIENT=@@CHARACTER_SET_CLIENT */;
/*!40101 SET @OLD_CHARACTER_SET_RESULTS=@@CHARACTER_SET_RESULTS */;
/*!40101 SET @OLD_COLLATION_CONNECTION=@@COLLATION_CONNECTION */;
/*!50503 SET NAMES utf8mb4 */;
/*!40103 SET @OLD_TIME_ZONE=@@TIME_ZONE */;
/*!40103 SET TIME_ZONE='+00:00' */;
/*!40014 SET @OLD_UNIQUE_CHECKS=@@UNIQUE_CHECKS, UNIQUE_CHECKS=0 */;
/*!40014 SET @OLD_FOREIGN_KEY_CHECKS=@@FOREIGN_KEY_CHECKS, FOREIGN_KEY_CHECKS=0 */;
/*!40101 SET @OLD_SQL_MODE=@@SQL_MODE, SQL_MODE='NO_AUTO_VALUE_ON_ZERO' */;
/*!40111 SET @OLD_SQL_NOTES=@@SQL_NOTES, SQL_NOTES=0 */;

--
-- Table structure for table `salary`
```

图 12-18 恢复数据库

12.1.5 Apache 使用 MySQL 进行网站的认证

(1) 以 CentOS6.5 系统为例,创建认证数据库,添加认证用户 jack,并设置其密码。

```
[root@localhost ~]# rpm -qa | grep mysql*
mysql-libs-5.1.71-1.el6.x86_64
mod_auth_mysql-3.0.0-11.el6_0.1.x86_64
mysql-server-5.1.71-1.el6.x86_64
mysql-devel-5.1.71-1.el6.x86_64
mysql-5.1.71-1.el6.x86_64
[root@localhost ~]# yum install -y mod_auth_mysql
```

```
[root@localhost ~]# service mysqld restart
[root@localhost ~]# mysql -u root -p
Enter password:
mysql> use company;
Reading table information for completion of table and column names
You can turn off this feature to get a quicker startup with -A
Database changed
mysql> show tables;
+------------------+
| Tables_in_company |
+------------------+
| renzheng         |
+------------------+
1 row in set (0.00 sec)
mysql> describe renzheng;
+----------+-------------+------+-----+---------+-------+
| Field    | Type        | Null | Key | Default | Extra |
+----------+-------------+------+-----+---------+-------+
| id       | char(9)     | NO   | PRI | NULL    |       |
| name     | varchar(10) | YES  |     | NULL    |       |
| password | varchar(20) | YES  |     | NULL    |       |
+----------+-------------+------+-----+---------+-------+
3 rows in set (0.00 sec)
mysql> select * from renzheng;
+------+------+----------+
| id   | name | password |
+------+------+----------+
| 1001 | jack | 123456   |
+------+------+----------+
1 row in set (0.00 sec)
```

（2）配置 httpd 服务，建立网站。

```
[root@localhost ~]# mkdir /web
[root@localhost ~]# echo "Welcome" > /web/index.html
[root@localhost ~]# vim /etc/httpd/conf/httpd.conf
……
<VirtualHost 192.168.1.1:80>
  DocumentRoot /web
</VirtualHost>
……
[root@localhost ~]# service httpd restart
[root@localhost ~]# setenforce 0
```

（3）在 hpptd 服务中启用 MySQL 认证。

```
[root@localhost~]# vim /etc/httpd/conf/httpd.conf
<VirtualHost 192.168.1.1:80>
DocumentRoot /web
<Directory /web>
AuthName Web Renzheng                    //认证的名字
AuthType Basic                           //认证的类型,设置为基本认证
AuthMYSQLEnable on                       //开启 mysql 认证
AuthMYSQLUser root                       //mysql 的用户为 root
AuthMYSQLPassword 123456                 //用户 root 的密码
AuthMYSQLDB company                      //登录的数据库
AuthMYSQLUserTable renzheng               //进行用户查询的数据表
AuthMYSQLNameField name                  //httpd 验证的用户名字段
AuthMYSQLPasswordField password          //httpd 验证的密码字段
AuthMySQLPwEncryption none               //不设置密码加密方式
Require valid-user                       //每个用户都可以访问
</Directory>
</VirtualHost>
[root@localhost~]# service httpd restart
```

（4）测试,这时需要输入用户名 jack 及其密码,才可以浏览网站,如图 12-19、图 12-20 所示。

```
[root@localhost~]# firefox http://192.168.1.1
```

图 12-19　输入用户名和密码

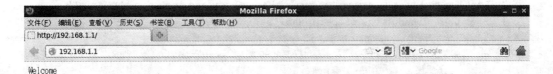

图 12-20　成功访问

任务 12.2　MariaDB 数据库服务器

12.2.1　MariaDB 数据库简介

MariaDB 数据库管理系统是 MySQL 的一个分支，主要由开源社区在维护。自从 MySQL 被 Oracle 公司收购之后，MySQL 就从开源软件转变为了非开源软件。MariaDB 数据库管理系统正是因为其开源软件的特性，并且在 API 和命令行等方面完全兼容 MySQL，从而很快就占领了市场。

12.2.2　安装 MariaDB 数据库

（1）为了避免数据库系统的冲突，先卸载 MySQL 数据库。

[root@localhost~]# yum -y remove mysql*

（2）安装 MariaDB 软件包。

[root@localhost~]# yum -y install mariadb mariadb-server
[root@localhost~]# rpm -qa|grep mariadb
mariadb-libs-5.5.56-2.el7.x86_64
mariadb-5.5.56-2.el7.x86_64
mariadb-server-5.5.56-2.el7.x86_64

（3）启动 MariaDB 服务。

[root@localhost~]# systemctl start mariadb

（4）初始化 MariaDB 数据库。

[[root@localhost~]# mysql_secure_installation

NOTE: RUNNING ALL PARTS OF THIS SCRIPT IS RECOMMENDED FOR ALLMariaDB
      SERVERS IN PRODUCTION USE!   PLEASE READ EACH STEP CAREFULLY!

In order to log intoMariaDB to secure it,we'll need the current
password for the root user.   If you've just installedMariaDB,and
you haven't set the root password yet,the password will be blank,
so you should just press enter here.

Enter current password for root (enter for none):   //直接敲回车，密码为空
OK,successfully used password,moving on…

Setting the root password ensures that nobody can log into theMariaDB
root user without the properauthorisation.

Set root password? [Y/n]y   //是否设置 root 用户密码
New password: //输入密码
Re-enter new password: //再次输入密码
Password updated successfully!

```
Reloading privilege tables..
...Success!

By default,aMariaDB installation has an anonymous user,allowing anyone
to log intoMariaDB without having to have a user account created for
them. This is intended only for testing,and to make the installation
go a bit smoother. You should remove them before moving into a
production environment.

Remove anonymous users? [Y/n]y        //删除匿名账户
...Success!

Normally,root should only be allowed to connect from 'localhost'.  This
ensures that someone cannot guess at the root password from the network.

Disallow root login remotely? [Y/n]y     //是否允许 root 用户远程连接数据库
...Success!
By default,MariaDB comes with a database named 'test' that anyone can
access. This is also intended only for testing,and should be removed
before moving into a production environment.

Remove test database and access to it? [Y/n]y      //是否删除 test 测试数据库
 -Dropping test database...
...Success!
 -Removing privileges on test database...
...Success!

Reloading the privilege tables will ensure that all changes made so far
will take effect immediately.

Reload privilege tables now? [Y/n]y   //是否重新加载授权表,让初始化后的配置立即生效
...Success!

Cleaning up...

All done! If you've completed all of the above steps,yourMariaDB
installation should now be secure.

Thanks for usingMariaDB!
```

(5) 配置防火墙,放行 MySQL 服务。

```
[root@localhost ~]# firewall-cmd --permanent --add-service=mysql
success
[root@localhost ~]# firewall-cmd --reload
success
[root@localhost ~]# firewall-cmd --list-all
public (active)
  target: default
icmp-block-inversion: no
```

```
        interfaces: ens33
        sources:
        services:ssh dhcpv6-client mysql
......
```

（6）登录 MariaDB 数据库。

```
[root@localhost~]# mysql-u root-p
Enter password:
Welcome to theMariaDB monitor.Commands end with;or\g.
YourMariaDB connection id is 10
Server version: 5.5.56-MariaDBMariaDB Server
Copyright (c) 2000,2017,Oracle,MariaDB Corporation Ab and others.
Type 'help;' or '\h' for help. Type '\c' to clear the current input statement.
MariaDB[(none)]>
```

12.2.3 数据库的创建与使用

这里，使用 MariaDB 数据库管理系统创建与管理数据库的方法，与 MySQL 数据库管理系统的方法是一致的。下面以创建表 12-5、表 12-6 的数据库为例，进行演示。

（1）查看已有的数据库，如图 12-21 所示。

```
MariaDB [(none)] > show databases;
+--------------------+
| Database           |
+--------------------+
| information_schema |
| mysql              |
| performance_schema |
+--------------------+
3 rows in set (0.00 sec)
```

图 12-21 查看已有的数据库

（2）创建数据库。

```
MariaDB[(none)]> create database company;
Query OK,1 row affected (0.00 sec)
```

（3）创建数据表，如图 12-22 所示。

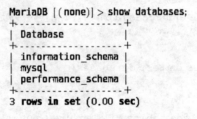

图 12-22 创建数据表

（4）显示数据库中的所有表，如图 12-23 所示。

```
MariaDB [company] > show tables;
+-------------------+
| Tables_in_company |
+-------------------+
| salary            |
| staff             |
+-------------------+
2 rows in set (0.00 sec)
```

图 12-23 显示数据库中的所有表

(5) 查看表结构,如图 12-24 所示。

```
MariaDB [company]> describe staff;
+------------+-------------+------+-----+---------+-------+
| Field      | Type        | Null | Key | Default | Extra |
+------------+-------------+------+-----+---------+-------+
| number     | char(9)     | NO   | PRI | NULL    |       |
| name       | varchar(10) | YES  |     | NULL    |       |
| sex        | char(4)     | YES  |     | NULL    |       |
| department | varchar(15) | YES  |     | NULL    |       |
+------------+-------------+------+-----+---------+-------+
4 rows in set (0.00 sec)

MariaDB [company]> describe salary;
+--------+---------------+------+-----+---------+-------+
| Field  | Type          | Null | Key | Default | Extra |
+--------+---------------+------+-----+---------+-------+
| number | char(9)       | NO   | PRI | NULL    |       |
| name   | varchar(10)   | YES  |     | NULL    |       |
| salary | decimal(18,2) | YES  |     | NULL    |       |
+--------+---------------+------+-----+---------+-------+
3 rows in set (0.00 sec)
```

图 12-24　查看表结构

(6) 修改表结构。

① 在员工信息表中增加一个字段 age(年龄),数据类型为 int(2),如图 12-25 所示。

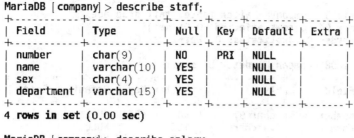

图 12-25　增加字段

② 将员工信息表中的字段 age 的类型改为 char(4),如图 12-26 所示。

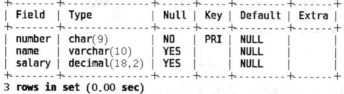

图 12-26　修改字段类型

③ 删除员工信息表中的字段 age，如图 12-27 所示。

```
MariaDB [company]> alter table staff drop age;
Query OK, 0 rows affected (0.01 sec)
Records: 0  Duplicates: 0  Warnings: 0

MariaDB [company]> describe staff;
+------------+-------------+------+-----+---------+-------+
| Field      | Type        | Null | Key | Default | Extra |
+------------+-------------+------+-----+---------+-------+
| number     | char(9)     | NO   | PRI | NULL    |       |
| name       | varchar(10) | YES  |     | NULL    |       |
| sex        | char(4)     | YES  |     | NULL    |       |
| department | varchar(15) | YES  |     | NULL    |       |
+------------+-------------+------+-----+---------+-------+
4 rows in set (0.01 sec)
```

图 12-27　删除字段

④ 将员工信息表的名称修改为 staffs，修改成功后再将该表名称修改为 staff，如图 12-28 所示。

```
MariaDB [company]> alter table staff rename to staffs;
Query OK, 0 rows affected (0.00 sec)

MariaDB [company]> describe staffs;
+------------+-------------+------+-----+---------+-------+
| Field      | Type        | Null | Key | Default | Extra |
+------------+-------------+------+-----+---------+-------+
| number     | char(9)     | NO   | PRI | NULL    |       |
| name       | varchar(10) | YES  |     | NULL    |       |
| sex        | char(4)     | YES  |     | NULL    |       |
| department | varchar(15) | YES  |     | NULL    |       |
+------------+-------------+------+-----+---------+-------+
4 rows in set (0.00 sec)

MariaDB [company]> alter table staffs rename to staff;
Query OK, 0 rows affected (0.00 sec)

MariaDB [company]> describe staff;
+------------+-------------+------+-----+---------+-------+
| Field      | Type        | Null | Key | Default | Extra |
+------------+-------------+------+-----+---------+-------+
| number     | char(9)     | NO   | PRI | NULL    |       |
| name       | varchar(10) | YES  |     | NULL    |       |
| sex        | char(4)     | YES  |     | NULL    |       |
| department | varchar(15) | YES  |     | NULL    |       |
+------------+-------------+------+-----+---------+-------+
4 rows in set (0.00 sec)
```

图 12-28　修改表的名称

（7）复制、删除表。

将 staff 表进行复制，并将表名命名为 sta，完成后删除表 sta，如图 12-29 所示。

（8）向员工信息表、工资表中插入记录，如图 12-30 所示。

（9）更新工资表中 jack 的工资为 4500，如图 12-31 所示。

```
MariaDB [company] > create table sta like staff;
Query OK, 0 rows affected (0.00 sec)

MariaDB [company] > show tables;
+-------------------+
| Tables_in_company |
+-------------------+
| salary            |
| sta               |
| staff             |
+-------------------+
3 rows in set (0.00 sec)

MariaDB [company] > drop table sta;
Query OK, 0 rows affected (0.00 sec)

MariaDB [company] > show tables;
+-------------------+
| Tables_in_company |
+-------------------+
| salary            |
| staff             |
+-------------------+
2 rows in set (0.00 sec)
```

图 12-29　删除表

```
MariaDB [company] > insert into staff values('1001','jack','male','jinglishi');
Query OK, 1 row affected (0.00 sec)

MariaDB [company] > insert into salary values('1001','jack','3500');
Query OK, 1 row affected (0.00 sec)

MariaDB [company] > select * from staff;
+--------+------+------+------------+
| number | name | sex  | department |
+--------+------+------+------------+
| 1001   | jack | male | jinglishi  |
+--------+------+------+------------+
1 row in set (0.00 sec)

MariaDB [company] > select * from salary;
+--------+------+---------+
| number | name | salary  |
+--------+------+---------+
| 1001   | jack | 3500.00 |
+--------+------+---------+
1 row in set (0.00 sec)
```

图 12-30　向表中插入记录

```
MariaDB [company] > update salary set salary='4500' where name='jack';
Query OK, 1 row affected (0.00 sec)
Rows matched: 1  Changed: 1  Warnings: 0

MariaDB [company] > select * from salary;
+--------+------+---------+
| number | name | salary  |
+--------+------+---------+
| 1001   | jack | 4500.00 |
+--------+------+---------+
1 row in set (0.00 sec)
```

图 12-31　更新表中数据

(10) 删除工资表中的信息,如图 12-32 所示。

```
MariaDB [company] > delete from staff;
Query OK, 1 row affected (0.00 sec)

MariaDB [company] > delete from salary;
Query OK, 1 row affected (0.00 sec)
```

图 12-32 删除表中的信息

12.2.4 备份与恢复数据库

1. 备份数据库

如图 12-33 所示,将数据库 company 备份到 /mysql 下,并命名为 company.sql。

```
[root@localhost ~]# mysqldump -u root -p company>/mysql/company.sql;
Enter password:
[root@localhost ~]# ls /mysql
company.sql
```

图 12-33 备份数据库

2. 恢复数据库

恢复数据库如图 12-34 所示(仅显示部分结果)。

```
[root@localhost ~]# mysqldump -u root -p company</mysql/company.sql;
Enter password:
-- MySQL dump 10.14  Distrib 5.5.56-MariaDB, for Linux (x86_64)
--
-- Host: localhost    Database: company
-- ------------------------------------------------------
-- Server version       5.5.56-MariaDB

/*!40101 SET @OLD_CHARACTER_SET_CLIENT=@@CHARACTER_SET_CLIENT */;
/*!40101 SET @OLD_CHARACTER_SET_RESULTS=@@CHARACTER_SET_RESULTS */;
/*!40101 SET @OLD_COLLATION_CONNECTION=@@COLLATION_CONNECTION */;
/*!40101 SET NAMES utf8 */;
/*!40103 SET @OLD_TIME_ZONE=@@TIME_ZONE */;
/*!40103 SET TIME_ZONE='+00:00' */;
/*!40014 SET @OLD_UNIQUE_CHECKS=@@UNIQUE_CHECKS, UNIQUE_CHECKS=0 */;
/*!40014 SET @OLD_FOREIGN_KEY_CHECKS=@@FOREIGN_KEY_CHECKS, FOREIGN_KEY_CHECKS=0 */;
/*!40101 SET @OLD_SQL_MODE=@@SQL_MODE, SQL_MODE='NO_AUTO_VALUE_ON_ZERO' */;
/*!40111 SET @OLD_SQL_NOTES=@@SQL_NOTES, SQL_NOTES=0 */;

--
-- Table structure for table `salary`
--

DROP TABLE IF EXISTS `salary`;
/*!40101 SET @saved_cs_client     = @@character_set_client */;
/*!40101 SET character_set_client = utf8 */;
```

图 12-34 恢复数据库

自我测试

一、填空题

1. 关系型数据库管理系统将数据保存在不同的_____中。
2. MySQL 数据库文件有".FRM"".MYD"和".MYI"三类。其中".FRM"是描述表_____的文件,".MYD"是表的_____文件,".MYI"是表数据文件中的_____文件。
3. MySQL 支持多个存储引擎作为对不同表的类型的处理器,存储引擎也称为_____。
4. _____是 MySQL 提供的自动保证数据完整性的一种机制,是数据库服务器强制用户必须遵从的业务逻辑。
5. _____是一种专门用来与数据库通信的语言,它可以帮助用户操作关系型数据库。

二、简答题

简述 MySQL 默认的四种数据库类型。

三、实训题

分别使用 MySQL 数据库、MariaDB 数据库实现以下操作:

(1) 创建一个学生数据库 students,包含学生表(student)、课程表(course)、成绩表(score),如表 12-7～表 12-9 所示。

表 12-7 学生表

字段名	数据类型	是否为空	长度	备注
学号	char	否	9	主键
姓名	varchar	是	10	
性别	char	是	2	
班级	Text	是		
出生日期	date	是		
民族	varchar	是	10	
政治面貌	varchar	是	8	

表 12-8 课程表

字段名	数据类型	是否为空	长度	备注
课程号	char	否	5	主键
课程名称	varchar	是	30	
课程简介	text	是		
课时	int	是		
学分	int	是		
开课学期	varchar	是	10	

表 12-9　成绩表

字段名	数据类型	是否为空	长度	小数位	备注
学号	char	否	9		外键
课程号	char	否	5		外键
成绩	decimal	是	4	1	

（2）使用 SQL 语言，向 3 个表中插入数据，如表 12-10～表 12-12 所示。

表 12-10　学生表数据

学号	姓名	性别	班级	出生日期	民族	政治面貌
202001001	李文	女	2020 大数据 1 班	2000-02-01	汉族	共青团员
202003002	张山	男	2020 财会 1 班	2001-10-20	汉族	共青团员
202005010	陈晨	女	2020 旅游 3 班	2002-11-14	苗族	群众
202005031	王雪	女	2020 旅游 3 班	2002-04-20	汉族	共青团员

表 12-11　课程表数据

课程号	课程名称	课程简介	课时	学分	开课学期
06001	计算机应用基础	计算机的基本操作	4	4	1
05003	大学英语	英语综合应用能力	4	4	1
07002	数据库应用技术	数据库系统的设计	2	2	1

表 12-12　成绩表数据

学号	课程号	成绩
202001001	06001	93
202001001	07002	88
202003002	05003	90
202003002	06001	88.5
202005010	06001	79
202005031	06001	86
202005031	05003	92.5

（3）使用 SQL 语言，查询李文的计算机应用基础课程的成绩、张山的大学英语课程的成绩、陈晨的计算机应用基础课程的成绩。

项目 13 无人值守安装系统

项目综述

公司办公使用的电脑，在使用过程中，难免需要重新安装系统，采用传统方式安装系统，较浪费时间，网络管理员将使用无人值守的方式安装系统，从而提高工作效率。

项目目标

- 了解无人值守安装系统相关知识；
- 掌握无人值守安装系统的配置方法；
- 掌握客户端无人值守安装系统的方法。

任务 13.1 无人值守安装系统简介

13.1.1 无人值守安装简介

无人值守安装是指在安装过程中，不需任何人工的操作，而是直接按照预先设置好的软件进行安装，通过这种方式安装系统将极大地提高工作效率，并大幅减小系统管理员的工作量。

在 Linux 系统中，使用 PXE＋TFTP＋FTP＋DHCP＋Kickstart 服务来建立无人值守安装系统。PXE(Preboot Execution Environment，预启动执行环境)是由 Intel 公司开发的技术，工作模式为 Client/Server 模式，支持工作站通过网络从远端服务器下载映像，并由此支持通过网络启动操作系统，在启动过程中，终端要求服务器分配 IP 地址，再用 TFTP 协议下载一个启动软件包到本机内存中执行，由这个启动软件包完成终端基本软件设置，从而引导预先安装在服务器中的终端操作系统。需要注意的是，要进行安装的计算机上的网卡必须支持 PXE 技术。

TFTP(Trivial File Transfer Protocol，简单文件传输协议)是 TCP/IP 协议族中的一个用来在客户端与服务器之间进行简单文件传输的协议。其采用的是 UDP 协议，端口号为 69。无人值守安装系统使用 TFTP 协议帮助客户端获取引导及驱动文件，使用 vsftpd 或者 httpd 服务程序将完整的系统安装镜像通过网络传输给客户端。

Kickstart 是一种无人值守的安装方式，其工作原理是预先把需要人工填写的各种参数保存成一个 ks.cfg 文件。如果在安装过程中需要填写参数，安装程序首先会去查找 Kickstart 生成的文件，如果找到合适的参数，就采用所找到的参数；如果没有找到合适的参数，就需要进行手工干预。

13.1.2 无人值守安装系统的工作流程

无人值守安装系统的工作流程如图 13-1 所示。

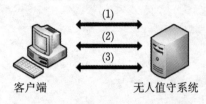

图 13-1 无人值守安装系统的工作流程

（1）客户端向无人值守系统服务器请求、分配 IP 地址；

（2）客户端向无人值守系统服务器请求、获取引导文件；

（3）客户端向无人值守系统服务器请求、下载应答文件。

◀ 任务 13.2 部署无人值守安装系统 ▶

13.2.1 配置服务器

按照表 13-1 来配置服务器和客户端的 IP 地址、网络设置。

表 13-1 服务器和客户端的 IP 地址

名称	IP 地址
服务器	192.168.1.1
客户端	无

实现步骤为：

（1）设置服务器的网络连接模式为 VMnet1，并关闭服务器自身的 DHCP 服务，避免虚拟机自带的 DHCP 服务影响手动部署的 DHCP 服务，如图 13-2 所示。

图 13-2 关闭虚拟机自带的 DHCP 服务

(2) 配置 DHCP 服务程序。

允许 BOOTP 引导程序协议(让局域网内暂时没有操作系统的主机可以获取静态 IP 地址),同时在配置文件的最下面加载引导驱动文件 pxelinux.0(让客户端主机获取到 IP 地址后主动获取引导驱动文件,自行进入下一步的安装过程)。

```
[root@localhost ~]# yum -y install dhcp
[root@localhost ~]# vim /etc/dhcp/dhcpd.conf
……
allow booting;
allow bootp;
ddns-update-style interim;
ignore client-updates;
subnet 192.168.1.0 netmask 255.255.255.0 {
option subnet-mask 255.255.255.0;
option domain-name-servers 192.168.1.1;
range dynamic-bootp 192.168.1.10 192.168.1.100;
default-lease-time 21600;
max-lease-time 43200;
next-server 192.168.1.1;
filename "pxelinux.0";
}……
[root@localhost ~]# systemctl restart dhcpd
```

(3) 配置 TFTP 服务程序,并配置防火墙,放行该服务。

```
[root@localhost ~]# yum -y install tftp-server
[root@localhost ~]# yum -y install xinetd*
[root@localhost ~]# vim /etc/xinetd.d/tftp
service tftp
{
socket_type=dgram
protocol=udp
wait=yes
user=root
server= /usr/sbin/in.tftpd
server_args= -s /var/lib/tftpboot
disable=no
per_source=11
cps=100 2
flags=IPv4
}
[root@localhost ~]# systemctl restart xinetd
[root@localhost ~]# firewall-cmd --permanent --add-port=69/udp
success
[root@localhost ~]# firewall-cmd --reload
success
```

(4) 配置 SYSLinux 服务程序。

SYSLinux 是一个用于提供引导加载的服务程序。安装好 SYSLinux 服务程序软件包后，在/usr/share/syslinux 目录中会出现很多引导文件。

```
[root@localhost ~]# yum -y install syslinux
[root@localhost ~]# ls /usr/share/syslinux
altmbr.bin      ethersel.c32    isohdpfx.bin        mbr_c.bin       sanboot.c32
altmbr_c.bin    gfxboot.c32     isohdpfx_c.bin      mbr_f.bin       sdi.c32
altmbr_f.bin    gptmbr.bin      isohdpfx_f.bin      memdisk         sysdump.c32
cat.c32         gptmbr_c.bin    isohdppx.bin        memdump.com     syslinux64.exe
chain.c32       gptmbr_f.bin    isohdppx_c.bin      meminfo.c32     syslinux.com
cmd.c32         gpxecmd.c32     isohdppx_f.bin      menu.c32        syslinux.exe
config.c32      gpxelinux.0     isolinux.bin        pcitest.c32     ver.com
cpuid.c32       gpxelinuxk.0    isolinux-debug.bin  pmload.c32      vesainfo.c32
cpuidtest.c32   hdt.c32         kbdmap.c32          poweroff.com    vesamenu.c32
diag            host.c32        linux.c32           pwd.c32         vpdtest.c32
disk.c32        ifcpu64.c32     ls.c32              pxechain.com    whichsys.c32
dmitest.c32     ifcpu.c32       lua.c32             pxelinux.0      zzjson.c32
dosutil         ifplop.c32      mboot.c32           reboot.c32
elf.c32         int18.com       mbr.bin             rosh.c32
```

接下来，将文件 pxelinux.0 以及系统光盘内的镜像文件拷贝到 TFTP 服务的默认目录/var/lib/tftpboot。

```
[root@localhost ~]# cd /var/lib/tftpboot/
[root@localhost tftpboot]# cp /usr/share/syslinux/pxelinux.0 .
[root@localhost tftpboot]# cp /media/images/pxeboot/{vmlinuz,initrd.img} .
[root@localhost tftpboot]# cp /media/isolinux/{vesamenu.c32,boot.msg} .
[root@localhost tftpboot]# ls
boot.msg  initrd.img  pxelinux.0  vesamenu.c32  vmlinuz
```

创建目录 pxelinux.cfg，将光盘内的文件 isolinux.cfg 拷贝到该目录下，并取名为 default。

```
[root@localhost tftpboot]# mkdir pxelinux.cfg
[root@localhost tftpboot]# cp /media/isolinux/isolinux.cfg pxelinux.cfg/default
[root@localhost tftpboot]# ls pxelinux.cfg/
default
```

修改文件 default 的第一行参数为 linux，第 64 行的参数为 FTP 传输方式，并指定光盘镜像的获取地址以及 Kickstart 应答文件的获取路径。

```
[root@localhost tftpboot]# vim pxelinux.cfg/default
1 default linux
……
64 append initrd=initrd.img inst.stage2=ftp://192.168.1.1 ks=ftp://192.168.1.1/pub/ks.cfg quiet
……
```

（5）配置 vsftpd 服务程序。

将系统光盘内的内容全部拷贝到 vsftpd 服务程序的默认根目录，并配置防火墙放行该服务以及在 SELinux 内放行 FTP 传输（或者直接关闭 SELinux 服务）。

```
[root@localhost~]# yum -y install vsftpd
[root@localhost~]# systemctl restart vsftpd
[root@localhost~]# cp -r /media/* /var/ftp/
[root@localhost~]# firewall-cmd --permanent --add-service=ftp
success
[root@localhost~]# firewall-cmd --reload
success
[root@localhost~]# setsebool -P ftpd_connect_all_unreserved=on
```

(6) 配置 Kickstart 应答文件。

应答文件默认在 root 管理员家目录下名为 anaconda-ks.cfg。需要将它复制到 vsftpd 服务程序的默认根目录下,并确保所有人都对其具有可读权限。

```
[root@localhost~]# cp ~/anaconda-ks.cfg /var/ftp/pub/ks.cfg
[root@localhost~]# chmod +r /var/ftp/pub/ks.cfg
```

然后,修改 ks.cfg 文件的第 7 行、第 34 行的参数。

```
[root@localhost~]# vim /var/ftp/pub/ks.cfg
……
7 url --url=ftp://192.168.1.1         //设置 FTP 服务器地址
……
34 clearpart --all --initlabel        //清空所有磁盘内容并初始化磁盘
……
```

13.2.2 配置客户端

实现步骤为:

(1) 创建客户端虚拟机,如图 13-3～图 13-8 所示。

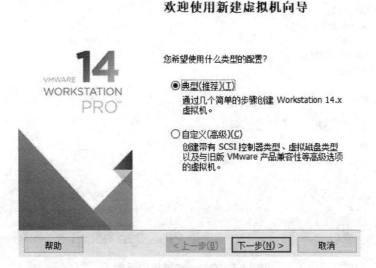

图 13-3 选择虚拟机配置类型

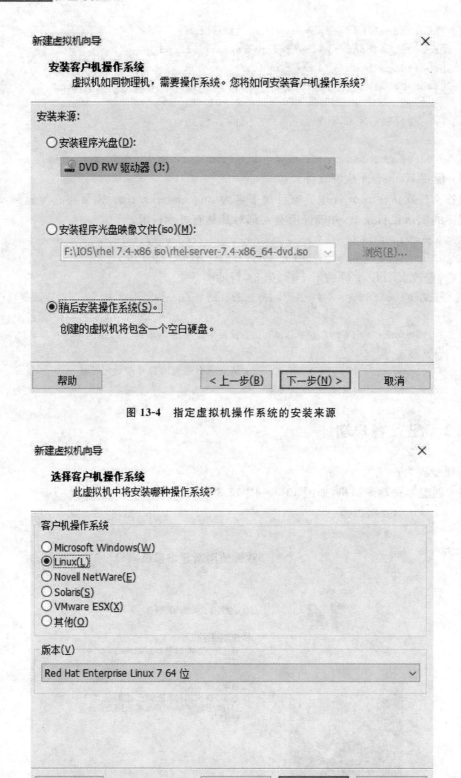

图 13-4　指定虚拟机操作系统的安装来源

图 13-5　选择操作系统的版本

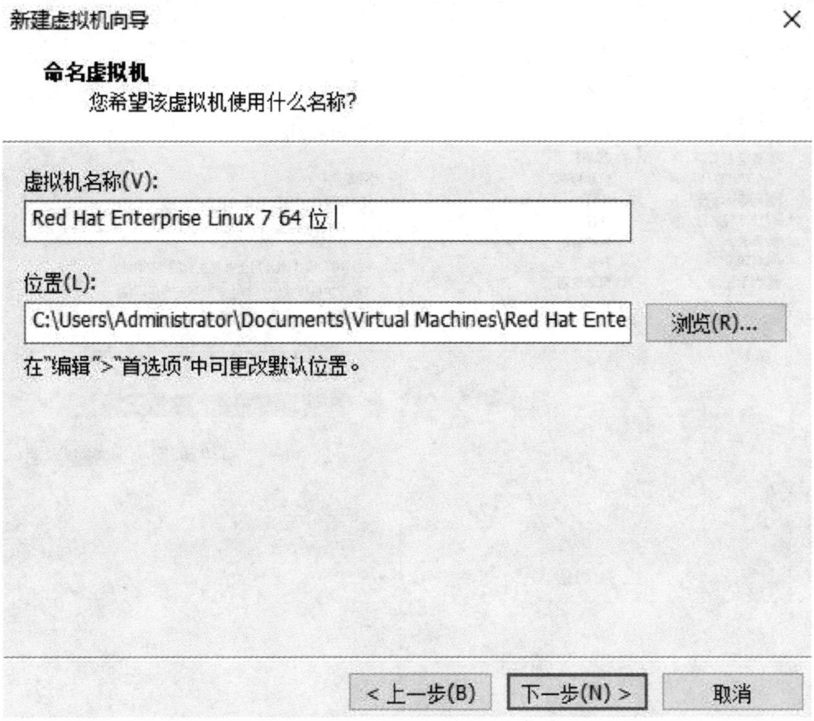

图 13-6　命名虚拟机以及指定安装路径

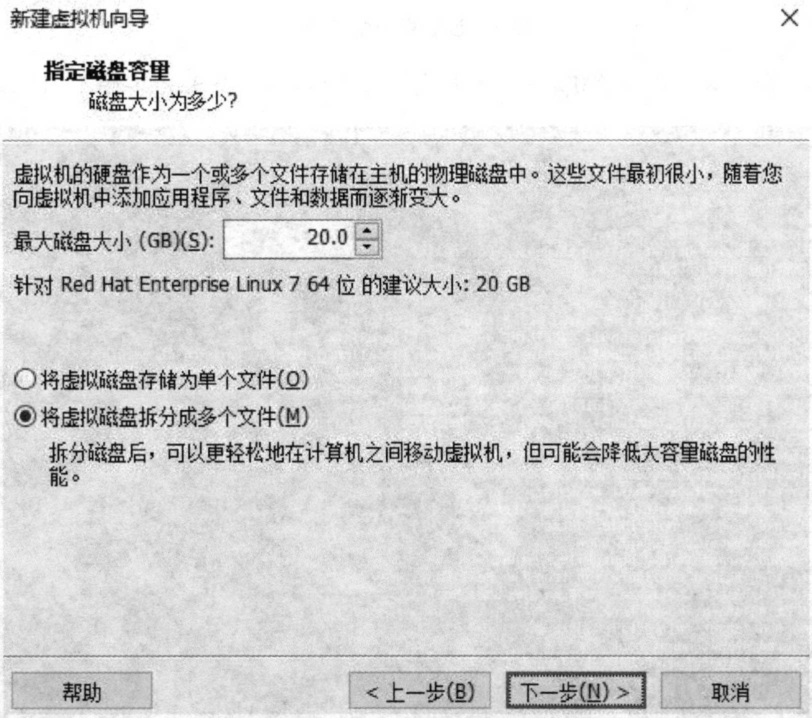

图 13-7　设置虚拟机磁盘大小

图 13-8　设置虚拟机网络模式

（2）启动客户端计算机，开始无人值守安装系统，如图 13-9～图 13-11 所示。

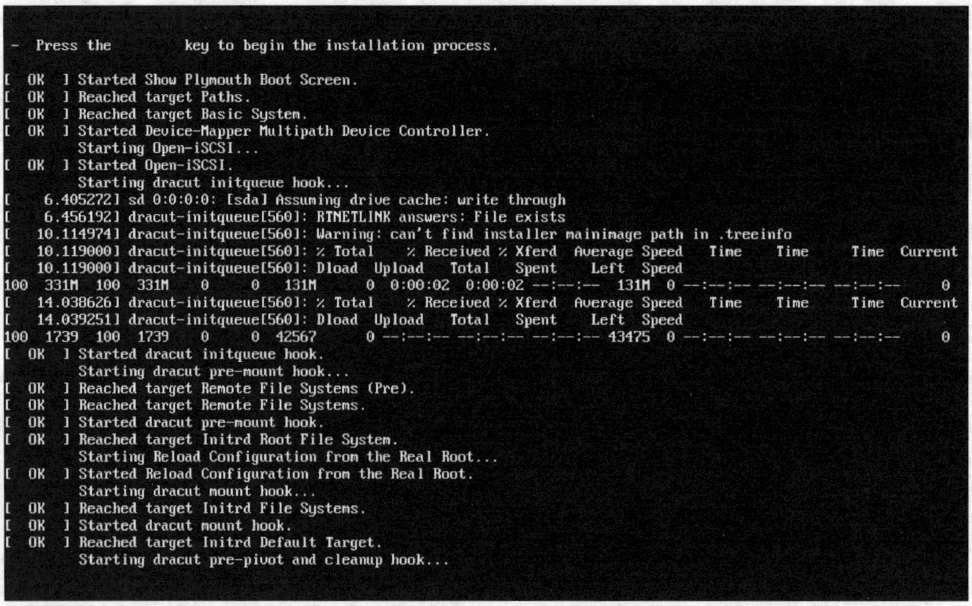

图 13-9　无人值守安装向导的初始化

图 13-10　自动传输光盘镜像文件并安装系统

图 13-11　无人值守自动安装系统

除此以外,也可以自行在系统中配置一个 Kickstart 应答文件(之前的 dhcpd、tptp、vsftpd 服务配置保持不变),如图 13-12～图 13-20 所示。

```
[root@localhost ~]# vim /etc/yum.repos.d/development.repo
[development]
name=development
baseurl=ftp://192.168.1.1
enabled=1
gpgcheck=0
[root@localhost ~]# yum -y install system-config-kickstart
[root@localhost ~]# system-config-kickstart
```

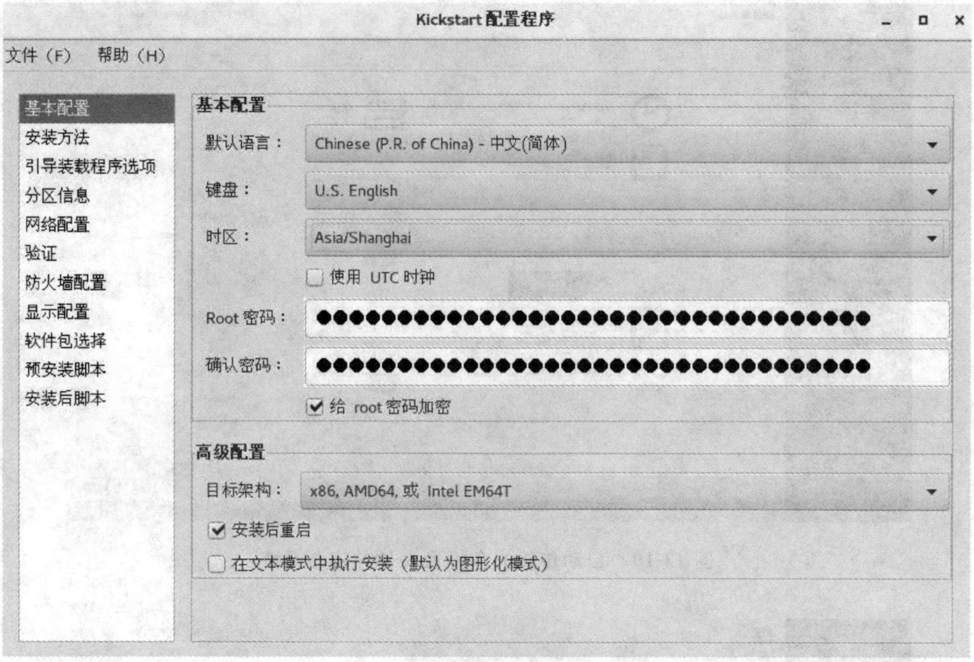

图 13-12　基本配置

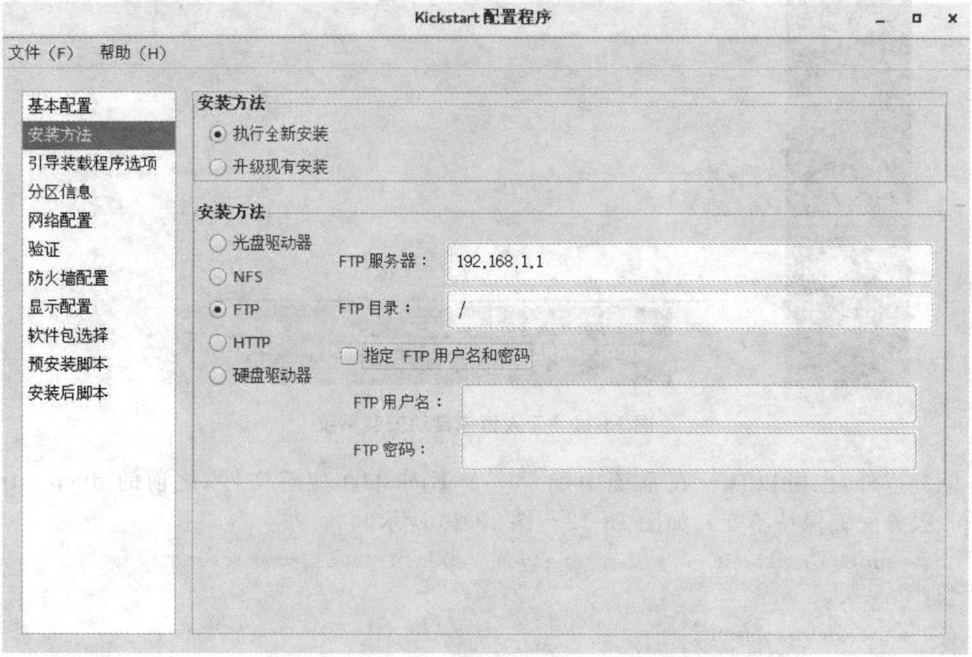

图 13-13　指定 FTP 服务器地址以及目录

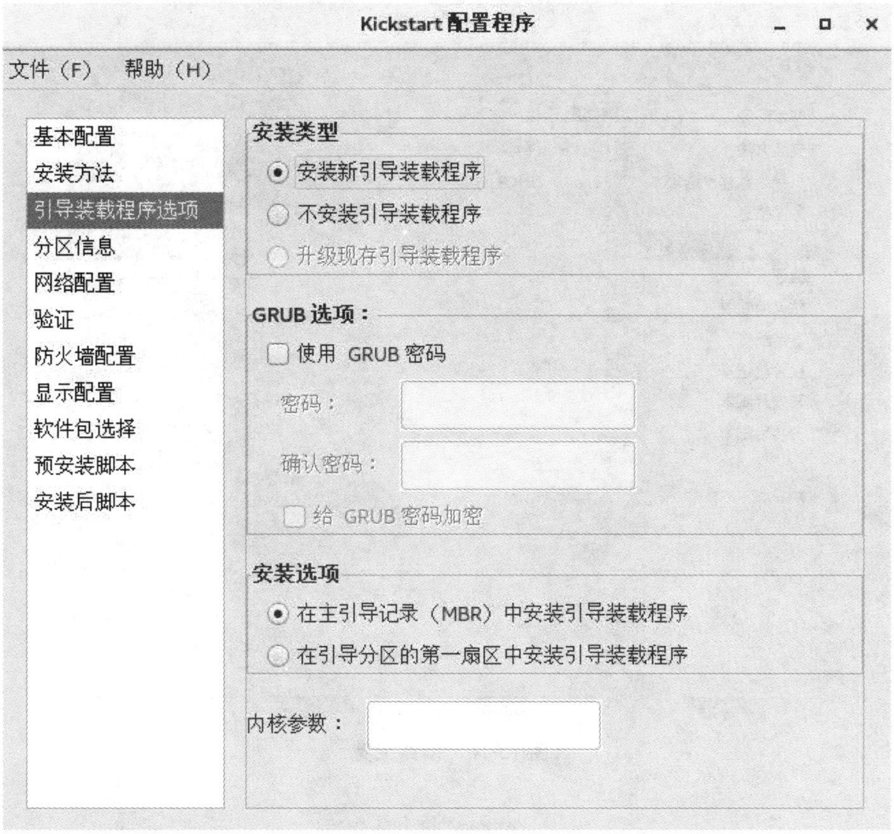

图 13-14　引导装载程序选项

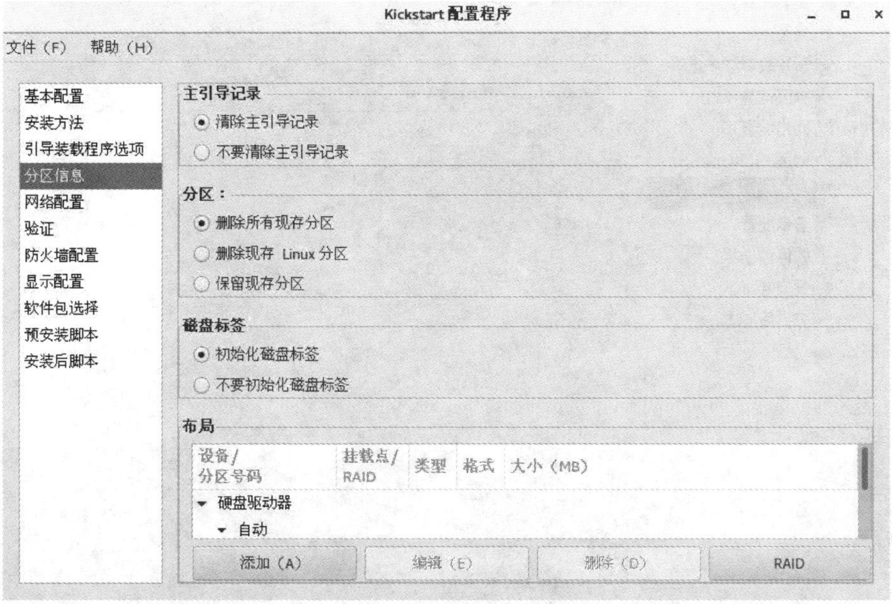

图 13-15　分区信息

图 13-16 网络配置

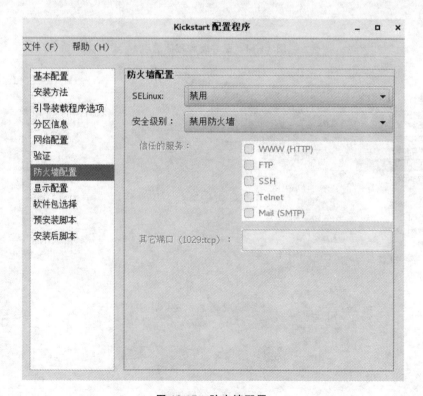

图 13-17 防火墙配置

项目 13　无人值守安装系统

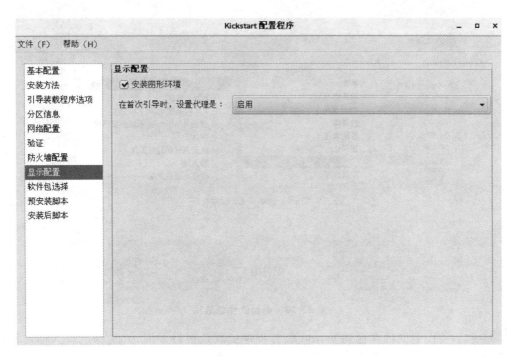

图 13-18　显示配置

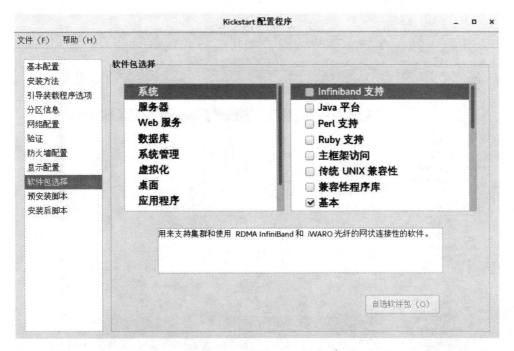

图 13-19　系统软件包选择

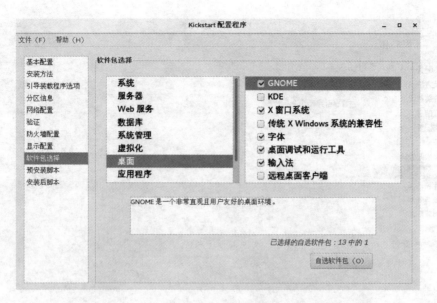

图 13-20　桌面软件包选择

完成后，将配置文件保存到 vsftpd 服务程序的默认根目录下，并确保所有人都对其具有可读权限，如图 13-21 所示。

```
[root@localhost~]# chmod +r /var/ftp/pub/ks.cfg
```

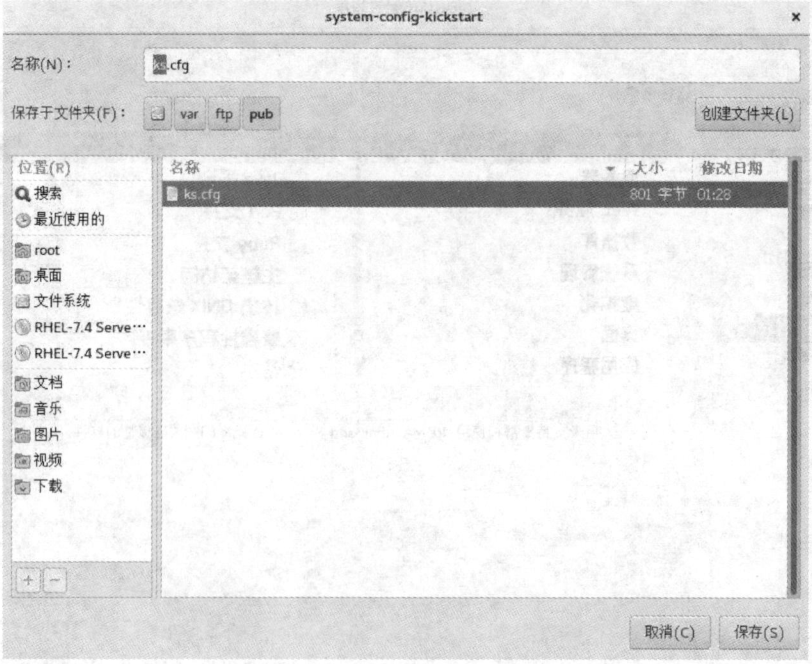

图 13-21　保存配置文件

接下来，客户端就可以进行无人值守安装系统（注意网卡模式为"仅主机模式"），如图 13-22～图 13-24 所示。

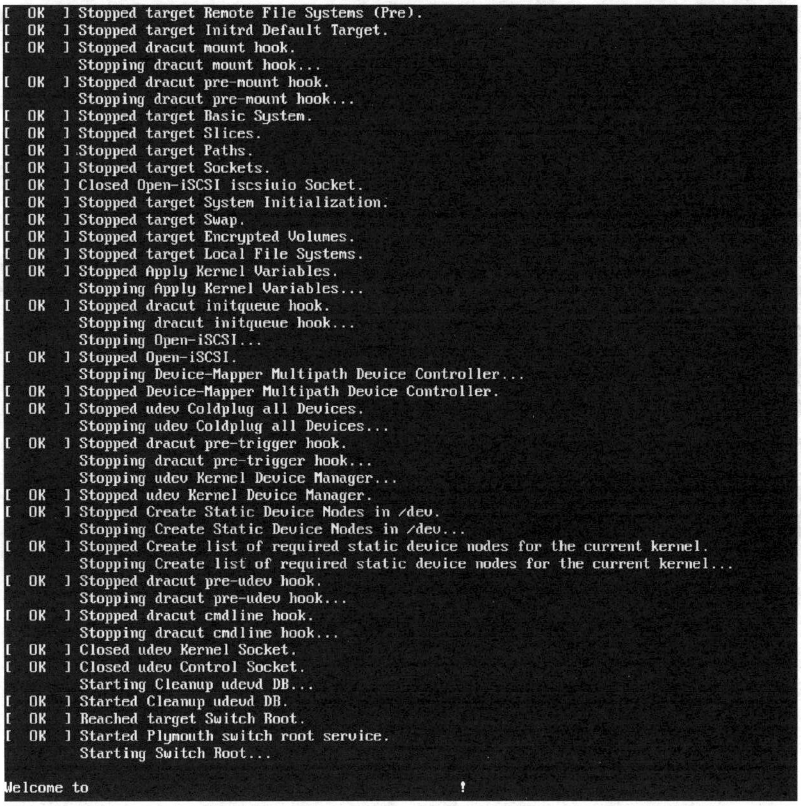

图 13-22　无人值守安装向导的初始化

图 13-23　从 FTP 服务器传输光盘镜像文件

图 13-24　无人值守自动安装系统

在现实工作环境中，只要服务器与客户端通过网络设备相连（如交换机），就能实现无人值守安装操作系统。

自我测试

一、填空题

1. 无人值守安装是指在安装过程中，不需任何_____的操作，而是直接按照预先设置好的软件进行安装。

2. TFTP 是 TCP/IP 协议族中的一个用来在客户端与服务器之间进行_____的协议。

3. Kickstart 是一种无人值守的安装方式，其工作原理是预先把需要人工填写的各种参数保存成一个_____文件。

二、简答题

简述无人值守安装系统的工作流程。

三、实训题

参照本项目内容，使用无人值守的方式安装系统。

项目 14　双机热备

项目综述

为确保公司的网站不因为服务器故障出现不能访问的情况,网络管理员将使用双机热备技术。

项目目标

- 了解双机热备相关知识;
- 掌握双机热备的配置方法。

任务 14.1　双机热备概述

14.1.1　集群基本概念

在互联网高速发展的今天,人们的生活已经离不开各类网络服务,如果在使用网络服务的过程中,出现了服务器故障(如设备故障、操作系统故障等),导致无法访问网络,将给人们的生活带去不便,更有甚者造成不可估量的损失。为了保证服务器能持续不断地为人们提供服务,可以通过双机热备来避免长时间的服务中断,保证系统长期、可靠地服务。

集群(Cluster)通过一组相对廉价的设备实现服务的可伸缩性,是提供相同网络资源的一组计算机系统。当服务请求急剧增加的时候,服务响应依然快速,且可以使用。集群内允许出现部分硬件或软件故障,然后通过集群管理软件将故障屏蔽,并将应用服务程序转移到集群中没有出现故障的设备(服务器)上,从而实现 24 小时不间断的服务。

高可用性集群(High Availability Cluster,HA Cluster),是指以减少服务中断(宕机)时间为目的的服务器集群技术。集群内的每台服务器称为节点(Node),它们之间通过物理硬件(电缆、光纤)和软件连接。如果集群中的任何一个节点出现故障,另外一个节点就会开始提供服务(这个过程被称为故障转移)。通过故障转移,用户几乎察觉不到服务已经中断,从而使用户能够继续使用服务而不被中断。如在 Windows 系统中,故障转移集群就是将多台服务器组合在一起,这些服务器相互协作,从而增强应用程序和服务的可用性。

双机热备属于集群中的一种,可以认为是集群的最小组成单位,就是将中心服务器安装成互为备份的两台服务器,并且在同一时间内只有一台服务器运行。

14.1.2 集群的分类

从工作方式出发，集群分为以下三类：

（1）主/主（Active/active）

这是最常用的集群模型，它提供了高可用性，并且必须保证在只有一个节点可以提供服务时，具有客户可以接受的性能。每个节点都通过网络对客户端提供网络服务，每个节点的容量被定义好，使得性能达到最优，并且每个节点都可以在故障转移时临时接管另一个节点的工作。所有的服务在故障转移后仍保持可用，而后端的实现客户端并不需要关心，所有后端的工作对客户端都是透明的。

（2）主/从（Active/passive）

为了提供最大的可用性，以及对性能最小的影响，主/从模型需要一个节点处于正常服务状态，另外一个节点处于备用状态。主节点处理客户端的请求，而备用节点处于空闲状态，当主节点出现故障时，备用节点会接管主节点的工作，继续为客户端提供服务，并且不会有任何性能上的影响。

（3）混合型（Hybrid）

混合型是上面两种模型的结合，可以只针对关键应用进行故障转移，这样可以在应用实现可用性的同时让非关键的应用在正常运行时也可以在服务器上运行。当出现故障时，出现故障的服务器上不太关键的应用就不可用了，但是那些关键应用会转移到另一个可用的节点上，从而达到性能和容错两方面的平衡。

◀ 任务 14.2 部署双机热备系统 ▶

14.2.1 使用 Pacemaker 部署双机热备系统

Pacemaker 是一个集群资源管理器，它可以利用管理员喜欢的集群基础构建提供的消息和成员管理能力来探测节点和资源故障，并从故障中恢复，从而实现集群资源的高可用性。与之前广泛使用的 Heartbeat 相比，Pacemaker 配置更为简单，并且支持的集群模式多样，资源管理方式更加灵活。

Pacemaker 的主要组件及作用如下：

① stonithd：心跳程序，主要用于处理与心跳相关的事件；
② lrmd：本地资源管理程序，直接调度系统资源；
③ pengine：政策引擎，依据当前集群状态计算下一步应该执行的操作等；
④ CIB：集群信息库，主要包含了当前集群中所有的资源及资源之间的关系等；
⑤ CRMD：集群资源管理守护进程。

Pacemaker 工作时会根据 CIB 中记录的资源，由 pengine 计算出集群的最佳状态及如何达到这个最佳状态，最后建立一个 CRMD 实例，由 CRMD 实例做出所有集群决策。这是

Pacemaker 的简要工作过程。

按照表 14-1 来配置双机热备系统。

表 14-1 双机热备信息

名称	IP	说明
Server1	192.168.1.10	node1(主节点)
Server2	192.168.1.20	node2(备用节点)
VIP	192.168.1.50	虚拟 IP

正常情况下由 Server1(192.168.1.10)提供服务,客户端可以根据主节点提供的 VIP 访问集群内的各种资源。当主节点出现故障时,备用节点可以自动接管主节点的 IP 资源(即 VIP 为 192.168.1.50)。

实现步骤为:

(1) 设置 server1、server2 的主机名以及主机名解析文件,配置 httpd 服务。

```
server1:
[root@localhost~]# vim /etc/hostname
node1
[root@node1~]# vim /etc/hosts
192.168.1.10 node1
192.168.1.20 node2
[root@node1~]# vim /etc/httpd/conf/httpd.conf
……
<Location/server-status>
SetHandler server-status
  Require all granted
</Location>
……
[root@node1~]# echo "Hello,192.168.1.50"> /var/www/html/index.html
[root@node1~]# systemctl stop firewalld
[root@node1~]# setenforce 0
[root@node1~]# systemctl restart httpd
[root@node1~]# systemctl enable httpd
server2:
[root@localhost~]# vim /etc/hostname
node2
[root@node2~]# vim /etc/hosts
192.168.1.10 node1
192.168.1.20 node2
[root@node2~]# vim /etc/httpd/conf/httpd.conf
……
<Location /server-status>
SetHandler server-status
  Require all granted
```

```
                </Location>
        ......
        [root@node2~]# echo "Hello,192.168.1.50">/var/www/html/index.html
        [root@node2~]# systemctl stop firewalld
        [root@node2~]# setenforce 0
        [root@node2~]# systemctl restart httpd
        [root@node2~]# systemctl enable httpd
```

(2) 在 server1、server2 之间设置 ssh 免密，如图 14-1 所示。

```
[root@node1 ~]# ssh-keygen
Generating public/private rsa key pair.
Enter file in which to save the key (/root/.ssh/id_rsa):
Created directory '/root/.ssh'.
Enter passphrase (empty for no passphrase):
Enter same passphrase again:
Your identification has been saved in /root/.ssh/id_rsa.
Your public key has been saved in /root/.ssh/id_rsa.pub.
The key fingerprint is:
SHA256:N3YNpADXzt7YGkYx1Ku95bXggQl6V8EvczcOFUkc30E root@node1
The key's randomart image is:
+---[RSA 2048]----+
|         ..o+. +E*|
|          o=oo =+|
|          oo+.o. +|
|         o +.+oo..|
|         .S*+B.+..|
|         . ooXo* .|
|            o B + |
|                 .|
|                  |
+----[SHA256]-----+
[root@node1 ~]# ssh-copy-id node2
The authenticity of host 'node2 (192.168.1.20)' can't be established.
ECDSA key fingerprint is SHA256:MIBn/OAwLAHfkVP9lmaNjpEQIinIuOucT/6lk7BqDnU.
ECDSA key fingerprint is MD5:b8:39:e1:29:c4:7f:bc:8d:11:62:cd:43:9e:15:f1:e0.
Are you sure you want to continue connecting (yes/no)? yes
/usr/bin/ssh-copy-id: INFO: attempting to log in with the new key(s), to filter out any that are already installed
/usr/bin/ssh-copy-id: INFO: 1 key(s) remain to be installed -- if you are prompted now it is to install the new keys
root@node2's password:

Number of key(s) added: 1

Now try logging into the machine, with:   "ssh 'node2'"
and check to make sure that only the key(s) you wanted were added.
```

图 14-1 设置 ssh 免密

(3) 在 server1、server2 上安装 pacemaker、corosync 应用程序，以及集群管理工具 pcs 和依赖包 psmisc policycoreutils-python。

在 RHEL7 中，系统镜像自带的资源有额外的高可用套件及存储套件，都存放在镜像文件的 addons 目录下。

```
server1:
[root@node1~]# vim /etc/yum.repos.d/dd.repo
[mydvd]
name=mydvd
baseurl=file:///media
enabled=1
gpgcheck=0

[High]      //高可用
name=High
baseurl=file:///media/addons/HighAvailability
enabled=1
gpgcheck=0


[Resi]      //存储
name=Resi
baseurl=file:///media/addons/ResilientStorage
```

```
enabled=1
gpgcheck=0
[root@node1~]# yum -y install pacemaker corosync
[root@node1~]# yum -y install pcs psmis policycoreutils-python
[root@node1~]# systemctl start pcsd
[root@node2~]# systemctl enable pcsd
```
server2:
```
[root@node2~]# vim /etc/yum.repos.d/dd.repo
[mydvd]
name=mydvd
baseurl=file:///media
enabled=1
gpgcheck=0
[High]
name=High
baseurl=file:///media/addons/HighAvailability
enabled=1
gpgcheck=0

[Resi]
name=Resi
baseurl=file:///media/addons/ResilientStorage
enabled=1
gpgcheck=0
[root@node2~]# yum -y install pacemaker corosync
[root@node2~]# yum -y install pcs psmis policycoreutils-python
[root@node2~]# systemctl start pcsd
[root@node2~]# systemctl enable pcsd
```

（4）在 servre1、server2 上为认证用户 hacluster 设置密码。然后在 server1 上建立认证节点，并创建集群（server2 上不需要配置，所有操作会自动同步到 node2 上）。

server1：
```
[root@node1~]# passwd hacluster
更改用户 hacluster 的密码。
新的密码:
无效的密码:密码少于 8 个字符
重新输入新的密码:
passwd:所有的身份验证令牌已经成功更新。
```
server2：
```
[root@node2~]# passwd hacluster
更改用户 hacluster 的密码。
新的密码:
无效的密码:密码少于 8 个字符
重新输入新的密码:
passwd:所有的身份验证令牌已经成功更新。
```

接下来，在server1上建立认证节点，并创建集群mycluster。

```
[root@node1~]# pcs clusterauth node1 node2
Username:hacluster
Password:
node1: Authorized
node2: Authorized
[root@node1~]# pcs cluster setup --namemycluster node1 node2
Destroying cluster on nodes: node1,node2…
node1: Stopping Cluster (pacemaker)…
node2: Stopping Cluster (pacemaker)…
node1: Successfully destroyed cluster
node2: Successfully destroyed cluster
Sending 'pacemaker_remote authkey' to 'node1','node2'
node2: successful distribution of the file 'pacemaker_remote authkey'
node1: successful distribution of the file 'pacemaker_remote authkey'
Sending clusterconfig files to the nodes…
node1: Succeeded
node2: Succeeded
Synchronizingpcsd certificates on nodes node1,node2…
node1: Success
node2: Success
Restartingpcsd on the nodes in order to reload the certificates…
node2: Success
node1: Success
[root@node1~]# pcs cluster start --all
node1: Starting Cluster…
node2: Starting Cluster…
```

进行集群状态的校验及查看。

```
[root@node1~]# corosync-cfgtool-s    //验证Corosync配置
Printing ring status.
Local node ID 1
RING ID 0
id=192.168.1.10
status=ring 0 active with no faults
[root@node1~]# pcs cluster status    //查看与集群相关的信息
Cluster Status:
Stack:corosync
Current DC: node1 (version 1.1.16-12.el7-94ff4df) - partition with quorum
Last updated: Tue Sep  1 04:58:56 2020
Last change: Tue Sep  1 04:58:18 2020 by hacluster via crmd on node1
2 nodes configured
0 resources configured
```

```
PCSD Status:
    node1: Online
    node2: Online
[root@node1~]# pcs status        //查看集群状态
Cluster name:mycluster
WARNING: no stonith devices and stonith-enabled is not false
Stack:corosync
Current DC: node1 (version 1.1.16-12.el7-94ff4df)-partition with quorum
Last updated: Tue Sep  1 04:59:12 2020
Last change: Tue Sep  1 04:58:18 2020 by hacluster via crmd on node1
2 nodes configured
0 resources configured
Online:[ node1 node2 ]
No resources
Daemon Status:
    corosync: active/disabled
    pacemaker: active/disable
    pcsd: active/enabled
[root@node2~]# corosync-cfgtool-s
Printing ring status.
Local node ID 2
RING ID 0
id=192.168.1.20
status=ring 0 active with no faults
[root@node2~]# pcs cluster status
Cluster Status:
    Stack:corosync
    Current DC: node1 (version 1.1.16-12.el7-94ff4df)-partition with quorum
    Last updated: Tue Sep  1 05:02:02 2020
    Last change: Tue Sep  1 04:58:18 2020 byhacluster via crmd on node1
    2 nodes configured
    0 resources configured
PCSD Status:
    node1: Online
    node2: Online
[root@node2~]# pcs status
Cluster name:mycluster
WARNING: no stonith devices and stonith-enabled is not false
Stack:corosync
Current DC: node1 (version 1.1.16-12.el7-94ff4df)-partition with quorum
Last updated: Tue Sep  1 05:02:32 2020
Last change: Tue Sep  1 04:58:18 2020 by hacluster via crmd on node1
2 nodes configured
```

```
    0 resources configured
Online:[ node1 node2 ]
No resources
Daemon Status:
   corosync: active/disabled
   pacemaker: active/disabled
   pcsd: active/enabled
[root@node1~]# pcs property setstonith-enabled=false    //关闭fence
[root@node1~]# crm_verify-LV    //校验集群状态
```

（5）配置集群资源。

向集群添加虚拟 IP 地址、httpd 服务，并让 pacemaker 自动检测 httpd 服务是否可用，同时可以在节点 node1、node2 之间相互切换（只需要在 server1 上配置）。

```
[root@node1~]# pcs resource create vip ocf:heartbeat:IPaddr2 ip=192.168.1.50 op monitor interval=30s
```

添加一个名为 vip 的 IP 地址资源，ocf 资源分类，使用 heartbeat 作为心跳检测，集群每隔 30s 检查该资源一次（op：option，选项；monitor：监控；interval：频率）。

```
[root@node1~]# pcs resource create apache systemd:httpd op monitor interval=30s
```

添加一个名为 apache 的 http 资源，调动 systemd 的 httpd 脚本，集群每隔 30s 检查该资源一次。然后，启动 pacemaker、corosync 服务，重新查看集群状态，确认资源已经加入。

```
[root@node1~]# systemctl start corosync
[root@node1~]# systemctl enable corosync
[root@node1~]# systemctl start pacemaker
[root@node1~]# systemctl enable pacemaker
[root@node1~]# pcs cluster status
Cluster Status:
   Stack:corosync
   Current DC: node1 (version 1.1.16-12.el7-94ff4df)- partition with quorum
   Last updated: Tue Sep  1 05:11:41 2020
   Last change: Tue Sep  1 05:09:53 2020 by root via cibadmin on node1
   2 nodes configured
   2 resources configured
PCSD Status:
   node1: Online
   node2: Online
[root@node1~]# pcs status
Cluster name:mycluster
Stack:corosync
Current DC: node1 (version 1.1.16-12.el7-94ff4df)-partition with quorum
Last updated: Tue Sep  1 05:11:50 2020
Last change: Tue Sep  1 05:09:53 2020 by root via cibadmin on node1
2 nodes configured
```

```
2 resourcesconfiguredOnline:[ node1 node2 ]
Full list of resources:
  vip (ocf::heartbeat:IPaddr2): Started node2
  apache (systemd:httpd): Started node1
Daemon Status:
  corosync: active /enabled
  pacemaker: active /enabled
```

pcsd：active /enabled 接下来将两个资源（vip、web）进行捆绑，避免出现单个资源只在一个节点上运行。

```
[root@node1~]# pcs resource group add my vip      //将 vip 加入组 my
[root@node1~]# pcs resource group add my apache   //将 apache 加入组 my
```

（6）配置节点优先级。

通过配置节点优先级（location）实现资源有限运行在优先级高的节点上（node1）。当 node1 失效后，再到优先级低的节点上（node2）运行。

```
[root@node1~]# pcs constraint locationapache prefers node1=10
[root@node1~]# pcs constraint locationapache prefers node2=5
[root@node1~]# crm_simulate-sL
Current cluster status:
Online:[ node1 node2 ]
  vip (ocf::heartbeat:IPaddr2): Started node2
  apache (systemd:httpd): Started node1
Allocation scores:
native_color: vip allocation score on node1: 0
native_color: vip allocation score on node2: 0
native_color: apache allocation score on node1: 10
native_color: apache allocation score on node2: 5
Transition Summary:
```

完成后重启集群。

```
[root@node1~]# pcs cluster stop --all
node1: Stopping Cluster (pacemaker)…
node2: Stopping Cluster (pacemaker)…
node1: Stopping Cluster (corosync)…
node2: Stopping Cluster (corosync)…
[root@node1~]# pcs cluster start --all
node2: Starting Cluster…
node1: Starting Cluster…
[root@node1~]# pcs status
Cluster name:mycluster
Stack:corosync
Current DC: node2 (version 1.1.16-12.el7-94ff4df) - partition with quorum
Last updated: Tue Sep  1 05:18:43 2020
Last change: Tue Sep  1 05:16:34 2020 by root via cibadmin on node1
```

```
2 nodes configured
2 resources configured
Online:[ node1 node2 ]
Full list of resources:
  vip (ocf::heartbeat:IPaddr2): Started node1
  apache (systemd:httpd): Started node1
Daemon Status:
  corosync: active /enabled
  pacemaker: active /enabled
  pcsd: active /enabled
```

这时,VIP 在主节点 node1。

```
[root@node1~]# ip addr
……
2: ens33: <BROADCAST,MULTICAST,UP,LOWER_UP> mtu 1500 qdisc pfifo_fast state UP qlen 1000
    link/ether 00:0c:29:9a:14:00brd ff:ff:ff:ff:ff:ff
    inet 192.168.1.10 /24 brd 192.168.1.255 scope global ens33
       valid_lft forever preferred_lft forever
    inet 192.168.1.50 /24 brd 192.168.1.255 scope global secondary ens33
       valid_lft forever preferred_lft forever
……
```

(7) 进行测试。

在主节点 node1 上,通过 VIP 访问网页。

```
[root@node1~]# curl http://192.168.1.50
Hello,192.168.1.50
```

如果此时主节点 node1 出现故障,比如系统重启、关闭集群等,备用节点 node2 将自动接管主节点的 IP 资源。

```
[root@node1~]# pcs cluster stop node1
node1: Stopping Cluster (pacemaker)…
node1: Stopping Cluster (corosync)…
[root@node2~]# pcs status
Cluster name:mycluster
Stack:corosync
Current DC: node2 (version 1.1.16-12.el7-94ff4df) -partition with quorum
Last updated: Tue Sep  1 05:34:44 2020
Last change: Tue Sep  1 05:18:15 2020 by root viacibadmin on node1
2 nodes configured
2 resources configured
Online:[ node2 ]
OFFLINE:[ node1 ]
Full list of resources:
  vip(ocf::heartbeat:IPaddr2): Started node2
```

```
        apache (systemd:httpd): Started node2
Daemon Status:
    corosync: active/disabled    //因为没有在 node2 上配置该服务开机启动
    pacemaker: active/disabled   //因为没有在 node2 上配置该服务开机启动
    pcsd: active/enabled
[root@node2~]# ip addr show
……
2: ens33: <BROADCAST,MULTICAST,UP,LOWER_UP> mtu 1500 qdisc pfifo_fast state UP qlen 1000
    link/ether 00:0c:29:ff:16:34brd ff:ff:ff:ff:ff:ff
    inet 192.168.1.20/24 brd 192.168.1.255 scope global ens33
        valid_lft forever preferred_lft forever
    inet 192.168.1.50/24 brd 192.168.1.255 scope global secondary ens33
        valid_lft forever preferred_lft forever
……
[root@node2~]# curl http://192.168.1.50
Hello,192.168.1.50
```

14.2.2 使用 Keepalive 部署双机热备系统

Keepalive 使用 vrrp（虚拟路由冗余协议）实现高可用，通过检查后端 TCP 服务的状态，如果一台提供 TCP 服务的后端节点出现故障，Keepalive 检测到后就将故障节点（主节点）及时从系统中剔除，将资源转移到其他节点。当故障节点（主节点）恢复后，就将资源重新转移到该节点上。下面将使用 Keepalive 配置表 14-1 的双机热备系统。

实现步骤为：

（1）在 server1、server2 上安装 keepalive 服务程序。

```
[root@node1~]# yum -y install keepalived
[root@node2~]# yum -y install keepalived
```

（2）在主节点、备用节点上配置 keepalive 服务，修改为以下内容，并启动 keepalive 服务。

```
[root@node1~]# vim /etc/keepalived/keepalived.conf
! Configuration File forkeepalived

global_defs {
    notification_email {    //设置 keepalived 在发生服务故障时,需要发送 email 地址
        acassen@firewall.loc
        failover@firewall.loc
        sysadmin@firewall.loc
    }
    notification_email_from Alexandre.Cassen@firewall.loc   //指定发件人
    smtp_server 192.168.200.1         //指定 SMTP 服务器地址
    smtp_connect_timeout 30           //指定连接 SMTP 的超时时间
    router_id LVS_DEVEL               //标识信息,在局域网内是唯一的
```

```
        vrrp_skip_check_adv_addr    //默认不跳过检查,如果报告与接收的上一个通告来自相
                                    同的 master 路由器,则不执行检查(跳过检查)
        # vrrp_strict     //严格遵守 VRRP 协议,注释掉,避免无法 PING 通 VIP 地址
        vrrp_garp_interval 0  //一个网卡上每组免费 ARP 消息之间的延迟时间
        vrrp_gna_interval 0   //一个网卡上每组 na 消息之间的延迟时间
    }

vrrp_instance VI_1 {
    state MASTER         //指定该节点为主节点,备用节点为 BACKUP
    interface ens33      //绑定虚拟 IP 的网络接口
    virtual_router_id 51 // VRRP 组名,主备节点需一致
    priority 100         //节点优先级,数字越大,实例优先级越高
    advert_int 1         //同步检查间隔,默认为 1 秒
    authentication {     //设置认证
        auth_type PASS   //认证方式,为 PASS 认证类型
        auth_pass 1111   //认证密码
    }
    virtual_ipaddress {  //设置 VIP
        192.168.1.50
    }
}


virtual_server 192.168.1.50 80 {        //设置 VIP 服务
    delay_loop 6          //设置健康检查时间间隔
    lb_algo rr            //设置 lvs 调度算法 rr|wrr|lc|wlc|lblc|sh|dh
    lb_kind NAT           //设置 lvs 模式
    persistence_timeout 50   //设置会话保持时间
    protocol TCP          //设置转发协议为 TCP
}
[root@node1~]# systemctl restart keepalived

[root@node2~]# vim /etc/keepalived/keepalived.conf
! Configuration File forkeepalived
global_defs {
    notification_email {
        acassen@firewall.loc
        failover@firewall.loc
        sysadmin@firewall.loc
    }
    notification_email_from Alexandre.Cassen@firewall.loc
    smtp_server 192.168.200.1
    smtp_connect_timeout 30
    router_id LVS_DEVEL
    vrrp_skip_check_adv_addr
      # vrrp_strict
    vrrp_garp_interval 0
```

```
        vrrp_gna_interval 0
    }
    vrrp_instance VI_1 {
        state BACKUP
        interface ens33
        virtual_router_id 51
        priority 50
        advert_int 1
        authentication {
            auth_type PASS
            auth_pass 1111
        }
        virtual_ipaddress {
            192.168.1.50
        }
    }

    virtual_server 192.168.1.50 80 {
        delay_loop 6
        lb_algo rr
        lb_kind NAT
        persistence_timeout 50
        protocol TCP
    }
    [root@node2~]# systemctl restart keepalived
```

这时,VIP 在主节点 node1。

```
    [root@node1~]# ip addr show
    ……
    2: ens33: <BROADCAST,MULTICAST,UP,LOWER_UP> mtu 1500 qdisc pfifo_fast state
UP qlen 1000
        link/ether 00:0c:29:9a:14:00 brd ff:ff:ff:ff:ff:ff
        inet 192.168.1.10/24 brd 192.168.1.255 scope global ens33
            valid_lft forever preferred_lft forever
        inet 192.168.1.50/32 scope global ens33
            valid_lft forever preferred_lft forever
    ……
```

(3) 进行测试。

在主节点 node1 上,通过 VIP 访问网页。

```
    [root@node1~]# curl http://192.168.1.50
    Hello,192.168.1.50
```

如果此时主节点 node1 出现故障,比如系统重启、关闭集群等,备用节点 node2 将自动接管主节点的 IP 资源。

```
[root@node1~]# systemctl stop  keepalived
[root@node2~]# ip addr show
……
2: ens33: <BROADCAST,MULTICAST,UP,LOWER_UP> mtu 1500 qdisc pfifo_fast state UP qlen 1000
    link /ether 00:0c:29:ff:16:34brd ff:ff:ff:ff:ff:ff
    inet 192.168.1.20 /24 brd 192.168.1.255 scope global ens33
       valid_lft forever preferred_lft forever
    inet 192.168.1.50 /32 scope global ens33
       valid_lft forever preferred_lft forever
……
[root@node2~]# curl http://192.168.1.50
Hello,192.168.1.50
```

自我测试

一、填空题

1. 高可用性集群是指以减少服务中断（宕机）时间为目的的服务器_____技术。
2. 集群内的每台服务器称为_____，它们之间通过物理硬件（电缆、光纤）和软件连接。
3. Keepalive 使用_____实现高可用。
4. 独立冗余磁盘阵列，简称_____。
5. 在服务器上配置 RAID，有_____、_____两种配置方式。。

二、简答题

1. 简述集群的分类。
2. 简述 Pacemaker 的主要组件及作用。

三、实训题

参照本项目内容，实操两种双机热备的方法。

参 考 文 献

1. 孙亚南,李勇. Red Hat Enterprise Linux 7 高薪运维入门[M]. 北京:清华大学出版社,2016.
2. 潘中强,王刚. Red Hat Enterprise Linux 7.3 系统管理实践[M]. 北京:清华大学出版社,2018.
3. 鸟哥. 鸟哥的 Linux 私房菜:基础学习篇[M]. 4 版. 北京:人民邮电出版社,2018.
4. 鸟哥. 鸟哥的 Linux 私房菜:服务器架设篇[M]. 3 版. 北京:机械工业出版社,2012.
5. 老男孩. 跟着老男孩学 Linux 运维:高性能 Web 集群实践(上)[M]. 北京:机械工业出版社,2019.
6. 老男孩,张耀. 跟着老男孩学 Linux 运维:核心系统命令实践[M]. 北京:机械工业出版社,2018.
7. 刘遄. Linux 就该这么学[M]. 北京:人民邮电出版社,2017.